Pipefitting

Level One

Annotated Instructor's Guide
Third Edition

Upper Saddle River, New Jersey
Columbus, Ohio

National Center for Construction Education and Research

President: Don Whyte
Director of Product Development and Revision: Daniele Stacey
Pipefitting Project Manager: Daniele Stacey
Production Manager: Jessica Martin
Product Maintenance Supervisor: Debie Ness
Editors: Bethany Harvey and Brendan Coote
Desktop Publisher: Jennifer Jacobs

NCCER would like to acknowledge the contract service provider for this curriculum:
Topaz Publications, Liverpool, New York.

10 9 8 7 6 5
ISBN 0-13-227312-8

PREFACE

There are some who may consider pipefitting synonymous with plumbing, but these are really two very distinct trades. Plumbers install and repair the water, waste disposal, drainage and gas systems in homes and commercial and industrial buildings. Pipefitters, on the other hand, install and repair both high- and low-pressure pipe systems used in manufacturing, in the generation of electricity, and in the heating and cooling of buildings.

If you're trying to imagine a setting involving pipefitters, think of large power plants that create and distribute energy throughout the nation; think of manufacturing plants, chemical plants, and piping systems that carry all kinds of liquids, gaseous, and solid materials.

If you're trying to imagine a job in pipefitting, picture a job that won't go away for a long time. As the U.S. government reports, the demand for skilled pipefitters continues to outpace the supply of workers trained in this craft. And high demand typically means 'higher pay' making pipefitters among the highest paid construction occupations in the nation.

While pipefitters and plumbers perform different tasks, the aptitudes involved in these crafts are comparable. Attention to detail, spatial and mechanical abilities, and the ability to work efficiently with the tools of their trade are key.

There are more than a half-million people employed in this work in the United States, and as most of them can tell you, there are many opportunities awaiting those with the skills and desire to move forward in the construction industry.

CONTREN® LEARNING SERIES

The National Center for Construction Education and Research (NCCER) is a not-for-profit 501(c)(3) education foundation established in 1995 by the world's largest and most progressive construction companies and national construction associations. It was founded to address the severe workforce shortage facing the industry and to develop a standardized training process and curricula. Today, NCCER is supported by hundreds of leading construction and maintenance companies, manufacturers, and national associations. The Contren® Learning Series was developed by NCCER in partnership with Prentice Hall, the world's largest educational publisher.

Some features of NCCER's Contren® Learning Series are as follows:

- An industry-proven record of success
- Curricula developed by the industry for the industry
- National standardization, providing portability of learned job skills and educational credits
- Compliance with Apprenticeship, Training, Employer, and Labor Services (ATELS) requirements for related classroom training (*CFR 29:29*)
- Well-illustrated, up-to-date, and practical information

NCCER also maintains a National Registry that provides transcripts, certificates, and wallet cards to individuals who have successfully completed modules of NCCER's Contren® Learning Series. *Training programs must be delivered by an NCCER Accredited Training Sponsor in order to receive these credentials.*

Contents

Contren® Curricula

NCCER's training programs comprise more than 40 construction, maintenance, and pipeline areas and include skills assessments, safety training, and management education.

Boilermaking
Carpentry
Carpentry, Residential
Cabinetmaking
Concrete Finishing
Construction Craft Laborer
Construction Technology
Core Curriculum: Introductory Craft Skills
Currículum Básico
Electrical
Electrical, Residential
Electrical Topics, Advanced
Electronic Systems Technician
Exploring Careers in Construction
Fundamentals of Mechanical and Electrical Mathematics
Heating, Ventilating, and Air Conditioning
Heavy Equipment Operations
Highway/Heavy Construction
Instrumentation
Insulating
Ironworking
Maintenance, Industrial
Masonry
Millwright
Mobile Crane Operations
Painting
Painting, Industrial
Pipefitting
Pipelayer
Plumbing
Reinforcing Ironwork
Rigging
Scaffolding
Sheet Metal
Site Layout
Sprinkler Fitting
Welding

Pipeline

Control Center Operations, Liquid
Corrosion Control
Electrical and Instrumentation
Field Operations, Liquid
Field Operations, Gas
Maintenance
Mechanical

Safety

Field Safety
Orientación de Seguridad
Safety Orientation
Safety Technology

Management

Introductory Skills for the Crew Leader
Project Management
Project Supervision

Acknowledgments

This curriculum was revised as a result of the farsightedness and leadership of the following sponsors:

Becon
Cianbro
Flint Hills Resources/Koch Industries
Fluor Global Craft Services
Kellogg Brown & Root
Lee College
Zachry Construction Corporation

This curriculum would not exist were it not for the dedication and unselfish energy of those volunteers who served on the Authoring Team. A sincere thanks is extended to the following:

Glynn Allbritton
Tom Atkinson
Ned Bush
Adrian Etie
Daniel Gomez
Tina Goode
Ron Harper
Ed LePage
Toby Linden

NCCER PARTNERING ASSOCIATIONS

American Fire Sprinkler Association
API
Associated Builders & Contractors, Inc.
Associated General Contractors of America
Association for Career and Technical Education
Association for Skilled and Technical Sciences
Carolinas AGC, Inc.
Carolinas Electrical Contractors Association
Construction Industry Institute
Construction Users Roundtable
Design-Build Institute of America
Electronic Systems Industry Consortium
Merit Contractors Association of Canada
Metal Building Manufacturers Association
National Association of Minority Contractors
National Association of Women in Construction
National Insulation Association
National Ready Mixed Concrete Association
National Systems Contractors Association
National Technical Honor Society
National Utility Contractors Association
North American Crane Bureau
North American Technician Excellence
Painting & Decorating Contractors of America
Portland Cement Association
SkillsUSA
Steel Erectors Association of America
Texas Gulf Coast Chapter ABC
U.S. Army Corps of Engineers
University of Florida
Women Construction Owners & Executives, USA

Product Supplements

Windows/Macintosh-Based TestGen

ISBN 0-13-229130-4 $30

Ensure test security with NCCER's windows-based computerized testing software. This software allows instructors to scramble the module exam questions and answer keys in order to print multiple versions of the same test, customize tests to suit the needs of their training units*, add questions, or easily create a final exam.

**Due to NCCER's Accreditation Guidelines, instructors may not delete existing questions from exams. Doing so may seriously jeopardize either the accreditation status or the training program sponsor or any recognition of trainees, instructors, and trainers through the NCCER National Registry.*

Transparency Masters

ISBN 0-13-229151-7 $25

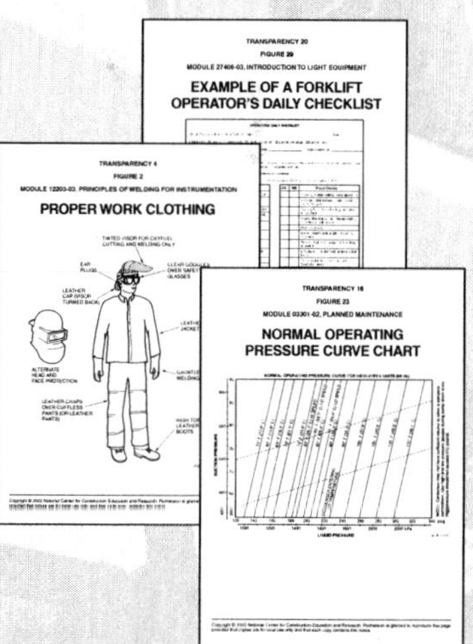

Spend more time training and less time at the copier. In response to instructor feedback NCCER offers loose, reproducible copies of the overhead transparencies referenced in the Instructor's Guides. The transparency masters package includes most of the Trainee Guide graphics, enlarged for projection and printed on loose sheets* for easy copying onto transparency film using your photocopier.

* *Transparency masters are provided on regular, loose sheets of paper, not acetates.*

To order any of these supplements, contact Prentice Hall Publishing at

800-922-0579

Module 08101-06

Orientation to the Trade

NCCER STANDARDIZED CRAFT TRAINING PROGRAM

The National Center for Construction Education and Research (NCCER) provides a standardized national program of accredited craft training. Key features of the program include instructor certification, competency-based training, and performance testing. The program provides trainees, instructors, and companies with a standard form of recognition through the National Registry. The program is described in full in the *Guidelines for Accreditation*, published by NCCER. For more information on standardized craft training, contact NCCER by writing to P.O. Box 141104, Gainesville, FL 32614-1104; calling 352-334-0911; or e-mailing info@nccer.org. More information is available at www.nccer.org.

HOW TO USE THIS ANNOTATED INSTRUCTOR'S GUIDE

Each page presents two sections of information. The larger section displays each page exactly as it appears in the Trainee Module. The narrow column ties suggested trainee and instructor actions to each page and provides icons (detailed below) to call your attention to material, safety, audiovisual, or testing requirements. The bottom of each page includes space for your notes.

The **Audiovisual** icon indicates an appropriate time to show a transparency or other audiovisual aid.

The **Classroom** icon prompts you to define a term, stress a point, ask trainees to explain a concept, or give examples.

The **Demonstration** icon directs you to show trainees how to perform tasks.

The **Examination** icon tells you to administer the written module examination.

The **Homework** icon is placed where you may wish to assign reading for the next class, assign a project, or advise trainees to prepare for an examination.

The **Laboratory** icon is used when trainees are to practice performing tasks.

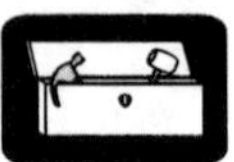

The **Materials** icon is a reminder for you to gather materials needed for classes, laboratories, and testing.

The **Performance Testing** icon tells you to administer a performance test or a portion thereof.

The **Safety** icon is used to emphasize safety issues. It is often keyed to *Caution* and *Warning!* statements in the Trainee Module.

The **Teaching Tip** icon indicated additional guidance is available, such as how to conduct an exercise, get the most educational value from a field trip, or encourage class participation. Teaching Tips may expand on a feature (*Think About It*, *Did You Know?*) or provide *Quick Quizzes* or similar exercises. You will be referred to the Teaching Tips section at the back of the module if there is additional material.

The **Combination** icon indicates that the laboratory listed corresponds with a performance task. If desired, you can note the proficiency of the trainees during the laboratory, and use it to satisfy performance testing requirements.

PREPARATION

Before teaching this module, you should review the Objectives, Performance Tasks, Materials and Equipment List, and Module Outline. Be sure to allow ample time to prepare your own training or lesson plan and gather all required materials and equipment.

MODULE OVERVIEW

This module provides the trainee with an overview of pipefitting, pipefitter responsibilities, and career opportunities. The module also covers basic principles of safety.

PREREQUISITES

Prior to training with this module, it is recommended that the trainee shall have successfully completed *Core Curriculum.*

OBJECTIVES

Upon completion of this module, the trainee will be able to do the following:

1. Describe the types of work performed by pipefitters.
2. Identify career opportunities available to pipefitters.
3. Explain the purpose and objectives of an apprentice training program.
4. Explain the responsibilities and characteristics of a good pipefitter.
5. Explain the importance of safety in relation to pipefitting.

PERFORMANCE TASKS

There are no performance tasks for this module.

MATERIALS AND EQUIPMENT LIST

Overhead projector and scree
Transparencies
Blank acetate sheets
Transparency pens
Whiteboard/chalkboard
Markers/chalk
Pencils and scratch paper
Appropriate personal protective equipment
Sample pipe
Common pipe wrenches
Copy of an employee manual
Job announcements for pipefitting from local newspapers (want ads)
NCCER Apprentice Training Recognition Forms
OSHA Safety and Health Standards for the Construction Industry
Module Examinations*

* Located in the Test Booklet.

SAFETY CONSIDERATIONS

Ensure that the trainees are equipped with appropriate personal protective equipment and know how to use it properly. Emphasize basic site safety.

ADDITIONAL RESOURCES

This module is intended to present thorough resources for task training. The following reference works are suggested for both instructors and motivated trainees interested in further study. These are optional materials for continued education rather than for task training.

The Pipefitters Blue Book, Latest Edition. W.V. Graves. Webster, TX: Graves Publishing Company.

The Pipefitters Handbook, 3rd Edition. Forrest R. Lindsey. New York, NY: Industrial Press.

TEACHING TIME FOR THIS MODULE

An outline for use in developing your lesson plan is presented below. Note that each Roman numeral in the outline equates to one session of instruction. Each session has a suggested time period of 2½ hours. This includes 10 minutes at the beginning of each session for administrative tasks and one 10-minute break during the session. Approximately 5 hours are suggested to cover *Orientation to the Trade*. You will need to adjust the time required for hands-on activity and testing based on your class size and resources.

Topic	Planned Time
Session I. Orientation to the Trade	
A. Introduction	__________
B. Pipefitting Work	__________
C. Opportunities in the Trade	__________
D. Your Training Program	__________
E. Responsibilities of the Employee	__________
Session II. Human Relations, Safety Roles, Review, and Module Examination	
A. Human Relations	__________
B. Employer and Employee Safety Obligations	__________
C. Review	__________
D. Module Examination	__________

1. Trainees must score 70 percent or higher to receive recognition from NCCER.
2. Record the testing results on Craft Training Report Form 200, and submit the results to the Training Program Sponsor.

Assign reading of Module 08101-06.

08101-06

Orientation to the Trade

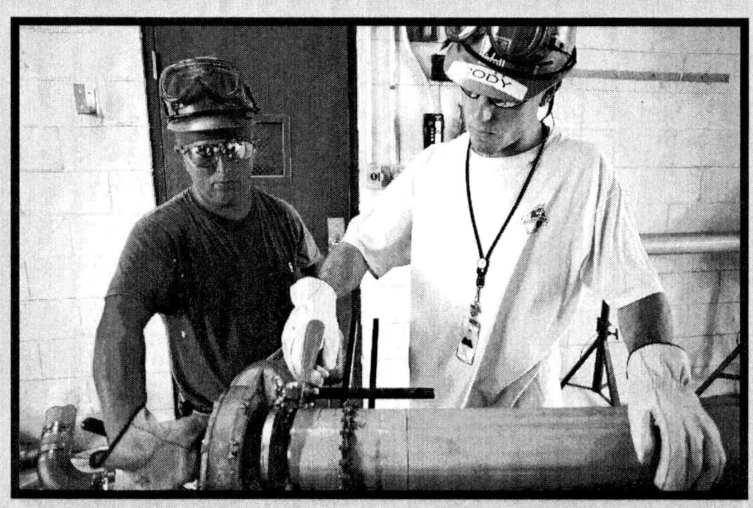

08101-06
Orientation to the Trade

Topics to be presented in this module include:

Overview

Although both work with pipe, there is a major difference between pipefitters and plumbers. Pipefitters lay out and install piping systems primarily for industrial facilities, such as chemical plants, oil refineries, food processing plants, and paper mills. Plumbers install and service water distribution and waste systems, primarily in residential and commercial applications.

A pipefitter will work with many kinds of pipe, ranging from small half-inch piping to piping that is three or four feet in diameter. A pipefitter must know how to work with threaded, grooved, and welded piping systems and must be able to master a variety of tools and equipment.

Instructor's Notes:

Objectives

When you have completed this module, you will be able to do the following:

1. Describe the types of work performed by pipefitters.
2. Identify career opportunities available to pipefitters.
3. Explain the purpose and objectives of an apprentice training program.
4. Explain the responsibilities and characteristics of a good pipefitter.
5. Explain the importance of safety in relation to pipefitting.

Trade Terms

Apprenticeship Training, Employer and Labor Services (ATELS)
Bevel
On-the-job training (OJT)
Occupational Safety and Health Administration (OSHA)
Sweat

Required Trainee Materials

1. Pencil and paper
2. Gloves and eye protection

Prerequisites

This course map shows all of the modules in the first level of the *Pipefitting* curriculum. The suggested training order begins at the bottom and proceeds up. Skill levels increase as you advance on the course map. The local Training Program Sponsor may adjust the training order.

Before you begin this module, it is recommended that you successfully complete *Core Curriculum.*

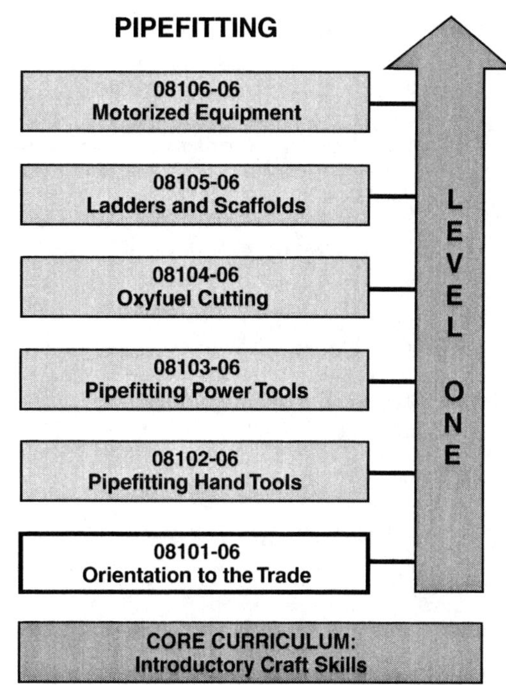

101CMAP.EPS

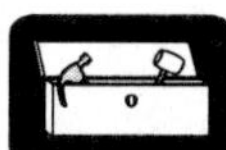

Ensure that you have everything required to teach the course. Check the Materials and Equipment list at the front of this module.

See the general Teaching Tip at the end of this module.

Show Transparency 1, Objectives. Review the goals of the module, and explain what will be expected of the trainee.

Explain that terms shown in bold are defined in the Glossary at the back of this module.

Review the modules covered in Level One and explain how this module fits in.

Explain what will be covered in the training program.

Discuss the difference between plumbers and pipefitters.

Describe the tasks performed by pipefitters.

Describe the different types of pipe that pipefitters work with.

Provide some sample pipe for trainees to examine.

1.0.0 ◆ INTRODUCTION

Piping systems carry water, gases, liquid chemicals, solids, and fuels in a variety of environments (*Figure 1*). Such environments include oil refineries, chemical plants, power plants, food processing plants, paper mills, ships, factories, and a host of other facilities. This training program covers the tools, materials, and techniques used by pipefitters to lay out, fabricate, and install piping systems.

Although pipefitters and plumbers both work with piping systems, these crafts are not the same. Plumbers work primarily with water distribution, drainage, and waste systems in residential and commercial environments. Pipefitters work with piping systems that carry all kinds of liquid, gaseous, and solid materials, primarily in an industrial environment.

2.0.0 ◆ PIPEFITTING WORK

A pipefitter must learn many skills, including working with a variety of specialized hand tools and power tools. Among the tasks performed by pipefitters are the following:

- Read and interpret blueprints and specifications
- Plan and sketch piping systems
- Lay out and fabricate fittings
- Fit-up pipe for welding
- Measure, cut, thread, and assemble pipe
- Bend pipe
- Install pipe, fittings, and valves using various joining techniques
- Inspect and test piping systems
- Perform routine maintenance and repairs on piping systems

Pipefitters must learn to work with several types of piping, including carbon steel, copper, plastic (PVC), stainless steel, and aluminum. In some environments, most of the pipe joints are welded, rather than threaded. This is especially true of aluminum and stainless steel pipe. Therefore, an important part of the pipefitter's work is to prepare and fit-up piping for welding.

Piping is likely to change direction several times between the source and the destination. Therefore, a pipefitter must know how to select and apply fittings or bend the pipe in order to

(A) POWER STATION

(B) INDUSTRIAL PIPING SYSTEM

101F01.EPS

Figure 1 ◆ Piping system examples.

Instructor's Notes:

1.2 PIPEFITTING ◆ LEVEL ONE

redirect the piping run. Some types of pipe, such as copper piping up to 2 inches in diameter, are fairly easy to bend. Because of its wall thickness, carbon steel pipe is difficult to bend, so fittings such as tees and elbows are generally used to change the direction of carbon steel pipe. However, carbon steel pipe can be bent using hydraulic benders. In some instances, such as high-pressure steam applications, bending of carbon steel pipe is required by the job specifications.

Valves are used to control the flow of material in a process. A pipefitter must therefore understand the operation, application, and installation of the many types of manual and automatic valves used in process control (*Figure 2*).

A pipefitter may work on a wide variety of piping systems:

- Cooling, heating, and refrigeration systems
- Compressed air systems
- High-pressure systems
- Steam systems
- Hydraulic systems
- Nuclear power systems
- Chemical storage and processing systems
- Oil, gasoline, and natural gas storage and processing systems
- Fire protection systems

These systems all have one thing in common: they require properly aligned and correctly assembled piping using materials that are compatible with the product being transported. This means that attention to detail is an essential characteristic of a successful pipefitter.

Figure 2 ◆ Piping section with valve.

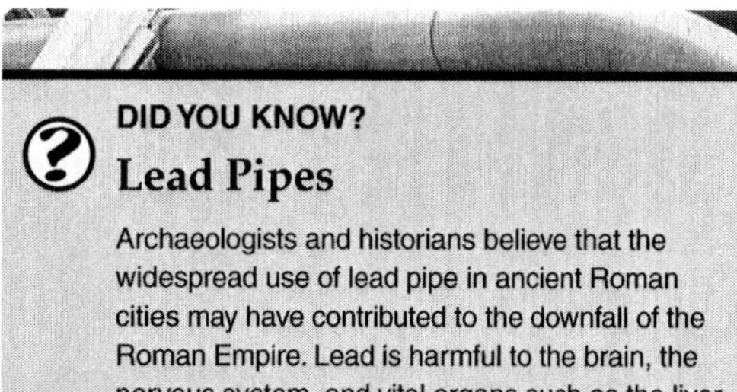

DID YOU KNOW?

Lead Pipes

Archaeologists and historians believe that the widespread use of lead pipe in ancient Roman cities may have contributed to the downfall of the Roman Empire. Lead is harmful to the brain, the nervous system, and vital organs such as the liver. Water carried through lead pipes may have gradually poisoned many people in the Roman Empire.

2.1.0 Tools of the Trade

Pipefitters work with a variety of hand and power tools. These tools are used primarily to cut, thread, and join pipe. A variety of wrenches are used to join pipe *(Figure 3)*. The pipefitter must know when to use each of these wrenches, how to select the right size, and how to use them correctly. Power saws and other types of cutting tools are used to cut pipe to the right size. Reamers are used to remove burrs from pipe. Grinding tools are used to **bevel** pipe for welding. In some cases, an oxyacetylene cutting torch is used to cut and bevel pipe prior to welding *(Figure 4)*.

Threading tools *(Figure 5)* are used to apply threads to pipe so that sections of pipe can be joined with threaded fittings. Carbon steel pipe is usually threaded, but other types of pipe such as copper, plastic, and steel alloys can be threaded as well. Both manual and powered threading tools are used.

Copper piping is generally bent to the desired shape and connected using **sweat** joints. Different types of pipe benders are used, depending on the size of the pipe.

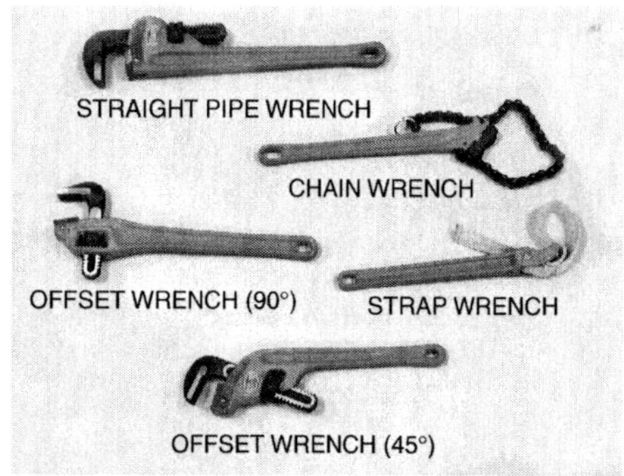

Figure 3 ◆ Common pipe wrenches.

Explain how different types of pipe are bent.

Describe various types of valves used in piping systems.

Explain that a pipefitter may work on various types of piping systems.

Describe pipe wrenches and other hand tools used by pipefitters.

Provide some common pipe wrenches for trainees to examine.

MODULE 08101-06 ◆ ORIENTATION TO THE TRADE 1.3

Discuss common hazards pipefitters face, including lifting, burns, trenching, falls, and hazardous chemicals.

Review common power tool safety practices.

101F04.EPS

Figure 4 ◆ Pipe cutting and beveling fixture.

Stainless steel piping is used in food processing and other specialized applications, such as medical facilities, where cleanliness is essential. Stainless steel is often joined by welding.

(A) HAND THREADER

(B) POWER THREADING MACHINE

101F05.EPS

Figure 5 ◆ Pipe threading tools.

2.2.0 Safety on the Job

There are many safety issues to be considered on any construction site. In addition to the hazards normally found on a construction site, pipefitters encounter specific hazards because of the nature of their work.

A single large pipe or a bundle of smaller pipe is too heavy to lift by hand, so a machine such as a crane must be used to move the pipe. The pipefitter participates in the rigging process by helping to attach slings to the load. Tools used for flame cutting and beveling of pipe create a potential burn hazard, as well as hazards from flammable and explosive gases. When working with underground piping systems, the pipefitter may be exposed to trenching hazards and possibly confined space hazards. Pipefitters may also have to work at elevations in order to install and maintain piping systems. Therefore, fall protection is a key consideration. Power tools such as automatic pipe threaders also present cutting and electrical hazards.

Another hazard that concerns pipefitters involves the materials carried in piping systems. Some piping systems carry corrosive chemicals or hot liquids. Other processes operate under high pressure. It is important to know what material is being carried by the system and to take the appropriate precautions to avoid injury.

Power Tool Safety Precautions

- Wear heavy protective clothing, a hard hat, goggles, and heavy work gloves to guard against injury from flying debris.
- Use adequate hearing protection.
- Use the appropriate respiratory protection when dust hazards are produced.
- To avoid shock, ensure that electrically powered tools are adequately grounded and used in a dry environment. For added protection, use tools that are double insulated and equipped with ground fault circuit interrupters.
- Do not use power tools in confined spaces where sparks could cause explosions.
- Use nonsparking tools around combustible and flammable materials.
- Do not exceed the rated speed of the tool.
- Use recommended tool guards.
- Inspect equipment regularly, and repair or replace it as necessary.

Instructor's Notes:

1.4 PIPEFITTING ◆ LEVEL ONE

3.0.0 ◆ OPPORTUNITIES IN THE TRADE

Pipefitters are employed all over the U.S. and the world in refineries, power plants, and industrial facilities. This creates an opportunity to travel while earning an excellent income. A journeyman-level pipefitter will earn as much as, and sometimes more than, a college-educated office worker. Because of the demand, pipefitters also have job security.

A person starting in the trade will first learn to properly handle and use the tools of the trade. They may be asked to bend, cut, and thread pipe and assist journey-level pipefitters in performing various other tasks. As their training progresses and their knowledge grows, apprentices will take on greater responsibilities.

Because pipes are joined by welding in many environments, pipefitters often work closely with welders. It is a natural progression then, for a pipefitter to learn welding. Pipe welding is very demanding, however, and thus requires a great deal of training and practice. But, once the pipefitter is certified as a pipe welder, the opportunities for jobs, income, and advancement increase significantly.

Career growth for a pipefitter can occur in many ways *(Figure 6)*. If you have leadership qualities, you can become a foreman of a pipefitting crew. Some experienced pipefitters become instructors at contractor, union, or vocational training schools. Others become quality control inspectors, estimators, or piping designers. With further education and training, a pipefitter can become a piping system designer.

Within a company that does pipefitting work, there are opportunities to move into supervisory and management positions such as quality control manager, project superintendent, and plant superintendent.

Another avenue for growth is the inspection of pipe welds. Various technical methods, including x-ray techniques, are used to inspect and certify critical pipe welds. The inspector must be able to perform specialized tests and analyze the results.

4.0.0 ◆ YOUR TRAINING PROGRAM

The Department of Labor's (DOL) Office of **Apprenticeship Training, Employer and Labor Services (ATELS)** sets the minimum standards for training programs across the country. ATELS programs rely on mandatory classroom instruction and **on-the-job training (OJT)**. ATELS requires 144 hours of classroom instruction per year and 2,000 hours of OJT per year. In a typical ATELS pipefitter apprentice program, trainees spend 576 hours in classroom instruction and 8,000 hours in OJT before receiving journey certificates.

To address the training needs of the professional communities, NCCER developed a four-year pipefitter training program. NCCER uses the minimum ATELS standards as a foundation for comprehensive curricula that provide trainees with in-depth classroom and OJT experience.

Explain that pipefitters are employed by many industries all over the world. Describe the initial tasks a pipefitter will perform. Explain that pipefitters often learn welding.

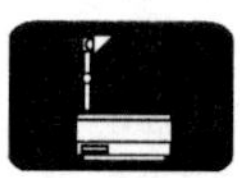

Show Transparency 2 (Figure 6). Discuss career opportunities for pipefitters.

See the Teaching Tip for Section 3.0.0 at the end of this module.

Ask a trainee to describe what ATELS does. Describe what is required in an ATELS apprenticeship program for pipefitters.

Provide copies of newspaper ads with job openings for pipefitters for the trainees to review.

CRAFT
APPRENTICESHIP (3–5 YEARS)
TECHNICAL COLLEGE (2 YEARS)
MANAGEMENT
COLLEGE APPRENTICESHIP (2–5 YEARS)
UNIVERSITY (4 YEARS)
JOURNEYMAN
PIPEFITTING INSTRUCTOR
QC INSPECTOR
FOREMAN
PIPEFITTING SUPERVISOR
PROJECT/ SAFETY MANAGER
ADMINISTRATOR/ BUSINESS MANAGEMENT
CEO/ COMPANY OWNER
BACHELOR'S DEGREE CONSTRUCTION MANAGEMENT ARCHITECTURE ENGINEERING
MASTER'S DEGREE CONSTRUCTION MANAGEMENT ARCHITECTURE ENGINEERING
DOCTORATE CONSTRUCTION MANAGEMENT ARCHITECTURE ENGINEERING
101F06.EPS

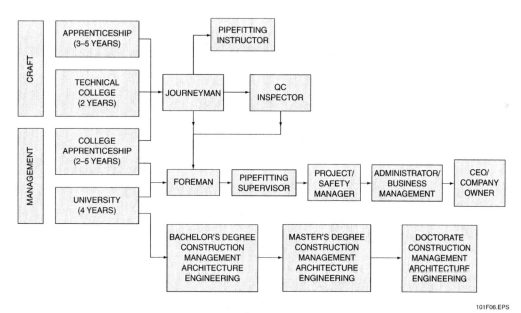

Figure 6 ◆ Career opportunities for pipefitters.

MODULE 08101-06 ◆ ORIENTATION TO THE TRADE 1.5

Explain that the NCCER provides competency-based training, which is recorded in the National Registry and is transferable.

Ask a trainee to describe their apprenticeship program. Discuss apprenticeship programs.

Discuss work-related on-the-job training requirements.

Explain some of the variances in classroom training.

Provide samples of NCCER Apprentice Training Recognition to the trainees for their examination or refer to the *Appendix*.

This NCCER curriculum provides trainees with industry-driven training and education. It adopts a purely competency-based teaching philosophy. This means that trainees must demonstrate to the instructor that they possess the understanding and the skills necessary to perform the hands-on tasks that are covered in each module before they can advance to the next stage of the curriculum.

When the instructor is satisfied that a trainee has successfully demonstrated the required knowledge and skills for a particular module, that information is sent to NCCER and kept in the National Registry. The National Registry can then confirm training and skills for workers as they move from state to state, company to company, or even within a company (see *Appendix*).

Whether you enroll in an NCCER program or another ATELS-approved program, ensure that you work for an employer or sponsor who supports a nationally standardized training program that includes credentials to confirm your skill development.

4.1.0 Apprenticeship Program

Apprentice training goes back thousands of years, and its basic principles have not changed in that time. First, it is a means for individuals entering the craft to learn from those who have mastered the craft. Second, it focuses on learning by doing; real skills versus theory. Although some theory is presented in the classroom, it is always presented in a way that helps the trainee understand the purpose behind the skill that is to be learned.

All apprenticeship standards prescribe certain work-related or on-the-job training. This on-the-job training is broken down into specific tasks in which the apprentice receives hands-on training during the period of the apprenticeship. In addition, a specified number of hours is required in each task. The total number of on-the-job training hours for the pipefitter apprenticeship program is traditionally 8,000, which amounts to about four years of training. In a competency-based program, it may be possible to shorten this time by testing out of specific tasks through a series of performance exams.

In a traditional program, the required on-the-job training may be acquired in increments of 2,000 hours per year. Layoff or illness may affect the duration.

The apprentice must log all work time and turn it in to the apprenticeship committee so that accurate time control can be maintained. After each 1,000 hours of related work, the apprentice will receive a pay increase as prescribed by the apprenticeship standards.

The classroom instruction and work-related training will not always run concurrently due to such reasons as layoffs, type of work needed to be done in the field, etc. Furthermore, apprentices with special job experience or coursework may obtain credit toward their classroom requirements. This reduces the total time required in the classroom while maintaining the total 8,000-hour on-the-job training requirement. These special cases will depend on the type of program and the regulations and standards under which it operates.

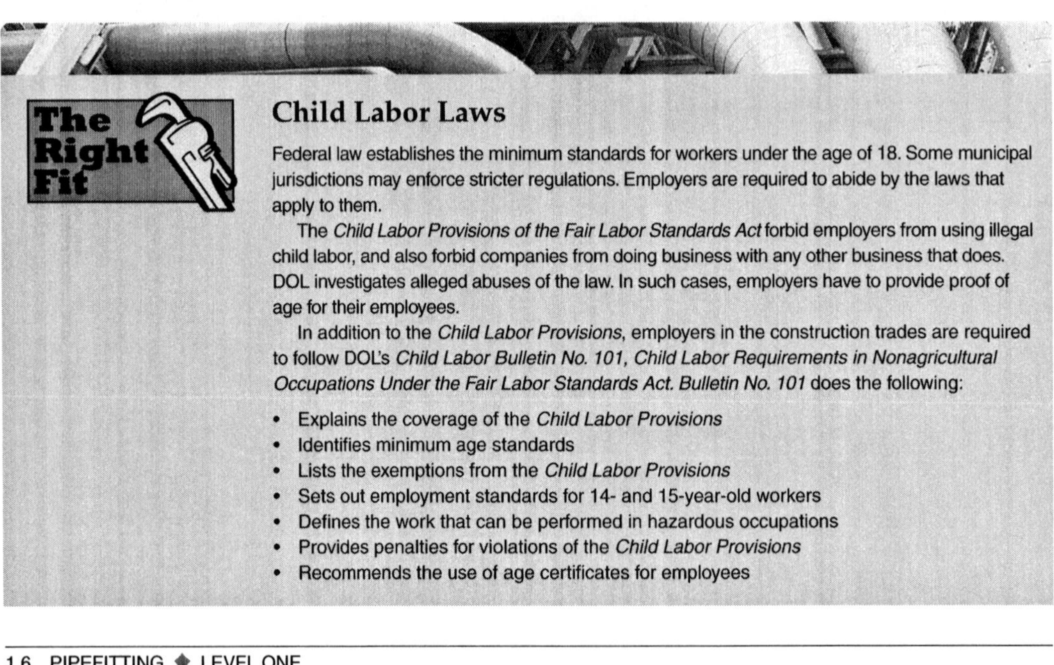

Child Labor Laws

Federal law establishes the minimum standards for workers under the age of 18. Some municipal jurisdictions may enforce stricter regulations. Employers are required to abide by the laws that apply to them.

The *Child Labor Provisions of the Fair Labor Standards Act* forbid employers from using illegal child labor, and also forbid companies from doing business with any other business that does. DOL investigates alleged abuses of the law. In such cases, employers have to provide proof of age for their employees.

In addition to the *Child Labor Provisions*, employers in the construction trades are required to follow DOL's *Child Labor Bulletin No. 101, Child Labor Requirements in Nonagricultural Occupations Under the Fair Labor Standards Act. Bulletin No. 101* does the following:

- Explains the coverage of the *Child Labor Provisions*
- Identifies minimum age standards
- Lists the exemptions from the *Child Labor Provisions*
- Sets out employment standards for 14- and 15-year-old workers
- Defines the work that can be performed in hazardous occupations
- Provides penalties for violations of the *Child Labor Provisions*
- Recommends the use of age certificates for employees

Instructor's Notes:

1.6 PIPEFITTING ◆ LEVEL ONE

Informal on-the-job training provided by employers is usually less thorough than that provided through a formal apprenticeship program. The degree of training and supervision in this type of program often depends on the size of the employing firm. A small contractor may provide training in only one area, while a large company may be able to provide training in several areas.

For those entering an apprenticeship program, a high school or technical school education is desirable, as are courses in shop, mechanical drawing, and general mathematics. Manual dexterity, good physical condition, and quick reflexes are important. The ability to solve problems quickly and accurately and to work closely with others is essential. You must have a high concern for safety.

The prospective apprentice must submit certain information to the apprenticeship committee. This may include the following:

- Aptitude test (General Aptitude Test Battery or GATB Form Test) results (usually administered by the local Employment Security Commission)
- Proof of educational background (candidate should have school transcripts sent to the committee)
- Letters of reference from past employers and friends
- Results of a physical examination
- Proof of age
- If the candidate is a veteran, a copy of Form DD214
- A record of technical training received that relates to the construction industry and/or a record of any pre-apprenticeship training
- High school diploma or General Equivalency Diploma (GED)

The apprentice must do the following:

- Wear proper safety equipment on the job
- Purchase and maintain tools of the trade as needed and required by the contractor
- Submit a monthly on-the-job training report to the committee
- Report to the committee if a change in employment status occurs
- Attend classroom-related instruction and adhere to all classroom regulations such as attendance requirements

5.0.0 ◆ RESPONSIBILITIES OF THE EMPLOYEE

In order to be successful, the professional must be able to use current trade materials, tools, and equipment to finish the task quickly and efficiently. A pipefitter must be adept at adjusting methods to meet each situation. The successful pipefitter must continuously train to remain knowledgeable about technical advancements and to gain the skills to use them. A professional never takes chances with regard to personal safety or the safety of others.

5.1.0 Professionalism

The word professionalism is a broad term that describes the desired overall behavior and attitude expected in the workplace. Professionalism is too often absent from the construction site and the various trades. Most people would argue that professionalism must start at the top in order to be successful. It is true that management support of professionalism is important to its success in the workplace, but it is more important that individuals recognize their own responsibility for professionalism.

Professionalism includes honesty, productivity, safety, civility, cooperation, teamwork, clear and concise communication, being on time and prepared for work, and regard for one's impact on one's co-workers. It can be demonstrated in a variety of ways every minute you are in the workplace. Most important is that you do not tolerate the unprofessional behavior of co-workers. This is not to say that you shun the unprofessional worker; instead, you work to demonstrate the benefits of professional behavior.

Professionalism is both a benefit to the employer and the employee. It is a personal responsibility. Our industry is what each individual chooses to make of it; choose professionalism and the industry image will follow.

5.2.0 Honesty

Honesty and personal integrity are important traits of the successful professional. Professionals pride themselves in performing a job well, and in being punctual and dependable. Each job is completed in a professional way, never by cutting corners or reducing materials. A valued professional

Describe the preliminary requirements and information that a prospective apprentice must submit to the committee.

Explain that professionalism is a broad term that describes the desired overall behavior and attitude expected in the workplace.

Ask a trainee to list qualities that demonstrate professionalism. Discuss individual responsibility for professionalism.

MODULE 08101-06 ◆ ORIENTATION TO THE TRADE 1.7

Explain that honesty, personal integrity, and loyalty are important traits of a successful professional. Discuss the need for a professional to be willing to learn, take responsibility, and cooperate with others.

Provide a copy of an employee manual for the trainees to examine.

Discuss tardiness and absenteeism.

Discuss various real-life situations that demonstrate the qualities discussed in this section.

maintains work attitudes and ethics that protect property such as tools and materials belonging to employers, customers, and other trades from damage or theft at the shop or job site.

Honesty and success go hand-in-hand. It is not simply a choice between good and bad, but a choice between success and failure. Dishonesty will always catch up with you. Whether you are stealing materials, tools, or equipment from the job site or simply lying about your work, it will not take long for your employer to find out. Of course, you can always go and find another employer, but this option will ultimately run out on you.

If you plan to be successful and enjoy continuous employment, consistency of earnings, and being sought after as opposed to seeking employment, then start out with the basic understanding of honesty in the workplace and you will reap the benefits.

Honesty means more, however, than just not taking things that do not belong to you. It means giving a fair day's work for a fair day's pay. It means carrying out your side of a bargain. It means that your words convey true meanings and actual happenings. Our thoughts as well as our actions should be honest. Employers place a high value on an employee who is strictly honest.

5.3.0 Loyalty

Employees expect employers to look out for their interests, to provide them with steady employment, and to promote them to better jobs as openings occur. Employers feel that they, too, have a right to expect their employees to be loyal to them—to keep their interests in mind, to speak well of them to others, to keep any minor troubles strictly within the plant or office, and to keep absolutely confidential all matters that pertain to the business. Both employers and employees should keep in mind that loyalty is not something to be demanded; rather, it is something to be earned.

5.4.0 Willingness to Learn

Every office and plant has its own way of doing things. Employers expect their workers to be willing to learn these ways. Adapting to change and being willing to learn new methods and procedures as quickly as possible are key. Sometimes, a change in safety regulations or the purchase of new equipment makes it necessary for even experienced employees to learn new methods and operations. Employees often resent having to accept improvements because of the retraining that is involved. However, employers will no doubt think they have a right to expect employees to put forth the necessary effort. Methods must be kept up to date in order to meet competition and show a profit. It is this profit that enables the owner to continue in business and provide jobs for the employees.

5.5.0 Willingness to Take Responsibility

Most employers expect their employees to see what needs to be done, then go ahead and do it. It is very tiresome to have to ask again and again that a certain job be done. It is obvious that having been asked once, an employee should assume the responsibility from then on. Once the responsibility has been delegated, the employee should continue to perform the duties without further direction. Every employee has the responsibility for working safely.

5.6.0 Willingness to Cooperate

To cooperate means to work together. In our modern business world, cooperation is the key to getting things done. Learn to work as a member of a team with your employer, supervisor, and fellow workers in a common effort to get the work done efficiently, safely, and on time.

5.7.0 Rules and Regulations

People can work together well only if there is some understanding about what work is to be done, when and how it will be done, and who will do it. Rules and regulations are a necessity in any work situation and should be so considered by all employees.

5.8.0 Tardiness and Absenteeism

Tardiness means being late for work, and absenteeism means being off the job for one reason or another. Consistent tardiness and frequent absences are an indication of poor work habits, unprofessional conduct, and a lack of commitment.

We are all creatures of habit. What we do once we tend to do again unless the results are too unpleasant. The habit of always being late may have begun back in our school days when we found it hard to get up in the morning. This habit can get us into trouble at school, and it can go right on getting us into trouble when we are through with school and go to work.

Our work life is governed by the clock. We are required to be at work at a definite time. So is everyone else. Failure to get to work on time results in confusion, lost time, and resentment on the part of those who do come on time. In addition, it may lead to penalties, including dismissal. Although it may be true that a few minutes out of a day are not very important, you must remember that a principle is involved. It is your obligation to be at work at

Instructor's Notes:

the time indicated. We agree to the terms of work when we accept the job. Perhaps it will help us to see things more clearly if we try to look at the matter from the point of view of the boss. Supervisors cannot keep track of people if they come in any time they please. It is not fair to others to ignore tardiness. Failure to be on time may hold up the work of fellow workers. Better planning of your morning routine will often keep you from being delayed and so prevent a breathless, late arrival. In fact, arriving a little early indicates your interest and enthusiasm for your work, which is appreciated by employers. The habit of being late is another one of those things that stand in the way of promotion.

It is sometimes necessary to take time off from work. No one should be expected to work when sick or when there is serious trouble at home. However, it is possible to get into the habit of letting unimportant and unnecessary matters keep us from the job. This results in lost production and hardship on those who try to carry on the work with less help. Again, there is a principle involved. The person who hires us has a right to expect us to be on the job unless there is some very good reason for staying away. Certainly, we should not let some trivial reason keep us home. We should not stay up nights until we are too tired to go to work the next day. If we are ill, we should use the time at home to do all we can to recover quickly. This, after all, is no more than most of us would expect of a person we had hired to work for us, and on whom we depended to do a certain job.

If it is necessary to stay home, then at least phone the office early in the morning so that the boss can find another worker for the day. Time and again, employees have remained home without sending any word to the employer. This is the worst possible way to handle the matter. It leaves those at work uncertain about what to expect. They have no way of knowing whether you have merely been held up and will be in later, or whether immediate steps should be taken to assign your work to someone else. Courtesy alone demands that you let the boss know if you cannot come to work.

The most frequent causes of absenteeism are illness, death in the family, accidents, personal business, and dissatisfaction with the job. Here we see that some of the causes are legitimate and unavoidable, while others can be controlled. One can usually plan to carry on most personal business affairs after working hours. Frequent absences will reflect unfavorably on a worker when promotions are being considered.

Employers sometimes resort to docking pay, demotion, and even dismissal in an effort to control tardiness and absenteeism. No employer likes to impose restrictions of this kind. However, in fairness to those workers who do come on time and who do not stay away from the job, an employer is sometimes forced to discipline those who will not follow the rules.

6.0.0 ◆ HUMAN RELATIONS

Most people underestimate the importance of working well with others. There is a tendency to pass off human relations as nothing more than common sense. What exactly is involved in human relations? One response would be to say that part of human relations is being friendly, pleasant, courteous, cooperative, adaptable, and sociable.

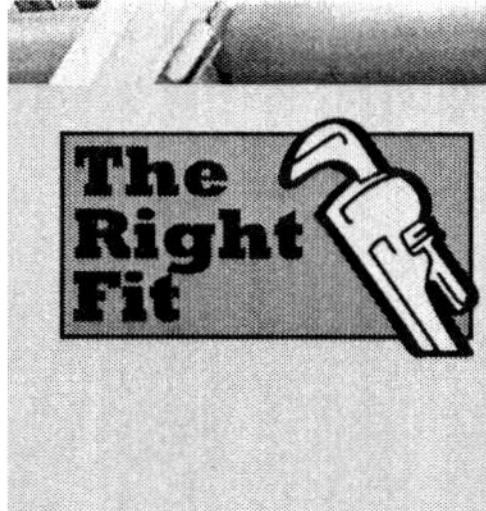

Ethical Principles for Members of the Construction Trades

Honesty: Be honest and truthful in all dealings. Conduct business according to the highest professional standards. Faithfully fulfill all contracts and commitments. Do not deliberately mislead or deceive others.
Integrity: Demonstrate personal integrity and the courage of your convictions by doing what is right even where there is pressure to do otherwise. Do not sacrifice your principles because it seems easier.
Loyalty: Be worthy of trust. Demonstrate fidelity and loyalty to companies, employers and sponsors, co-workers, and trade institutions and organizations.
Fairness: Be fair and just in all dealings. Do not take undue advantage of another's mistakes or difficulties. Fair people are open-minded and committed to justice, equal treatment of individuals, and tolerance for and acceptance of diversity.
Respect for others: Be courteous and treat all people with equal respect and dignity.
Obedience: Abide by laws, rules, and regulations relating to all personal and business activities.
Commitment to excellence: Pursue excellence in performing your duties, be well informed and prepared, and constantly try to increase your proficiency by gaining new skills and knowledge.
Leadership: By your own conduct, seek to be a positive role model for others.

Explain the importance of notifying your boss if you will not be able to work that day. Discuss causes of absenteeism and methods employers may use to control it.

Have trainees review Sections 6.0.0–7.0.0.

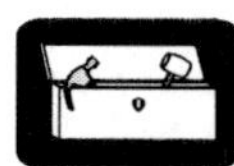

Ensure that you have everything required for teaching this session.

Explain that human relations is more than getting people to like you, it is also the ability to handle difficult situations.

Discuss how human relations and productivity are related. Ask trainees to give examples of how good working relations can improve productivity and poor working relations can impair productivity.

Discuss the importance of a positive attitude on safety and productivity. Ask trainees to give examples of how a negative attitude will decrease your chances for being successful on the job.

Discuss ways of maintaining a positive attitude.

Have trainees work in pairs and role play different situations that can cause conflict in the workplace. Have a group discussion afterwards on which methods made the situation better or worse.

6.1.0 Making Human Relations Work

As important as the previously noted characteristics are for personal success, they are not enough. Human relations is much more than just getting people to like you. It is also knowing how to handle difficult situations as they arise.

Human relations is knowing how to work with supervisors who are often demanding and sometimes unfair. It involves understanding your own personality traits and those of the people you work with. Building sound working relationships in various situations is important. If working relationships have deteriorated for one reason or another, restoring them is essential. Human relations is learning how to handle frustrations without hurting others.

6.2.0 Human Relations and Productivity

Effective human relations is directly related to productivity. Productivity is the key to business success. Every employee is expected to produce at a certain level. Employers quickly lose interest in an employee who has a great attitude but is able to produce very little. There are work schedules to be met and jobs that must be completed.

All employees, both new and experienced, are measured by the amount of quality work they can safely turn out. The employer expects every employee to do his or her share of the workload.

However, doing one's share in itself is not enough. If you are to be productive, you must do your share (or more than your share) without antagonizing your fellow workers. You must perform your duties in a manner that encourages others to follow your example. It makes little difference how ambitious you are or how capably you perform. You cannot become the kind of employee you want to be or the type of worker management wants you to be without learning how to work with your peers.

Employees must do everything they can to build strong, professional working relationships with fellow employees, supervisors, and clients.

6.3.0 Attitude

A positive attitude is essential to a successful career. First, being positive means being energetic, highly motivated, attentive, and alert. A positive attitude is essential to safety on the job. Second, a positive employee contributes to the productivity of others. Both negative and positive attitudes are transmitted to others on the job. A persistent negative attitude can spoil the positive attitudes of others. It is very difficult to maintain a high level of productivity while working next to a person with a negative attitude. Third, people favor a person who is positive. Being positive makes a person's job more interesting and exciting. Fourth, the kind of attitude transmitted to management has a great deal to do with an employee's future success in the company. Supervisors can determine a subordinate's attitude by their approach to the job, reactions to directives, and the way they handle problems.

6.4.0 Maintaining a Positive Attitude

A positive attitude is far more than a smile, which is only one example of an inner positive attitude. As a matter of fact, some people transmit a positive attitude even though they seldom smile. They do this by the way they treat others, the way they look at their responsibilities, and the approach they take when faced with problems. Here are a few suggestions that will help you to maintain a positive attitude:

- Remember that your attitude follows you wherever you go. If someone makes a greater effort to be a more positive person in their social and personal lives, it will automatically help them on the job. The reverse is also true. One effort will complement the other.
- Negative comments are seldom welcomed by fellow workers on the job. Neither are they welcome on the social scene. The solution: talk about positive things and be complimentary. Constant complainers do not build healthy and fulfilling relationships.

The Right Fit

Tips for a Positive Attitude

Here is a short checklist of things that you should keep in mind to help you develop and maintain a positive attitude:

- Remember that your attitude follows you wherever you go.
- Helpful suggestions and compliments are much more effective than negative ones.
- Look for the positive characteristics of your teammates and supervisors.

Instructor's Notes:

1.10 PIPEFITTING ◆ LEVEL ONE

- Look for the good things in people on the job, especially your supervisor. Nobody is perfect, and almost everyone has a few worthwhile qualities. If you dwell on people's good features, it will be easier to work with them.
- Look for the good things where you work. What are the factors that make it a good place to work? Is it the hours, the physical environment, the people, the actual work being done? Or is it the atmosphere? Keep in mind that you cannot expect to like everything. No work assignment is perfect, but if you concentrate on the good things, the negative factors will seem less important and bothersome.
- Look for the good things in the company. Just as there are no perfect assignments, there are no perfect companies. Nevertheless, almost all organizations have good features. Is the company progressive? What about promotional opportunities? Are there chances for self-improvement? What about the wage and benefit package? Is there a good training program? You cannot expect to have everything you would like, but there should be enough to keep you positive. In fact, if you decide to stick with a company for a long period of time, it is wise to look at the good features and think about them. If you think positively, you will act the same way.
- You may not be able to change the negative attitude of another employee, but you can protect your own attitude from becoming negative.

7.0.0 ◆ EMPLOYER AND EMPLOYEE SAFETY OBLIGATIONS

An obligation is like a promise or a contract. In exchange for the benefits of your employment and your own well-being, you agree to work safely. In other words, you are obligated to work safely. You are also obligated to make sure anyone you happen to supervise or work with is working safely. Your employer is obligated to maintain a safe workplace for all employees. Safety is everyone's responsibility.

Some employers will have safety committees. If you work for such an employer, you are then obligated to that committee to maintain a safe working environment. This means two things:

- Follow the safety committee's rules for proper working procedures and practices.
- Report any unsafe equipment and conditions directly to the committee or your supervisor.

On the job, if you see something that is not safe, report it! Do not ignore it. It will not correct itself. You have an obligation to report it.

In the long run, even if you do not think an unsafe condition affects you, it does. Do not mess around; report what is not safe. Do not think your employer will be angry because your productivity suffers while the condition is corrected. On the contrary, your employer will be more likely to criticize you for not reporting a problem. Your employer knows that the short time lost in making conditions safe again is nothing compared with shutting down the whole job because of a major disaster. If that happens, you are out of work anyway. So do not ignore an unsafe condition. In fact, the **Occupational Safety and Health Administration (OSHA)** regulations require you to report hazardous conditions.

This applies to every part of the construction industry. Whether you work for a large contractor or a small subcontractor, you are obligated to report unsafe conditions. The easiest way to do this is to tell your supervisor. If that person ignores the unsafe condition, report it to the next highest supervisor. If it is the owner who is being unsafe, let that person know your concerns. If nothing is done about it, report it to OSHA. If you are worried about your job being on the line, think about it in terms of your life, or someone else's, being at risk.

The U.S. Congress passed the *Occupational Safety and Health Act* in 1970. This act also created OSHA. It is part of the U.S. Department of Labor. The job of OSHA is to set occupational safety and health standards for all places of employment, enforce these standards, ensure that employers provide and maintain a safe workplace for all employees, and provide research and educational programs to support safe working practices.

OSHA was adopted with the stated purpose "to assure as far as possible every working man and woman in the nation safe and healthful working conditions and to preserve our human resources."

OSHA requires each employer to provide a safe and hazard-free working environment. OSHA also requires that employees comply with OSHA

Ask a trainee to define obligation. Discuss employer and employee obligations to maintain a safe workplace. Discuss the costs of accidents.

Explain that OSHA regulations require employees on all job sites to report unsafe conditions. Explain that OSHA is a part of the Department of Labor and responsible for ensuring a safe and healthful working environment for all workers.

MODULE 08101-06 ◆ ORIENTATION TO THE TRADE 1.11

Describe OSHA standards for the construction industry and explain how they are enforced by the states.

Discuss the general requirements OSHA places on employers in the construction industry.

Provide a copy of the *OSHA Safety and Health Standards for the Construction Industry* for the trainees to examine.

rules and regulations that relate to their conduct on the job. To gain compliance, OSHA inspect job sites, impose fines for violations, and even stop any more work from proceeding until the job site is safe.

According to OSHA standards, you are entitled to on-the-job safety training. As a new employee, you are entitled to the following:

- Being shown how to do your job safely
- Being provided with the required personal protective equipment
- Being warned about specific hazards
- Being supervised for safety while performing the work

The enforcement of this act of Congress is provided by the federal and state safety inspectors, who have the legal authority to make employers pay fines for safety violations. The law allows states to have their own safety regulations and agencies to enforce them, but they must first be approved by the U.S. Secretary of Labor. For states that do not develop such regulations and agencies, federal OSHA standards must be obeyed.

These standards are listed in *OSHA Safety and Health Standards for the Construction Industry* (29 CFR, Part 1926), sometimes called *OSHA Standards 1926*. Other safety standards that apply to construction are published in *OSHA Safety and Health Standards for General Industry* (29 CFR, Parts 1900 to 1910).

The most important general requirements that OSHA places on employers in the construction industry are as follows:

- The employer must perform frequent and regular job site inspections of equipment.
- The employer must instruct all employees to recognize and avoid unsafe conditions, and to know the regulations that pertain to the job so they may control or eliminate any hazards.
- No one may use any tools, equipment, machines, or materials that do not comply with *OSHA Standards 1926*.
- The employer must ensure that only qualified individuals operate tools, equipment, and machines.

Instructor's Notes:

Review Questions

1. Which of the following is a task *not* generally performed by pipefitters?
   a. Installing residential plumbing systems
   b. Inspecting piping systems
   c. Installing valves
   d. Planning piping systems

2. Which of the following is a correct statement about pipefitters?
   a. Pipefitters must learn how to weld because welding is an important part of their work.
   b. Pipefitters only work with carbon steel pipe.
   c. Pipefitters sometimes work in food processing plants.
   d. Pipefitting work is limited to the use of hand tools.

Figure 1

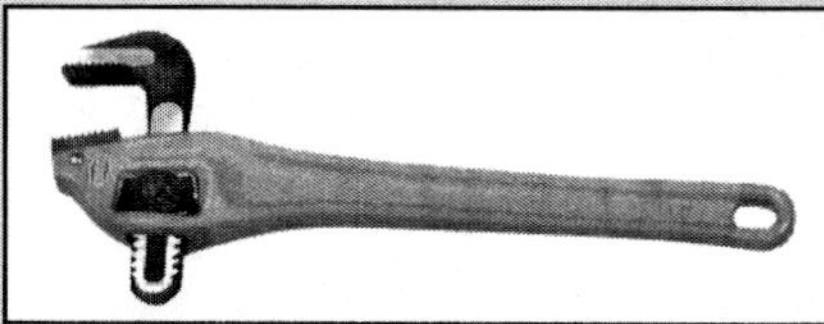

101RQ01.EPS

3. The tool shown in *Figure 1* is a(n) _____ wrench.
   a. straight pipe
   b. chain
   c. strap
   d. offset

Figure 2

101RQ02.EPS

4. The tool shown in *Figure 2* is used to _____.
   a. bend pipe
   b. measure pipe
   c. operate a valve
   d. thread pipe

5. Journey-level pipefitters often earn as much as, or more than, college-educated office workers.
   a. True
   b. False

6. Minimum standards for apprentice training programs are established by _____.
   a. OSHA
   b. ATELS
   c. NCCER
   d. your employer

7. If one of your co-workers does *not* act in a professional manner, the best approach is to _____.
   a. report the person to your supervisor
   b. advise other workers to shun the person
   c. demonstrate the benefits of professional behavior
   d. refuse to cooperate with the person

8. It is okay to be a little late for work as long as you make up the time.
   a. True
   b. False

9. The primary mission of OSHA is to _____.
   a. inspect job sites for safety violations
   b. fine companies that violate safety regulations
   c. distribute safety equipment to workers
   d. ensure that employers maintain a safe workplace

10. If you see a safety violation at your job site, you should _____.
    a. inform your supervisor
    b. ignore it unless it affects you directly
    c. report it to an OSHA inspector
    d. walk off the job until it is corrected

Have the trainees complete the Review Questions, and go over the answers prior to administering the Module Examination.

Summarize the major concepts presented in the module.

Summary

Pipefitters have many opportunities for career advancement and personal growth. If they choose to do so, they can advance into supervisory and management positions within a company.

Construction work is a team effort in which every person must do their job on time and in a professional manner. This means that all members of a construction crew must learn to work together and must develop interpersonal skills and good work habits.

By its very nature, construction work is potentially hazardous to the pipefitters and others working. This means that everyone must be particularly safety conscious when on a job site.

Notes

Instructor's Notes:

Trade Terms Quiz

Fill in the blank with the correct trade term that you learned from your study of this module.

1. You can solder copper pipe using the ________ method.
2. Nation-wide training standards are set by ________.
3. Learning new skills while working is called ________.
4. A cut made at an angle is called a(n) ________.
5. Job-site safety and health rules are set by the ________.

Trade Terms

Apprenticeship Training, Employer and Labor Services (ATELS)
Bevel
On-the-job Training (OJT)
Occupational Safety and Health Administration (OSHA)
Sweat

Have the trainees complete the Trade Terms Quiz, and go over the answers prior to administering the Module Examination.

Administer the Module Examination. Record the results on Craft Training Report Form 200, and submit the results to the Training Program Sponsor.

MODULE 08101-06 ◆ ORIENTATION TO THE TRADE 1.15

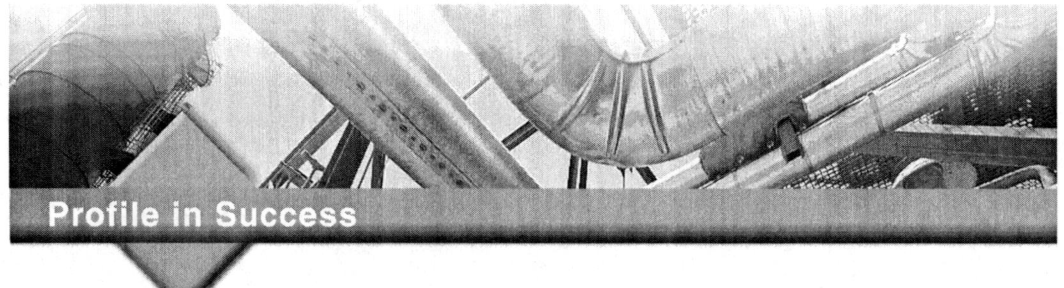

Edward LePage

Cianbro Corporation "The Constructors"
Project Superintendent and Northern New England Mechanical Training Coordinator

Ed LePage has always enjoyed new challenges. When new opportunities arise, he learns as much as he can and works hard at the new craft. He has mastered several trades and is now a subject matter expert for three NCCER committees: Pipefitter, Millwright, and Industrial Maintenance. He is also the Instrumentation Project Manager for Associate Builders and Contractors (ABC) National Craft Championships.

How did you choose a career in the pipefitting field?
Our company was mostly a civil, structural construction company that decided to venture into the heavy mechanical markets of New England. We needed mechanical craftsmen and there were not enough in our area to staff our projects. I was an ironworker-rigger and was asked to work with the pipers to install large bore piping, because most of the fitters were not riggers.

While working with the pipers, I found myself with spare time. I started installing flanges and elbows while the fitter was away from the fabrication areas. One day the piping superintendent came up to me with a toolbox full of pipefitter's tools and said that I was to start on Monday as a pipefitter. When I told him that I wasn't a fitter, he said that he was betting that I would become a top fitter and suggested that I buy a fitting book and start reading. On the following Monday I was on my way to becoming one of our top pipefitters.

What types of training have you been through?
I have gone through NCCER's Train the Trainer class and I am an instructor for Core, Millwright, Pipefitter, Welding, and Instrumentation training. I have attended Ludeca, Inc.'s training for laser alignment on Optalign® Plus and Rotalign® Pro and I train and certify all of our laser alignment people. I have attended Beloit's Engineering Maintenance School. I have also attended Falk Corporation's Bearings and Gearbox School.

What kinds of work have you done in your career?
I started my career in construction as a laborer after attending the University of Maine for two years in Civil Engineering. I worked my way up to ironworker, rebar rod buster, and rigger. I had the opportunity to enter the pipefitter craft and was also allowed to enter the millwright craft. I worked my way up from foreman of many crafts to general foreman, assistant mechanical superintendent, mechanical superintendent, and project superintendent.

I have been the construction manager on several extremely large projects. I have worked on bridges, recycled fiber projects, in nuclear power plants, biomass boilers, paper machine installation and rebuilds, heat recovery steam generator projects, water turbine and generator installations and rebuilds, steam turbine and generator installation and rebuilds, chemical plant installation and rebuilds, two semi-submersible oil drilling rigs, and various other mechanical projects.

Instructor's Notes:

Tell us about your present job.

I work for some of our customers as a consultant helping them to estimate, budget, and schedule their larger mechanical projects. I oversee most of our large mechanical shutdowns that we do in the pulp and paper mills in northern New England. I develop, coordinate, and teach our training programs to our pipefitting, millwright, and welding trainees.

What factors have contributed most to your success?

I had the opportunity to expand my career by working for this company. The people that I have worked for believed in me and allowed me the opportunity to challenge myself for the last 35 years. And a lot of hard work always helps.

What advice would you give to those new to the pipefitting field?

Do the very best that you can at whatever you try. At the end of the day if you can say I did the very best that I could do, you should feel good about yourself.

Ask questions if you don't know what to do or how to do something. The only dumb question is the one that you didn't ask.

MODULE 08101-06 ◆ ORIENTATION TO THE TRADE 1.17

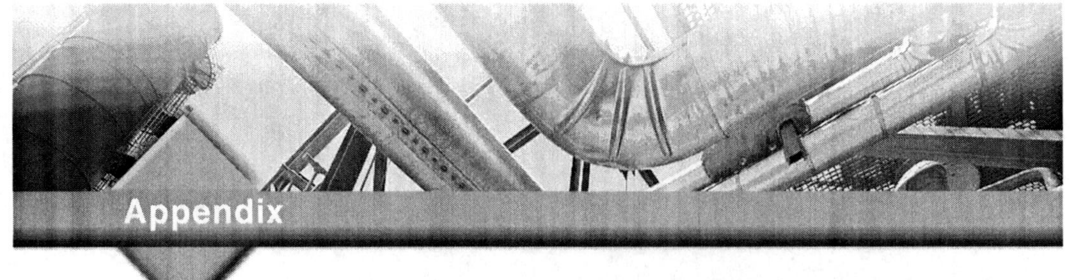

Samples of NCCER Apprentice Training Recognition

October 31, 2005

John Doe
NCCER
3600 NW 43rd St Bldg G
Gainesville, FL 32606

Dear John,

On behalf of the National Center for Construction Education and Research, I congratulate you for successfully completing the NCCER's standardized craft training program.

As the NCCER's most recent graduate, you are a valuable member of today's skilled construction and maintenance workforce. The skills that you have acquired through the NCCER craft training programs will enable you to perform quality work on construction and maintenance projects, promote the image of these industries and enhance your long-term career opportunities.

We encourage you to continue your education as you advance in your construction career. Please do not hesitate to contact us for information regarding our Management Education and Safety Programs or if we can be of any assistance to you.

Enclosed please find your certificate, transcript and wallet card. Once again, congratulations on your accomplishments and best wishes for a successful career in the construction and maintenance industries.

Sincerely,

Donald E. Whyte
Donald E. Whyte
President, NCCER

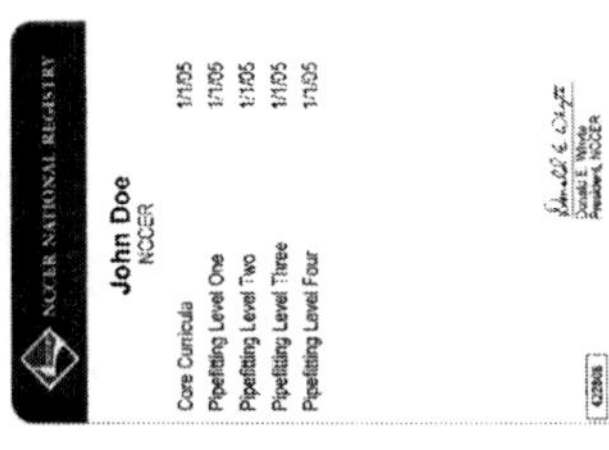

NATIONAL CENTER FOR CONSTRUCTION EDUCATION AND RESEARCH
P.O. Box 141104 ■ Gainesville, Florida ■ 32614-1104 ■ PH: 352.334.0911 ■ FX: 352.334.0932 ■ www.nccer.org

101A01.EPS

Instructor's Notes:

1.18 PIPEFITTING ◆ LEVEL ONE

National Center for Construction
Education and Research
This is to certify that
John Doe
has fulfilled the requirements for
Pipefitting Level One
in the NCCER's standardized training curriculum
this First day of January, 2005
Donald E. Whyte
President

101A02.EPS

MODULE 08101-06 ◆ ORIENTATION TO THE TRADE 1.19

NATIONAL CENTER FOR CONSTRUCTION EDUCATION AND RESEARCH
P.O. Box 141104 ■ Gainesville, Florida 32614-1104
PH: 352.334.0911 ■ FX: 352.334.0932 ■ www.nccer.org
Affiliated with the University of Florida

10/31/05
Page: 1

Official Transcript

John Doe
NCCER
3600 NW 43rd St Bldg G
Gainesville, FL 32606
R#: 000110490

Current Employer:

Course / Description	Instructor	Training Location	Date Compl.
00101-04 Basic Safety	Don Whyte	NCCER	1/1/05
00102-04 Introduction to Construction Math	Don Whyte	NCCER	1/1/05
00103-04 Introduction to Hand Tools	Don Whyte	NCCER	1/1/05
00104-04 Introduction to Power Tools	Don Whyte	NCCER	1/1/05
00105-04 Introduction to Blueprints	Don Whyte	NCCER	1/1/05
00106-04 Basic Rigging	Don Whyte	NCCER	1/1/05
00107-04 Basic Communication Skills	Don Whyte	NCCER	1/1/05
00108-04 Basic Employability Skills	Don Whyte	NCCER	1/1/05
08101-04 Orientation to the Trade	Don Whyte	NCCER	1/1/05
08102-04 Pipefitting Hand Tools	Don Whyte	NCCER	1/1/05
08103-04 Pipefitting Power Tools	Don Whyte	NCCER	1/1/05
08104-04 Oxyfuel Cutting	Don Whyte	NCCER	1/1/05
08105-04 Ladders and Scaffolds	Don Whyte	NCCER	1/1/05
08106-04 Motorized Equipment	Don Whyte	NCCER	1/1/05

Donald E. Whyte

101A03.EPS

Instructor's Notes:

Trade Terms Introduced in This Module

Apprenticeship Training, Employer and Labor Services (ATELS): The U.S. Department of Labor office that sets the minimum standards for training programs across the country.

Bevel: A cut made at an angle.

On-the-job training (OJT): Job-related learning acquired while working.

Occupational Safety and Health Administration (OSHA): The federal government agency established to ensure a safe and healthy environment in the workplace.

Sweat: A method of joining pipe in which solder is applied to the joint and heated until the solder flows into the joint.

MODULE 08101-06 ◆ ORIENTATION TO THE TRADE 1.21

Additional Resources

This module is intended to be a thorough resource for task training. The following reference work is suggested for further study. This is optional material for continued education rather than for task training.

The Pipefitters Blue Book, Latest Edition. W.V. Graves. Webster, TX: Graves Publishing Company.

The Pipefitters Handbook, 3rd Edition. Forrest R. Lindsey. New York, NY: Industrial Press.

Figure Credits

The Industrial Company, 101F01A

Ivey Mechanical Company, 101F01B

Topaz Publications, Inc., 101F02, 101F03, 101RQ01

H&M Pipe Beveling, 101F04, 101E01

Ridge Tool Company (RIDGID®), 101F05, 101RQ02, 101E02

Zachary Construction Corporation, Title Page

Instructor's Notes:

The following are suggested activities or instructional methods to help you teach the material in this AIG.

General

When you call on someone to answer a question, the rest of the class relaxes or even tunes out because they expect that the question and answer will take place only between you and the trainee you called on. Instead, use this technique to involve more the trainees in answering questions and to keep them on their toes.

1. Ask the trainees to define a term or explain a concept.
2. After one trainee has answered, ask a trainee seated nearby if the answer is right. Then ask whether a trainee in the back of the room agrees.
3. Ask the trainees to explain why they think an answer is right or wrong.
4. Use the session to clear up incorrect ideas and encourage the trainees to learn from their mistakes.

Section 3.0.0

Careers in Pipefitting

Pipefitters have opportunities within several industries. Trainees will need appropriate personal protective equipment, pencils, and paper. Arrange for a pipefitter to give a presentation on careers in pipefitting. Some areas to include in the presentation are travel, job opportunities, and challenging assignments. Allow 40 to 50 minutes for this exercise.

1. Introduce the speaker who will give a presentation on job opportunities in pipefitting. Ask the presenter to speak about the proper attitude needed to become a good pipefitter.
2. Have the trainees take notes and write down questions during the presentation.
3. Ask the presenter to spend some time after the presentation to answer any questions the trainees may have.

MODULE 08101-06 — ANSWERS TO REVIEW QUESTIONS

Answer		Section
1.	a	1.0.0
2.	c	1.0.0
3.	d	2.1.0; Figure 3
4.	d	2.1.0; Figure 5
5.	a	3.0.0
6.	b	4.0.0
7.	c	5.1.0
8.	b	5.8.0
9.	d	7.0.0
10.	a	7.0.0

Answers to Trade Terms Quiz

1. Sweat
2. Apprenticeship Training, Employer and Labor Services (ATELS)
3. On-the-job training (OJT)
4. Bevel
5. Occupational Safety and Health Administration (OSHA)

CONTREN® LEARNING SERIES — USER UPDATE

NCCER makes every effort to keep these textbooks up-to-date and free of technical errors. We appreciate your help in this process. If you have an idea for improving this textbook, or if you find an error, a typographical mistake, or an inaccuracy in NCCER's Contren® textbooks, please write us, using this form or a photocopy. Be sure to include the exact module number, page number, a detailed description, and the correction, if applicable. Your input will be brought to the attention of the Technical Review Committee. Thank you for your assistance.

Instructors – If you found that additional materials were necessary in order to teach this module effectively, please let us know so that we may include them in the Equipment/Materials list in the Annotated Instructor's Guide.

Write: Product Development and Revision
National Center for Construction Education and Research
P.O. Box 141104, Gainesville, FL 32614-1104

Fax: 352-334-0932

E-mail: curriculum@nccer.org

Craft ______________________ Module Name ______________________

Copyright Date __________ Module Number __________ Page Number(s) __________

Description

____

____

____

____

(Optional) Correction

____

____

____

(Optional) Your Name and Address

____

____

____

Module 08102-06

Pipefitting Hand Tools

NCCER STANDARDIZED CRAFT TRAINING PROGRAM

The National Center for Construction Education and Research (NCCER) provides a standardized national program of accredited craft training. Key features of the program include instructor certification, competency-based training, and performance testing. The program provides trainees, instructors, and companies with a standard form of recognition through the National Registry. The program is described in full in the *Guidelines for Accreditation*, published by NCCER. For more information on standardized craft training, contact NCCER by writing to P.O. Box 141104, Gainesville, FL 32614-1104; calling 352-334-0911; or e-mailing info@nccer.org. More information is available at www.nccer.org.

HOW TO USE THIS ANNOTATED INSTRUCTOR'S GUIDE

Each page presents two sections of information. The larger section displays each page exactly as it appears in the Trainee Module. The narrow column ties suggested trainee and instructor actions to each page and provides icons (detailed below) to call your attention to material, safety, audiovisual, or testing requirements. The bottom of each page includes space for your notes.

The **Audiovisual** icon indicates an appropriate time to show a transparency or other audiovisual aid.

The **Classroom** icon prompts you to define a term, stress a point, ask trainees to explain a concept, or give examples.

The **Demonstration** icon directs you to show trainees how to perform tasks.

The **Examination** icon tells you to administer the written module examination.

The **Homework** icon is placed where you may wish to assign reading for the next class, assign a project, or advise trainees to prepare for an examination.

The **Laboratory** icon is used when trainees are to practice performing tasks.

The **Materials** icon is a reminder for you to gather materials needed for classes, laboratories, and testing.

The **Performance Testing** icon tells you to administer a performance test or a portion thereof.

The **Safety** icon is used to emphasize safety issues. It is often keyed to *Caution* and *Warning!* statements in the Trainee Module.

The **Teaching Tip** icon indicated additional guidance is available, such as how to conduct an exercise, get the most educational value from a field trip, or encourage class participation. Teaching Tips may expand on a feature (*Think About It, Did You Know?*) or provide *Quick Quizzes* or similar exercises. You will be referred to the Teaching Tips section at the back of the module if there is additional material.

The **Combination** icon indicates that the laboratory listed corresponds with a performance task. If desired, you can note the proficiency of the trainees during the laboratory, and use it to satisfy performance testing requirements.

PREPARATION

Before teaching this module, you should review the Objectives, Performance Tasks, Materials and Equipment List, and Module Outline. Be sure to allow ample time to prepare your own training or lesson plan and gather all required materials and equipment.

MODULE OVERVIEW

This module covers general hand tool safety and procedures for identifying, selecting, inspecting, using, and caring for pipe vises and stands, pipe wrenches, levels, pipe fabrication tools, and pipe bending tools.

PREREQUISITES

Prior to training with this module, it is recommended that the trainee shall have successfully completed *Core Curriculum;* and *Pipefitting Level One*, Module 08101-06.

OBJECTIVES

Upon completion of this module, the trainee will be able to do the following:

1. Describe the safety requirements that apply to the use of pipefitter hand tools.
2. Explain how to properly care for selected pipefitter hand tools.
3. Demonstrate how to safely and properly use selected pipefitter hand tools.
4. Identify tools and state their uses.
5. Use selected hand tools.

PERFORMANCE TASKS

Under the supervision of the instructor, the trainee should be able to do the following:

1. Identify various pipefitting hand tools.
2. Secure a section of pipe in a vise and pipe stand.
3. Properly use:
   - Straight pipe wrenches
   - Offset pipe wrenches
   - Chain wrenches
   - Strap wrenches
4. Properly use:
   - Laser level
   - Torpedo and larger levels
   - Tubing water level
   - Center finder
5. Check square and level:
   - Turn tongue 180 degrees from where it was
   - Flip level to ensure it is level

MATERIALS AND EQUIPMENT LIST

Overhead projector and screen
Transparencies
Blank acetate sheets
Transparency pens
Whiteboard/chalkboard
Markers/chalk
Pencils and scratch paper

Appropriate personal protective equipment
Assorted diameters of pipe
Assorted diameters of tubing at various lengths
Conduit
Chain vises
Yoke vises
Strap vises

Various jacks, stands, rollers, and supports
Straight pipe wrenches
Offset pipe wrenches
Compound leverage wrenches
Chain wrenches
Pipe tongs
Strap wrenches
Open-end wrenches
Adjustable wrenches
Framing levels
Torpedo levels
Laser levels
Tubing water levels
Framing squares
Pipefitter's squares
Combination tri squares
Center finders
Straight butt welding clamps
Flange welding clamps
T-joint welding clamps
Elbow welding clamps
Shop-made aligning dogs
Hi-Lo gauges
Wraparounds
Drift pins
Two-hole pins
Flange spreaders
Hacksaws
Hacksaw blades
Soil pipe cutters
Tube and pipe cutters
Manual pipe reamers
Hand pipe and bolt threaders
Die heads
Thread gauges
Pipe extractors
Pipe taps
Spring tube benders
Lever compression tube benders
Manual benders
Hammer type flaring tools
Screw-in type flaring tools
Module Examinations*
Performance Profile Sheets*

*Located in the Test Booklet.

SAFETY CONSIDERATIONS

Ensure that the trainees are equipped with appropriate personal protective equipment and know how to use it properly. This module requires trainees to use hand tools. Emphasize basic hand tool safety.

ADDITIONAL RESOURCES

This module is intended to present thorough resources for task training. The following reference work is suggested for both instructors and motivated trainees interested in further study. This is optional material for continued education rather than for task training.

Tools and Their Uses, Latest Edition. Naval Education and Training Program Development Center. Washington, DC: U.S. Government Printing Office.

TEACHING TIME FOR THIS MODULE

An outline for use in developing your lesson plan is presented below. Note that each Roman numeral in the outline equates to one session of instruction. Each session has a suggested time period of 2½ hours. This includes 10 minutes at the beginning of each session for administrative tasks and one 10-minute break during the session. Approximately 20 hours are suggested to cover *Pipefitting Hand Tools*. You will need to adjust the time required for hands-on activity and testing based on your class size and resources. Because laboratories often correspond to Performance Tasks, the proficiency of the trainees may be noted during these exercises for Performance Testing purposes.

Topic	Planned Time
Session I. Introduction, Safety, Vises, and Stands	
A. Introduction	____________
B. Hand Tool Safety	____________
C. Vises and Stands	____________
D. Laboratory – Trainees practice securing a section of pipe in a stand. This laboratory corresponds to Performance Task 2.	____________
Sessions II and III. Pipe Wrenches and Levels	
A. Wrenches	____________
B. Laboratory – Trainees practice using various types of wrenches. This laboratory corresponds to Performance Task 3.	____________
C. Levels	____________
D. Laboratory – Trainees practice using various types of levels. This laboratory corresponds to Performance Task 4.	____________
Session IV. Pipe Fabrication Tools	
A. Squares and Center Finders	____________
B. Clamps	____________
C. Gauges and Wraparounds	____________
D. Pins	____________
E. Flange Spreaders	____________
Sessions V and VI. Pipe Cutting Tools	
A. Saws, Tube Cutters, and Pipe Cutters	____________
B. Reamers and Threaders	____________
C. Extractors and Taps	____________
Session VII. Benders and Flaring Tools	
A. Benders	____________
B. Flaring Tools	____________
C. Laboratory – Trainees practice fabricating pipe and checking square and level. This laboratory corresponds to Performance Task 5.	____________
Session VIII. Laboratory, Review, Module Examination, and Performance Testing	
A. Laboratory – Trainees identifying various pipefitting hand tools. This laboratory corresponds to Performance Task 1.	____________
B. Review	____________
C. Module Examination	____________
1. Trainees must score 70 percent or higher to receive recognition from NCCER.	
2. Record the testing results on Craft Training Report Form 200, and submit the results to the Training Program Sponsor.	
D. Performance Testing	____________
1. Trainees must perform each task to the satisfaction of the instructor to receive recognition from NCCER. If applicable, proficiency noted during laboratory exercises can be used to satisfy the Performance Testing requirements.	
2. Record the testing results on Craft Training Report Form 200, and submit the results to the Training Program Sponsor.	

Assign reading of Module 08102-06.

08102-06

Pipefitting Hand Tools

08102-06
Pipefitting Hand Tools

Topics to be presented in this module include:

Overview

Pipefitters use hand tools to grip, level, fabricate, cut, and bend pipe. Although hand tools do not present the immediate hazards of power tools, improper use of hand tools can cause serious injury. Use the correct tool for the job safely and properly. and you will be able to perform your job more efficiently.

Vises and stands hold the work to free the pipefitter to work with both hands. Various sizes and configurations of wrenches are available to grip and turn pipe. Always select the right size wrench for the job. Levels are used to ensure that a pipe is level or plumb. Tools used to fabricate pipe include squares, clamps, gauges, wraparounds, and pins. Specialized tools are used to cut, thread, bend, and flare pipe.

Pipefitters use their tools regularly and must learn not only how they are used but also how to take care of them. Properly maintained tools are more accurate and safer.

Instructor's Notes:

Objectives

When you have completed this module, you will be able to do the following:

1. Describe the safety requirements that apply to the use of pipefitter hand tools.
2. Explain how to properly care for selected pipefitter hand tools.
3. Demonstrate how to safely and properly use selected pipefitter hand tools.
4. Identify tools and state their uses.
5. Use selected hand tools.

Trade Terms

Burr	Outside diameter
Conduit	Pipe fitting
Die	Ratchet
Female threads	Tack welds
Flare	Thread gauge
Male threads	

Required Trainee Materials

1. Pencil and paper
2. Appropriate personal protective equipment

Prerequisites

Before you begin this module, it is recommended that you successfully complete *Core Curriculum*; and *Pipefitting Level One*, Module 08101-06.

This course map shows all of the modules in the first level of the *Pipefitting* curriculum. The suggested training order begins at the bottom and proceeds up. Skill levels increase as you advance on the course map. The local Training Program Sponsor may adjust the training order.

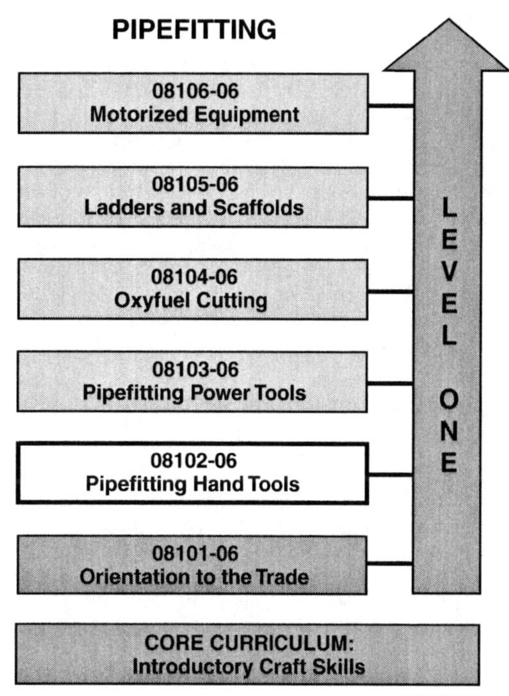

102CMAP.EPS

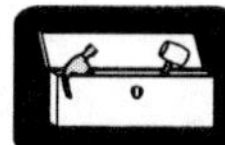

Ensure that you have everything required to teach the course. Check the Materials and Equipment list at the front of this module.

See the general Teaching Tip at the end of this module.

Show Transparency 1, Objectives, and Transparency 2, Performance Tasks. Review the goals of the module, and explain what will be expected of the trainee.

Explain that terms shown in bold are defined in the Glossary at the back of this module.

Review the modules covered in Level One and explain how this module fits in.

Discuss the general safety rules that must be followed when using hand tools.

Explain that pipefitters use vises and stands to support pipe while they are working on it.

Describe the different types of vises and their applications.

Explain how to use yoke, chain, and strap vises.

Emphasize the importance of securely mounting a vise.

1.0.0 ◆ INTRODUCTION

Pipefitting hand tools are generally categorized by their physical description and method of use. For example, a wraparound is a tool that a pipefitter wraps around a piece of pipe to lay out a straight line around the pipe. There is a variety of pipefitting hand tools. Some of them are so widely used that several sizes are made. In many cases, the tool selected to perform a job must be the exact size to perform the job, or personal injury or equipment damage can occur. The following sections list the descriptions, use, maintenance, and potential hazards of nonpowered pipefitting hand tools.

2.0.0 ◆ GENERAL HAND TOOL SAFETY

Although hand tools do not present the immediate hazards of power tools, improper use of hand tools can cause serious personal injury. Safe use of hand tools requires following some basic safety rules. Some hand tools are more hazardous than others and require additional personal protection and care when being used. For safe use of hand tools, follow these basic rules:

- Always choose the proper tool for the job.
- Never use a tool for anything other than its intended use. For example, do not use a screwdriver as a prybar.
- Never use dull or broken cutting tools. Dull cutting tools require greater force to do the job. A sharp cutting tool is a safe tool.
- Keep your hands and fingers away from sharp edges of cutting tools.
- Work away from your body when using cutting tools.
- Ensure that a tool is in good condition and that handles are tightly fastened to the tools.
- Keep tools clean and free of rust.
- Wear eye protection when using chisels, punches, or other tools that may produce flying particles and debris.
- Wear gloves when using tools that require them.
- Use tools in a safe and proper manner. An improperly held wrench can slip, causing injury to the hand or knuckles.
- Do not modify any hand tool.

3.0.0 ◆ VISES AND STANDS

Pipefitters use pipe vises and stands to temporarily support pipe or other materials being worked on. These support devices are specially designed to securely hold pipe and other round objects.

3.1.0 Pipefitter's Vises

Pipefitters use standard vises, as well as yoke, chain, and strap vises. All vises are designed to hold or clamp, but the yoke, chain, and strap vises are designed for pipefitting jobs. The yoke vise is similar to a standard vise, except that it is hinged to open vertically so the pipe may be laid into the vise. Chain or strap vises also open so the pipe may be laid in easily. Chain or strap vises allow larger objects to be held. *Figure 1* shows pipefitter's vises.

Follow these steps to use a pipefitter's vise:

Step 1 Obtain the pipe to be held.

Step 2 Identify where the work must be performed.

Step 3 Select a vise.

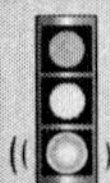

NOTE

The location of the work may determine the kind of vise needed. Standard bench vises may be used, but yoke, chain, or strap vises may do the job better.

Step 4 Inspect the selected vise for obvious damage, excessive jaw wear or screw play, worn-out chain links or strap, grease, rust, or excessive dirt. If any damage to the vise is found, fix or replace the vise. If grease or excessive dirt is found, clean and oil as necessary.

Step 5 Check to see that the vise is securely mounted.

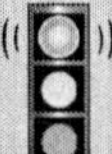

WARNING!

A vise must be securely mounted to prevent injury to the operator and damage to the vise, the pipe, and other equipment in the area.

Step 6 Loosen and open the vise.

Step 7 Place the pipe into the vise.

CAUTION

To prevent marring of the object held, insert material that is softer than the object under the vise jaws or chain, or use a strap vise. For long objects, use a jack stand or similar support to hold the section of the object not in the vise.

Instructor's Notes:

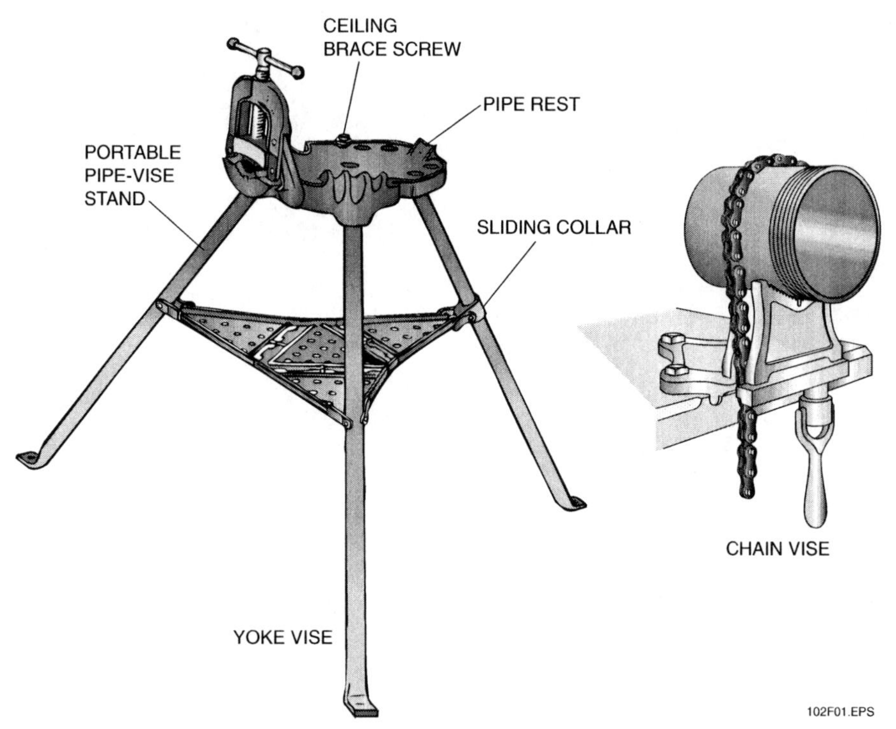

Figure 1 ◆ Pipefitter's vises.

Show trainees how to use each type of vise.

Describe the different types of jacks and stands that pipefitters use and their applications.

Provide some jacks and stands for trainees to examine.

Step 8 Close and secure the moveable side of the vise.

Step 9 Tighten the vise.

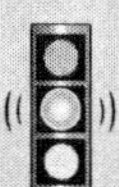

CAUTION

Tighten only enough to hold the object securely. Overtightening may crush the pipe in the vise.

Step 10 Perform the work on the pipe.

WARNING!

Wear safety equipment, such as glasses, gloves, apron, or face shield, as needed when working on the object being held.

CAUTION

When sawing an object in a vise, saw as close to the vise as possible. Do not use a vise as an anvil. Never use an extension bar for extra tightening.

Step 11 Hold the pipe in the vise, and loosen the vise grip until the moveable side of the vise can be fully opened.

Step 12 Remove the pipe from the vise.

Step 13 Inspect the vise for any obvious damage, and correct any problem with the vise.

Step 14 Clean the vise of any dust, dirt, grease, shavings, or chips, especially in the screw area and the chain links. Also, clean the general area around the vise.

Step 15 Close the moveable side of the vise and secure it.

Step 16 Store the vise if it is portable and is no longer needed.

3.2.0 Pipe Jack Stands

Pipe jack stands are support devices used in the repair or fabrication of pipe or other such objects. Some of these devices have rollers on the tops or yokes that allow the object supported to be easily maneuvered (rolled) and worked on safely. Three-legged jacks, which can be used for 6-inch pipe and smaller, are vertically adjustable; four-legged stands can hold pipe up to 24 inches in diameter.

Discuss the importance of using commercial jack stands and rollers.

Explain how to use the various jacks and stands used by pipefitters.

Emphasize the importance of properly positioning supports.

Identify the pinch points on support devices and explain the hazards.

The wheels on top of the four-legged stands can be adjusted horizontally to fit different diameters of pipe. All jack stands have a load rating stamped on them. This load rating must not be exceeded. *Figure 2* shows support devices.

WARNING!

Jack stands and roller assemblies should not be field-fabricated. Load capacities and other safety considerations are built into these devices by their manufacturers, so only commercial jack stands and rollers should be used. Field-fabricated jack stands and rollers can fail and cause personal injury or even death. For pipe over 10 inches in diameter, OSHA requires that four-legged stands be used.

Follow these steps to use jack stands:

Step 1 Select the kind of support device needed.

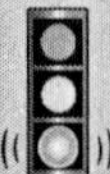

NOTE

For this exercise, assume that two 6-foot sections of 4-inch steel pipe must be set up to be welded into one 12-foot section. Since you will have to adjust the pipes with each other, four adjustable jacks are needed.

Step 2 Inspect the selected supports for obvious damage, such as bent, rusted, or weakened legs or supports; grease on the yokes; rusted wheels that will not turn easily; damaged screw threads; or collar adjustment handles that will not turn. If a support is damaged, replace it. If it needs cleaning and oiling, do it before the device is used. The oil should be placed on the threads only.

Step 3 Select a level, safe location to set the supports while they are being used.

WARNING!

These supports often hold heavy objects and must have safe footing. Position the supports out of the normal flow of work if possible. If using these supports on the ground, they should be set up on plywood.

Step 4 Position the supports in a straight line, and space them so that two will support each section of pipe.

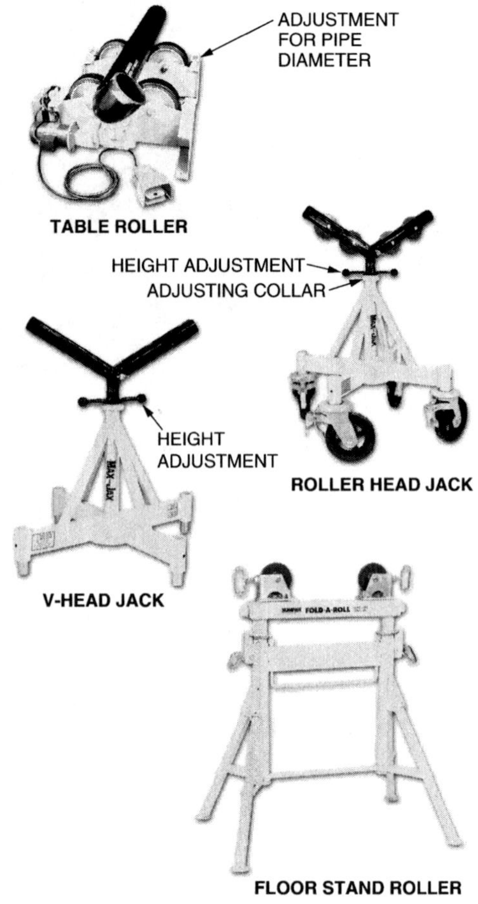

Figure 2 ◆ Support devices.

Step 5 Turn the collar adjustment on each support so that the yokes are fully lowered.

Step 6 Ensure that all the yokes are roughly level with each other.

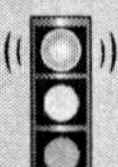

WARNING!

Beware of pinch points between the collar adjust screws and the safety lock rings.

NOTE

Unlock the vertical lock screws and the safety lock ring on each support, and adjust the main body (tube) of the support. When they are aligned, tighten the vertical lock screws so the tubes will not compress later.

Instructor's Notes:

2.4 PIPEFITTING ◆ LEVEL ONE

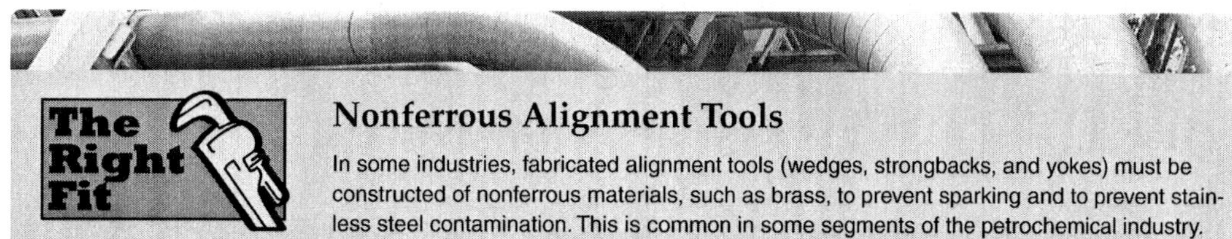

Nonferrous Alignment Tools

The Right Fit

In some industries, fabricated alignment tools (wedges, strongbacks, and yokes) must be constructed of nonferrous materials, such as brass, to prevent sparking and to prevent stainless steel contamination. This is common in some segments of the petrochemical industry.

Step 7 Position the first section of pipe onto one set of supports.

WARNING!

Check to see that the support is firmly under the pipe before the pipe is released. An object this heavy could injure a worker or damage equipment if it falls. Also ensure that the supports are located toward the ends of the pipe.

Step 8 Position the inside end support for the second section of pipe so that it does not touch the end support for the first section.

Step 9 Position the second section of pipe onto the second set of supports.

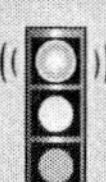

WARNING!

Ensure that the pipe is well seated and that it will not fall.

Step 10 Push the two sections of pipe together carefully until the ends meet.

Step 11 Adjust the main body tubes and the collar adjustments of the supports to make the two sections of pipe roughly level with each other.

4.0.0 ◆ PIPE WRENCHES

Many types and sizes of pipe wrenches are used in the pipefitting trade. Pipe wrenches are heavy-duty wrenches made of durable cast iron or aluminum that are used to grip and turn pipe or to grip and hold pipe in a stationary position. It is critical to select the right size pipe wrench for a given job. A pipe wrench that is too small will not hold the pipe firmly, and a handle that is too short will not provide enough leverage. A pipe wrench that is too large can strip the threads, break the pipe or fitting, or cause excessive marring or scratching of the pipe.

4.1.0 Pipe Wrench

The pipe wrench is used to grip and turn round pipe and tubing. Types include straight pipe wrenches, heavy-duty pipe wrenches, and 45-degree and 90-degree offset pipe wrenches (*Figure 3*). Wrenches come in a number of lengths, such as 6, 12, 14, 18, 24, 36, and 48 inches. The wrench's length determines the size of pipe that it can handle.

The straight pipe wrench is the most common type. When working in tight quarters, use offset wrenches; their angles (45 or 90 degrees) make it easier to reach pipe in cramped areas. The jaws of pipe wrenches always leave marks on the pipe; therefore, do not use them on pieces where appearance is important. When applying force to a wrench, always direct it toward the open side of the jaws. This technique will give you the best grip and leverage.

A variation on the straight pipe wrench is the compound-leverage pipe wrench (*Figure 4*). This design increases the leverage that can be applied on a pipe. This wrench is generally used on pipe joints that are frozen or are locked together.

Another variation of the pipe wrench is the chain wrench (*Figure 5*). This tool has a length of chain permanently attached to the wrench handle at one end. The chain is looped around the pipe to grip and secure it, and the other end of the chain can be secured to various positions on the wrench's handle. Oil the chain frequently to prevent it from becoming stiff or rusty.

4.2.0 Pipe Tongs

Pipe tongs are the wrenches typically used on large pipe (*Figure 6*). Pipe tongs also have chains, which need to be oiled frequently. They provide extra leverage needed for tougher, more heavy-duty jobs.

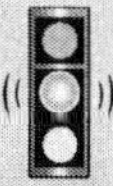

CAUTION

Never use cheaters (pipes used to extend wrenches) because they can cause equipment damage. Always use the right-sized wrench to avoid damaging the pipe.

Demonstrate how to secure a section of pipe in a vise and a pipe stand.

Have trainees practice securing a section of pipe in a vise and a pipe stand. Note the proficiency of each trainee. This laboratory corresponds to Performance Task 2.

Have trainees review Sections 4.0.0–5.5.0.

Identify the different types of wrenches used by pipefitters. Explain how to select the correct wrench.

Describe different types of pipe wrenches and explain their uses.

Provide various types of pipe wrenches to the trainees for their examination.

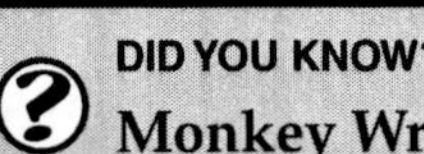

DID YOU KNOW?

Monkey Wrench

The monkey wrench is a general-purpose wrench with an adjustable jaw for turning nuts of various sizes. Charles Moncky invented this tool around 1858. "Monkey wrench" also has an informal meaning: to disrupt, as in, "He threw a monkey wrench into our plans." Monkey comes from the last name of the wrench's inventor. Wrench was a term that meant trick or deception long before the tool of the same name was developed.

DID YOU KNOW?

The Pipe Wrench

The pipe wrench was invented by a steamboat firefighter named Daniel Stillson. Stillson had suggested to Walworth, a heating and piping firm, that it design a wrench that could be used to join or separate pipes. The firm's owner told Stillson to take a crack at making a prototype that would twist off a stuck pipe. To the everlasting gratitude of future plumbers, Stillson was successful and his prototype twisted off the pipe. In 1870, Stillson's design was patented, and Walworth began manufacturing the wrench. Even today, pipe wrenches are sometimes called Stillson wrenches.

Source: Inventors website. Mary Bellis. "Pipe Wrench." http://inventors.about.com/library/inventors/blwrench.htm, reviewed January 22, 2004.

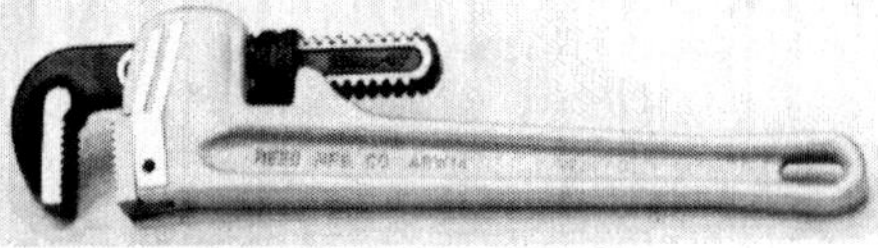

STRAIGHT PIPE WRENCH

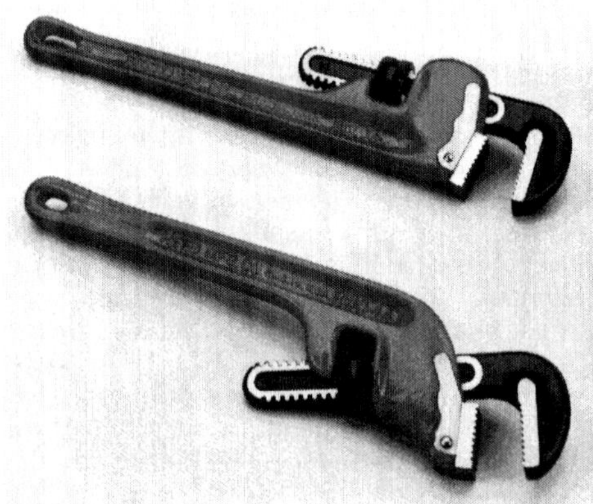

HEAVY-DUTY PIPE WRENCH
45-DEGREE OFFSET WRENCH

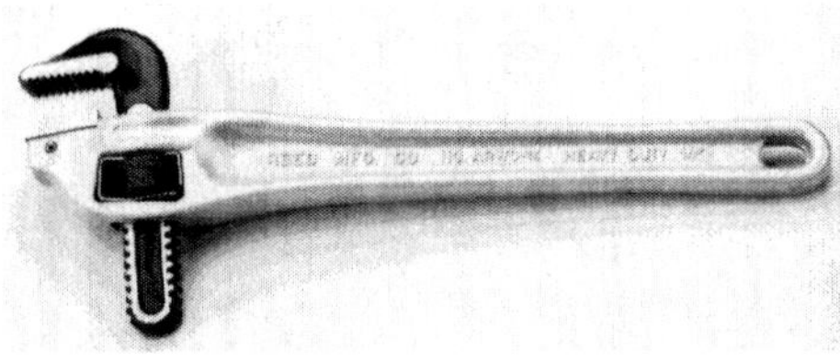

90-DEGREE OFFSET PIPE WRENCH

102F03.EPS

Figure 3 ◆ Pipe wrenches.

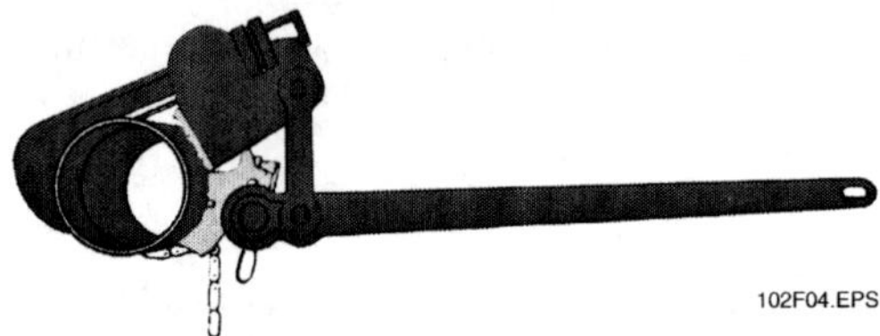

102F04.EPS

Figure 4 ◆ Compound-leverage pipe wrench.

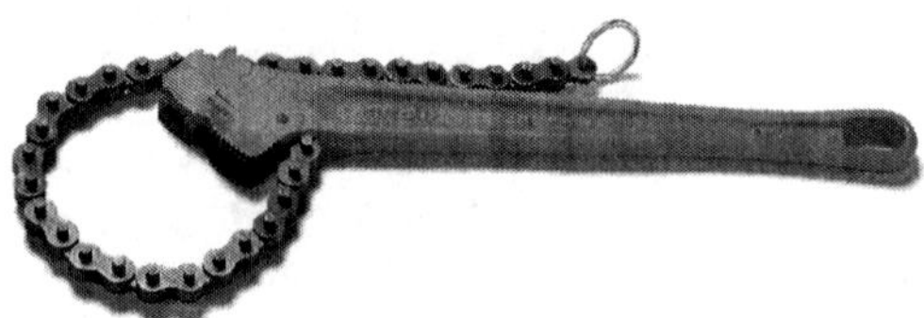

102F05.EPS

Figure 5 ◆ Chain wrench.

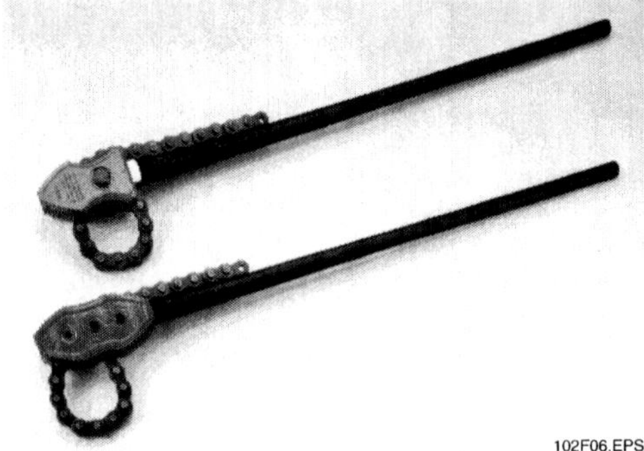

102F06.EPS

Figure 6 ◆ Pipe tongs.

Instructor's Notes:

4.3.0 Strap Wrench

The strap wrench (*Figure 7*) is used to hold chrome-plated or other types of finished pipe. The strap wrench does not leave jaw marks or scratches on the pipe because it grips with a strap rather than a jaw or chain. With some strap wrenches, rosin needs to be applied to the strap so that it does not slip. Other strap wrenches use vinyl straps that do not need rosin.

Describe strap, open-ended, and adjustable wrenches and explain their uses.

4.4.0 Open-End Wrench

The open-end wrench (*Figure 8*) has two different-sized openings, each on one end of the tool. The size of each opening is stamped on the handle. This wrench is usually used for assembling fittings and fasteners that are less than 1 inch across.

4.5.0 Adjustable Wrench

The adjustable wrench, like the Crescent® wrench in *Figure 9*, is similar to the open-end wrench, except that it has an adjustable jaw. A roller at the stem of the jaw adjusts the size of the opening. Because it is so easy to adjust, this wrench is found in virtually every toolbox. Various sizes of adjustable wrenches are available, from 6 to 18 inches in length.

Adjustable wrenches have smooth jaws, making these tools ideal for turning nuts, bolts, small **pipe fittings**, and chrome-plated pipe fittings. Using an adjustable wrench can save you considerable time, because you will not have to constantly switch wrench sizes as you work with different sizes of nuts and bolts.

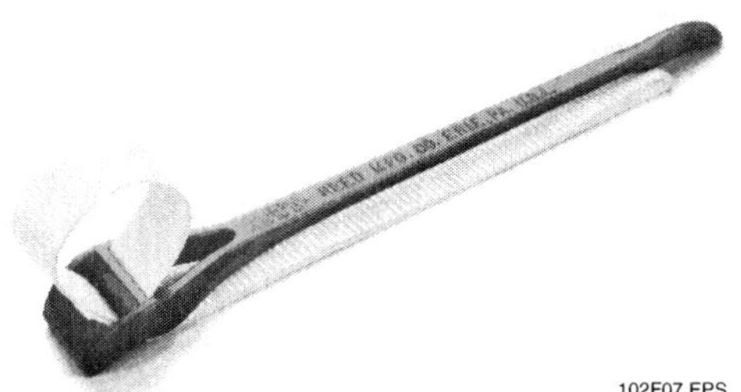

102F07.EPS

Figure 7 ◆ Strap wrench.

102F08.EPS

Figure 8 ◆ Open-end wrench.

4.6.0 Using a Pipe Wrench

When using a pipe wrench, either place the wrench on the pipe or fitting with the jaw opening facing you and pull on the handle, or place the wrench on the pipe or fitting with the jaw opening facing away from you and push on the handle.

Discuss the importance of bracing yourself when using a wrench.

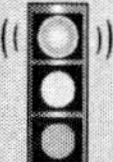

WARNING!

Be careful when pulling the handle toward you to keep the jaw from slipping off the pipe, causing the handle to hit you. Be sure to brace yourself when pushing on the handle so that you do not fall if the wrench slips off the pipe.

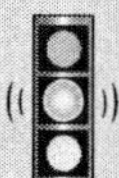

CAUTION

Never place the jaws of a pipe wrench onto the threaded end of a pipe because this will ruin the threads.

Explain how to use a pipe wrench to tighten a fitting.

Follow these steps to tighten a fitting onto the threaded end of a pipe:

Step 1 Select a pipe wrench suitable for the size of pipe you are using and the job you are performing.

Step 2 Inspect the selected wrench for excessively worn jaw teeth and corners and any noticeable damage to the handle. Replace the wrench if necessary.

Step 3 Secure the pipe in a chain vise, allowing the threaded end to extend from the vise about 8 inches.

Step 4 Tighten the fitting onto the end of the pipe by hand.

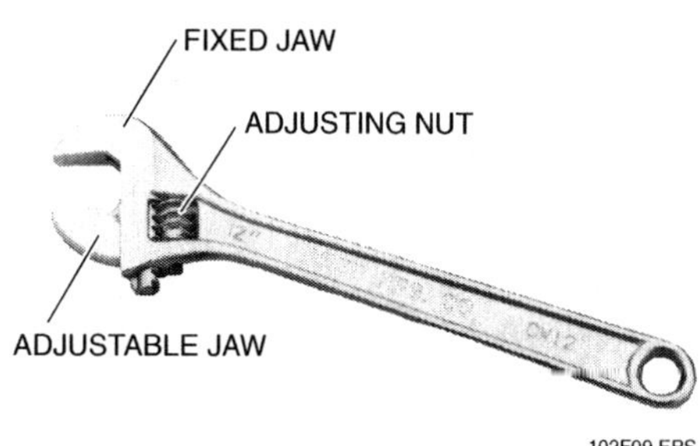

102F09.EPS

Figure 9 ◆ Adjustable Crescent® wrench.

MODULE 08102-06 ◆ PIPEFITTING HAND TOOLS 2.7

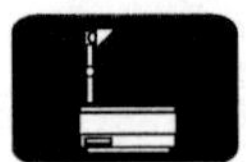

Show Transparency 3 (Figure 10). Explain how to use two pipe wrenches to install or remove a fitting.

Show trainees how to use various types of pipe wrenches.

Have trainees practice using various types of wrenches. Note the proficiency of each trainee. This laboratory corresponds to Performance Task 3.

Discuss the guidelines for using and caring for pipe wrenches.

Step 5 Position yourself so that the fitting is on either your left or right side.

Step 6 Turn the knurled nut on the pipe wrench until the jaws are open slightly wider than the diameter of the fitting.

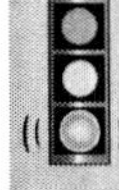

NOTE

Many straight pipe wrenches have pipe sizes marked on the hook jaw just above the housing. If your wrench has these marks, line up the correct mark with the top of the housing.

Step 7 Slide the jaws onto the fitting. Be sure the jaws fit snugly onto the fitting. The fitting should be centered inside the jaws, and there should be a gap between the fitting and the back of the hook jaw.

Step 8 Push or pull the handle as far as possible to tighten the fitting.

Step 9 Pull or push the handle in the opposite direction to release the jaws from the fitting.

Step 10 Regrip the fitting with the wrench and push or pull the wrench again.

Step 11 Repeat steps 9 and 10 until the fitting is tight on the pipe.

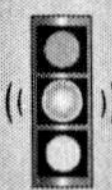

CAUTION

Do not overtighten the fitting on the pipe or the threads may strip. Three full threads should be exposed on the pipe at the end of the fitting.

Step 12 Remove the wrench from the pipe fitting.

Step 13 Clean the wrench, and inspect it for any damage.

Step 14 Close the jaws, and store the wrench.

You may have to install or remove a fitting from a piece of pipe when a vise is not available or the fitting is part of a piping system. In this case, use two pipe wrenches. Follow these steps to remove a fitting from a pipe without using a vise:

Step 1 Position yourself with the fitting on your left or right side.

Step 2 Place one pipe wrench on the pipe.

Step 3 Place the other pipe wrench on the fitting with the jaw opening facing in the opposite direction (*Figure 10*).

Step 4 Pull on the handle of the wrench on the pipe while pushing on the handle of the wrench on the fitting.

Step 5 Regrip the pipe and the fitting when you have turned the wrenches as far as you can.

Step 6 Repeat steps 2 through 5 until the fitting is loose enough to remove by hand.

Step 7 Remove the wrenches from the pipe and fitting.

Step 8 Clean the wrenches and inspect them for any damage.

Step 9 Close the jaws and store the wrenches.

4.7.0 Using and Caring for Pipe Wrenches

Wrenches will only work well if they are used correctly and properly cared for. Follow these guidelines to use and care for pipe wrenches:

- Use pipe wrenches only to turn pipe and fittings. Do not use a pipe wrench to bend, raise, or lift a pipe.
- Do not use the pipe wrench as a hammer. It is not designed for any sort of pounding.
- Do not drop or throw a pipe wrench. You may break or crack the wrench, and you may cause injury to yourself or others.
- Check the teeth of the wrench often. They should be kept clean and sharp to keep the wrench from slipping on the pipe.
- Apply penetrating oil to the wrench threads to prevent them from sticking.
- Never use a cheater or extender bar on a pipe wrench handle to increase leverage. This may cause the handle or the pipe to bend or break.

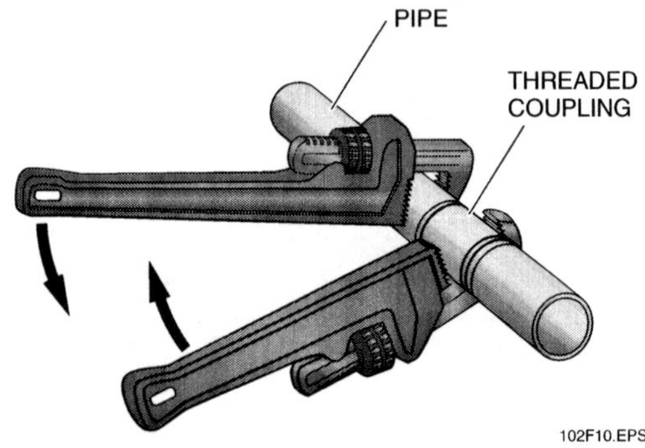

Figure 10 ◆ Using two pipe wrenches.

Instructor's Notes:

5.0.0 ◆ LEVELS

Levels are used to determine the plumb or levelness of pipe. Plumb refers to vertical alignment, and level refers to horizontal alignment. Several types of levels are used by pipefitters, including framing levels, torpedo levels, and tubing water levels. Most levels used by pipefitters are made of tough, lightweight metals, such as magnesium or aluminum. They generally have three vials, two to measure plumb and one to measure level. The amount of liquid each vial contains is not enough to fill the vial completely. This creates a bubble. Centering this bubble between the lines scribed on the outside of the vial produces plumb or level (*Figure 11*).

5.1.0 Framing Levels

Framing levels, also known as spirit levels, come in a variety of sizes. Longer levels are more accurate than shorter ones. The most common sizes for framing levels are 18, 24, 28, and 48 inches. Most pipefitters carry a 24-inch framing level. The frame of the framing level is milled and ground on both the top and bottom to be uniformly parallel and to ensure smoothness. Framing levels are available with or without a 45-degree vial. *Figure 12* shows a framing level.

5.2.0 Torpedo Levels

A torpedo level (*Figure 13*) is approximately 9 inches long and, as its name suggests, is tapered at both ends. Torpedo levels are best used in tight places or where accuracy is not critical. The top of the torpedo level has a groove running down the center from end to end that helps it sit on the round surface of a pipe. The bottom side of some torpedo levels has magnets encased in it so it will adhere to metal pipes. Torpedo levels have three vials and will measure level, plumb, and a true 45 degrees. All pipefitters should carry a torpedo level.

5.3.0 Measuring Using Spirit Levels

Pipefitters use framing and torpedo levels in the same way to measure plumb and level. Framing levels can be used to measure true 45-degree angles if they have a 45-degree vial.

5.3.1 Measuring Horizontal Level

Follow these steps to use a framing or torpedo level to measure horizontal level:

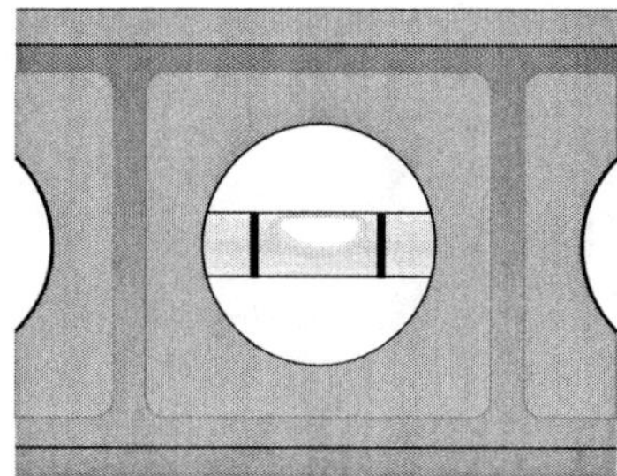

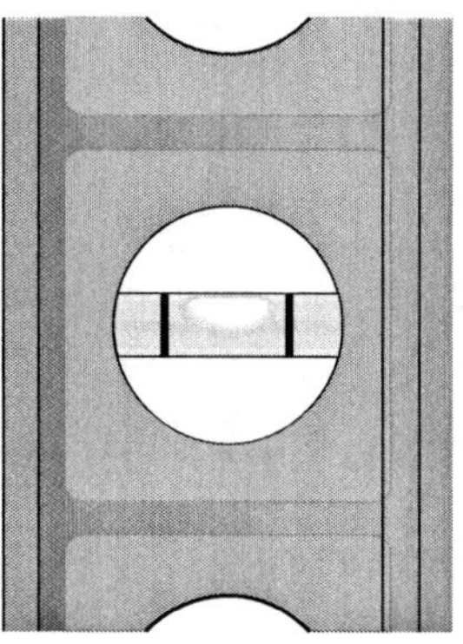

Figure 11 ◆ Correctly aligned spirit level vials.

Figure 12 ◆ Framing level.

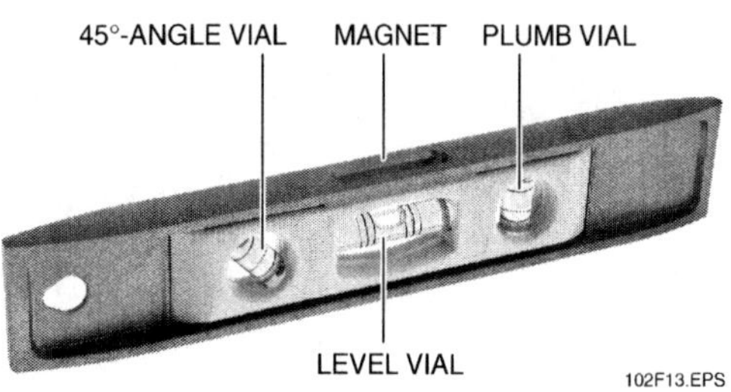

Figure 13 ◆ Torpedo level.

Ask a trainee to explain how pipefitters use levels. Identify different types of levels used by pipefitters.

Show Transparency 4 (Figure 11). Explain how to read plumb or level on a spirit level vial.

Describe framing and torpedo levels.

Explain how to measure horizontal level.

Provide a framing level and torpedo level for the trainees to examine.

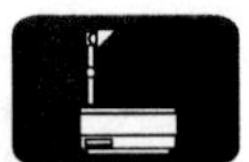

Show Transparency 5 (Figure 14). Explain how to use a framing or torpedo level to measure vertical plumb.

Explain how to use a torpedo level to measure a true 45-degree angle.

Checking a Spirit Level

You should periodically check your spirit level to make sure it has not gotten out of calibration. Position the level on a flat surface so that the bubble registers in the center of the vial, then use the length of the level as a straightedge to draw a line. Flip the level over from side to side and lay it against the line. If the level is good, the bubble will be centered in the vial.

Step 1 Make sure that the surface of the pipe where the level is to be placed is smooth and clean.

Step 2 Make sure that the sides of the level are clean.

Step 3 Place the level on the pipe to be leveled.

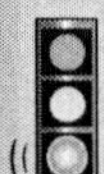

NOTE
The bubble inside the vial will move to the high end of the pipe.)

Step 4 Adjust the pipe so that the bubble lies between the two marks scribed on the vial.

5.3.2 Measuring Vertical Plumb

Follow these steps to use a framing or torpedo level to measure vertical plumb:

Step 1 Make sure that the surface of the pipe or object is smooth and clean.

Step 2 Make sure that the sides of the level are clean.

Step 3 Hold the level vertically against the side of the pipe or object to measure vertical plumb from side to side. Hold the level so that the vertical plumb vial is toward the top of the pipe or object. *Figure 14* shows measuring vertical plumb.

Step 4 Adjust the pipe or object until the bubble lies between the scribe marks on the vial.

Step 5 Move the level 90 degrees around the pipe or object to measure vertical plumb from front to back.

Step 6 Adjust the pipe or object until the bubble lies between the scribe marks on the vial.

Step 7 Recheck the pipe or object for vertical plumb from side to side.

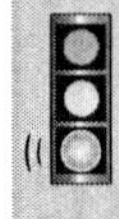

NOTE
You should recheck the vertical plumb on all sides of the pipe or object until no further adjustments are necessary.

5.3.3 Measuring for True 45-Degree Angle

Follow these steps to adjust a pipe to a true 45-degree angle, using a torpedo level:

Step 1 Make sure that the pipe is smooth and clean.

Step 2 Make sure that the sides of the level are clean.

Step 3 Place the torpedo level on the pipe so that the 45-degree vial is level.

Step 4 Adjust the pipe until the bubble lies between the marks scribed on the vial.

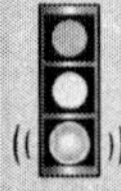

NOTE
As always, the bubble will move to the high side.

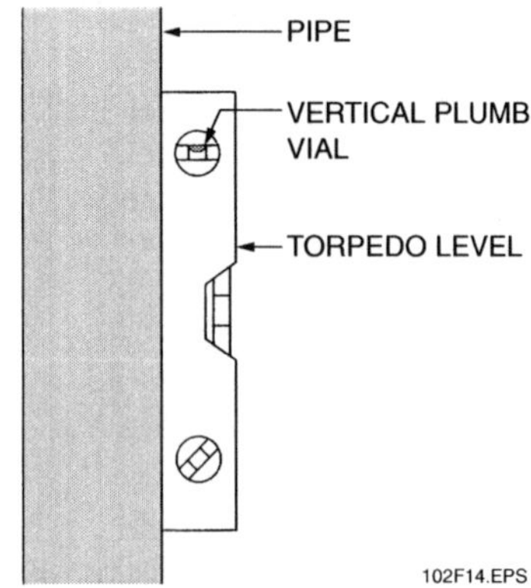

Figure 14 ◆ Measuring vertical plumb.

Instructor's Notes:

5.4.0 Laser Instruments

Laser instruments are widely used today to determine level, plumb, and grade. Depending on the type of laser diode being used, the laser beam produced may be invisible to the human eye. Visible laser beams have wavelengths in the visible light portion of the radiant energy spectrum. Invisible laser beams have wavelengths in the infrared portion of the radiant energy spectrum. Because these beams are invisible, they must be detected using electronic laser detectors. Electronic laser detectors are described in detail later in this section.

A helium-neon laser is basically an electron tube filled with a combination of helium and neon gas. It produces a very intense, narrow beam of red light that retains its shape over long distances. The beam can be manipulated, controlled, and detected quite easily. By expanding or focusing the beam with simple optical elements, the desired spot size can be achieved at almost any distance.

Modern laser instruments are typically battery-operated units that generate either a fixed laser beam, a rotating beam, or both. Fixed laser beam instruments can generate either single or multiple beams. One example of a fixed-beam device is the laser spirit level, which is a carpenter's level with a built-in visible beam laser (*Figure 15*). It can be used as an ordinary level or switched on to emit a laser beam, thereby extending the level's reference line typically between 200 and 300 feet. It can be mounted on a tripod or placed on a level surface. Another example of a single-beam device is a handheld distance meter (*Figure 16*) used to measure distances up to about 100 feet without a target, and even farther distances with a reflective target. The auto-leveling laser alignment tool is a good example of a multiple-use laser instrument. It can be used to provide simultaneous plumb, level, and square reference points.

102F15.EPS

Figure 15 ◆ Laser input level.

5.5.0 Tubing Water Level

The tubing water level (*Figure 17*) is a small diameter, clear plastic tube that is generally 50 feet long and partially filled with water. Pipefitters use the tubing water level to achieve an accurate reading of level between two distant points. It can also be used to transfer elevation bench marks between two locations. The tubing water level is preferred over the string line level when leveling a taut string or line because it is so accurate.

102F16.EPS

Figure 16 ◆ Fixed-beam laser instrument.

102F17.EPS

Figure 17 ◆ Tubing water level.

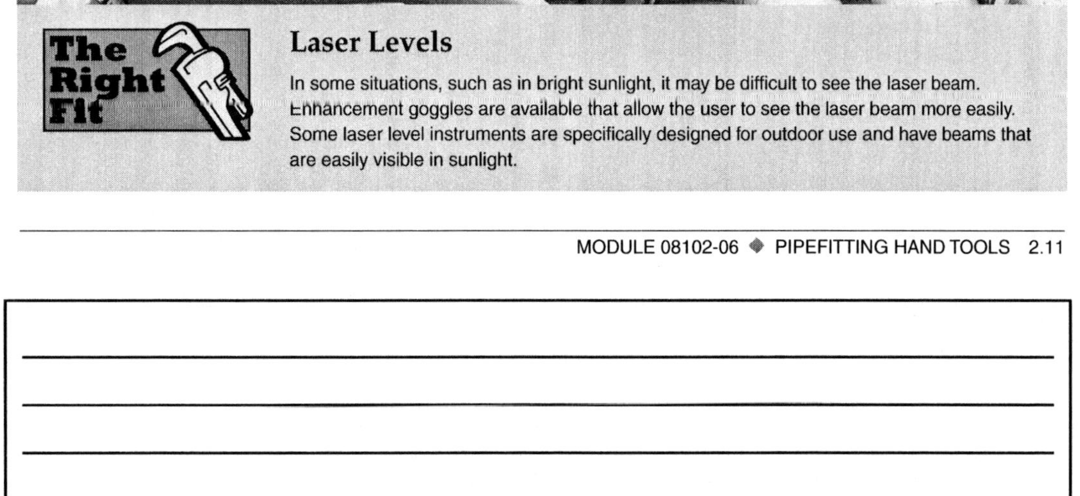

The Right Fit

Laser Levels

In some situations, such as in bright sunlight, it may be difficult to see the laser beam. Enhancement goggles are available that allow the user to see the laser beam more easily. Some laser level instruments are specifically designed for outdoor use and have beams that are easily visible in sunlight.

Explain how lasers are used to measure level, plumb, and grade. Describe laser instruments commonly used by pipefitters.

Describe and explain the uses of a tubing water level.

Provide a laser level and tubing water level to the trainees for their examination.

MODULE 08102-06 ◆ PIPEFITTING HAND TOOLS 2.11

Explain how to use a tubing water level.

Show trainees how to use various types of levels.

Have trainees practice using various types of levels. Note the proficiency of each trainee. This laboratory corresponds to Performance Task 4.

Have trainees review Sections 6.0.0–7.0.0.

Ensure that you have everything required for teaching this session.

Identify common squares used by pipefitters. Describe a framing square and explain how it is used.

To use the tubing water level, you will need a helper. One person holds the tube end vertically near the reference point, or one end of the line, and adjusts the tube until the water level is even with the reference point. Another person takes the other end of the tube and holds it vertically at the other end of the line. The water level in the tube will rest exactly level with the water level at the other end of the tube. Tie a string between these two points, and it will be perfectly level.

6.0.0 ◆ PIPE FABRICATION TOOLS

Pipe fabrication tools are used to lay out angles, find center lines of pipe, scribe straight lines around pipe, and plumb and square pipe sections together. The most common fabrication tools are the following:

- Framing squares
- Pipefitter's squares
- Combination tri squares
- Center finders
- Pipe line-up clamps
- Hi-Lo gauges
- Wraparounds
- Drift pins
- Two-hole pins
- Flange spreaders

6.1.0 Framing Squares

The framing square, also known as a carpenter's square or a steel square, can be used by pipefitters for a variety of tasks. The most common functions of a framing square are the following:

- Laying out guidelines for cutting steel and pipe, and then squaring up adjacent markings at right angles
- Using tables and scales on the square for measuring and fabricating
- Determining plumb or square fits in fabricating, with the use of a tape or rule
- Aligning two sections of pipe or a fitting to a pipe
- Finding and laying out the center lines of pipe
- Checking the squareness of the end of a pipe

The framing square (*Figure 18*) consists of two parts: the tongue and the body, or blade. The tongue is the shorter, narrower part, and the body is the longer, wider part. The point at which the tongue and the body meet on the outside edge is called the heel. The two most commonly used sizes are the 12 by 8 inch and the 24 by 16 inch.

6.1.1 Straight Pipe Alignment Using Framing Squares

Straight pipe alignment is one of the many applications of framing squares. To align two pieces of straight pipe, two framing squares are needed (see *Figures 19* and *20*). Check the framing squares for trueness by placing the blades and tongues side by side. Procedures for pipe alignment will be taught later.

6.2.0 Pipefitter's Squares

Built on the same principle as the framing square, the pipefitter's square is designed specifically for pipefitting methods and techniques. The main differences between the pipefitter's square and the framing square are tables and graduations. The pipefitter's square has takeoffs, dimensions, and other useful information printed on the square itself. The pipefitter's square can be used in all the same ways as the framing square, with the added advantage of having tables that eliminate many calculations. Special care should be taken when using this tool because it can easily be bent out of square. In addition, it will rust when exposed to the elements and it is several times more expensive than a framing square.

6.3.0 Combination Tri Squares

A combination tri square (*Figure 21*) is a versatile tool composed of four parts, giving it a variety of uses in layout and fabrication. These four components are a grooved steel rule graduated into scales of 1/8 inch, 1/16 inch, 1/32 inch, and 1/64 inch; a locking square head with a level glass; a locking center head; and a protractor with a revolving turret marked with degrees. All of these parts are removable and can be used independently.

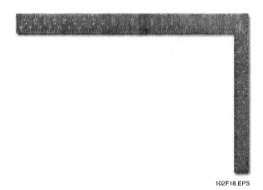

102F18.EPS

Figure 18 ◆ Framing square.

Instructor's Notes:

2.12 PIPEFITTING ◆ LEVEL ONE

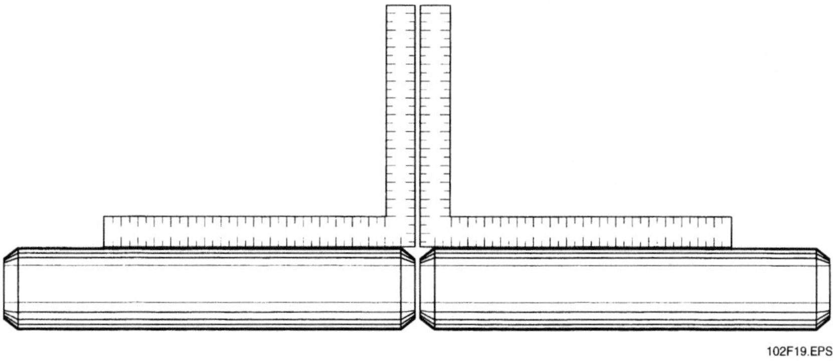

Figure 19 ◆ Positioning squares to align straight pipe.

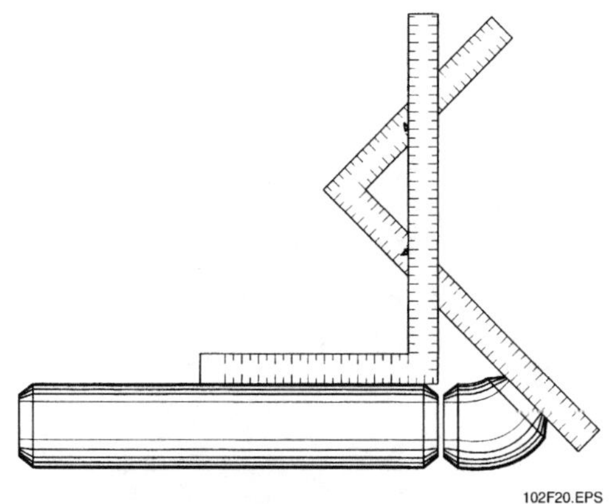

Figure 20 ◆ Aligning a 45-degree elbow to straight pipe.

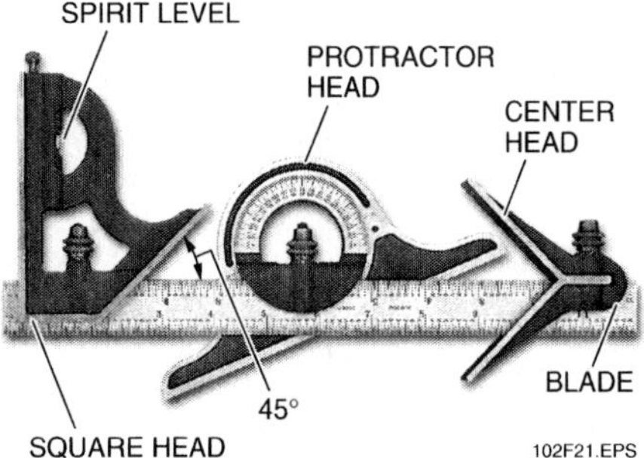

Figure 21 ◆ Combination tri square.

The components of the combination tri square allow numerous uses. It can be used as a height and depth gauge, bevel or angle protractor, level, plumb, square, center finder, ruler, and for many of the functions of a framing square. To attach any of the components, loosen the spring-loaded locking nut, allowing the part to be fitted into the groove on the steel rule. Firmly lock the nut after the desired position is obtained. To remove a component, loosen the locking nut, and slide the component off the steel rule.

The square head and rule are used for measuring height or depth, plumb or levelness, squareness, and for laying out angles. *Figure 22* shows uses of the square head.

The center head and rule and any type of level are used to find the center line of pipe and other cylindrical objects. The center line of pipe is found by adjusting the depth of the rule so that the protruding edge of the rule is approximately $\frac{1}{16}$ inch off the surface of the pipe when the center head is placed on the pipe. A level is placed against the long end of the rule, above the center head, and the tri square is moved until the glass vial reads level. The location of the rule on the pipe is then marked for center line. *Figure 23* shows using the center head.

The protractor head is used for laying out and marking angles in the same manner as the square head and rule by revolving the rotating turret to the desired degree (*Figure 24*). By placing a level on the steel rule, the protractor can be used to determine the degree of slope of a pipe. Place the edge of the protractor on the pipe, and then rotate the turret until the level on the rule reads true. The angle on the protractor is the degree of slope.

6.4.0 Center Finders

Another tool used to find the center of a pipe is the center finder (*Figure 25*). The center finder, also known as a centering head, consists of a Y-type head that straddles the pipe, a centering pin, and an adjustable dial bubble protractor for indicating center at some angle away from top dead center. To mark top dead center, set the center finder on top of the pipe, and adjust till the bubble indicates center. A light blow with a hammer to the centering pin makes an indentation on the surface to mark its center.

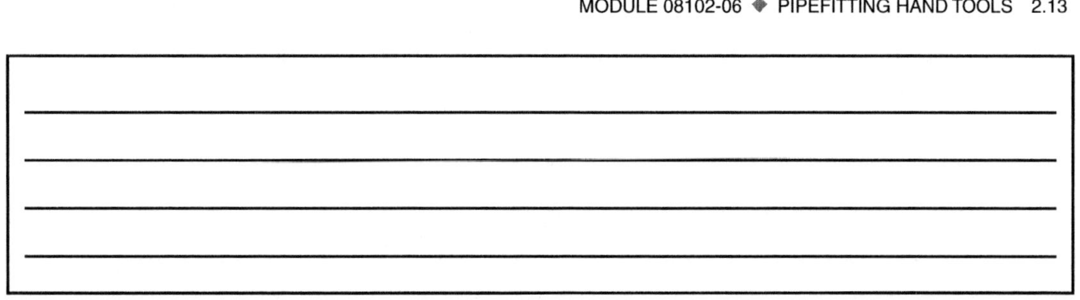

Show Transparencies 6 and 7 (Figures 19 and 20). Explain the straight pipe alignment procedure.

Describe a pipefitter's square and explain how it is used. Describe a combination tri square.

See the Teaching Tip for Section 6.0.0 at the end of this module.

Show Transparencies 8 through 10 (Figures 22 through 24). Explain how to use a combination square to measure various angles.

Describe a center finder and explain how it is used.

Show trainees how to use framing, pipefitter's, and combination tri squares and a center finder.

MODULE 08102-06 ◆ PIPEFITTING HAND TOOLS 2.13

Have trainees practice using framing, pipefitter's, and combination tri squares and a center finder. Note the proficiency of each trainee. This laboratory corresponds to Performance Task 4.

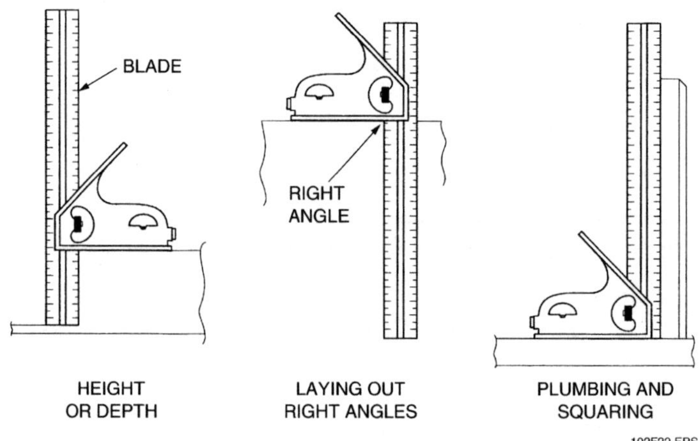

Figure 22 ◆ Uses of the square head.

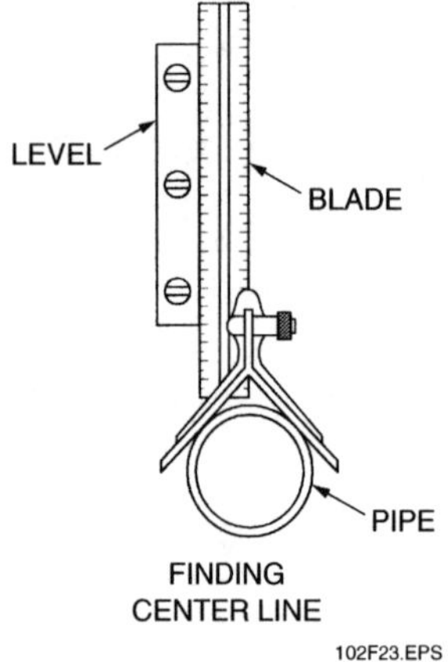

Figure 23 ◆ Using the center head.

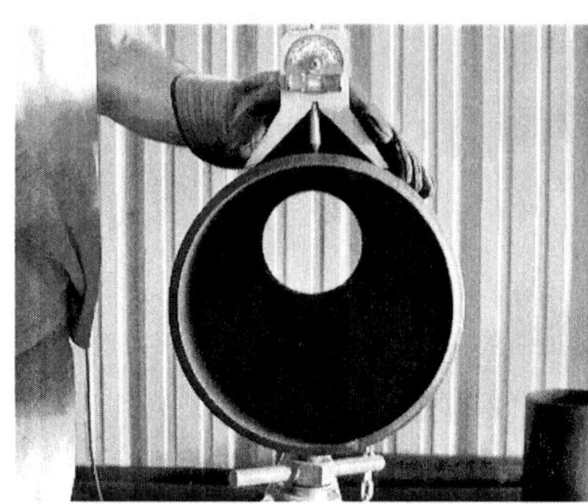

Figure 25 ◆ Center finder.

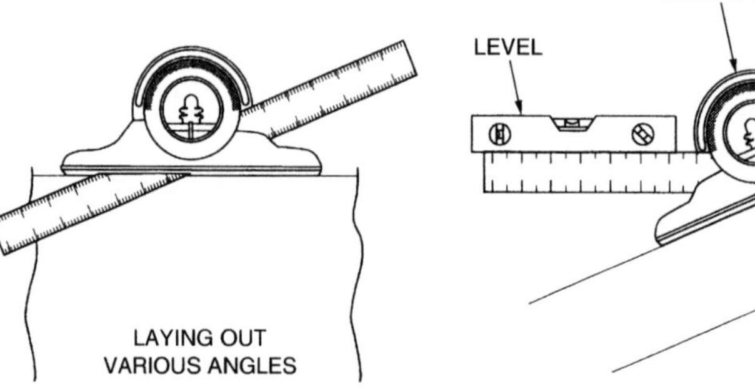

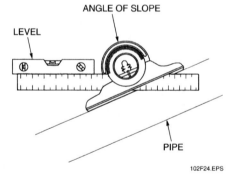

Figure 24 ◆ Uses of a protractor.

2.14 PIPEFITTING ◆ LEVEL ONE

Instructor's Notes:

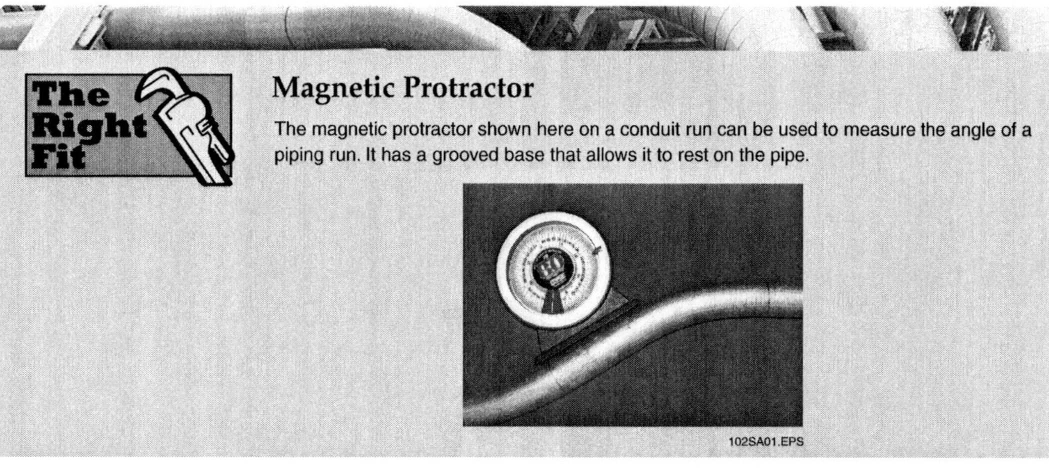

The Right Fit

Magnetic Protractor

The magnetic protractor shown here on a conduit run can be used to measure the angle of a piping run. It has a grooved base that allows it to rest on the pipe.

102SA01.EPS

Identify common clamps used by pipefitters to line up pipe. Describe straight butt welding clamps and explain how they are used.

Show trainees how to use straight butt welding clamps.

6.5.0 Pipe Line-Up Clamps

Pipe line-up clamps range from simple field-fabricated jigs to large hydraulic clamps capable of aligning pressure vessels with external diameters of over 30 feet. Even though there are many differences in the types of pipe line-up clamps, they all serve the same purpose: to keep the pipes square and level before and during welding. The most common pipe line-up clamps are the following:

- Straight butt welding clamps
- Flange welding clamps
- Other chain-type welding clamps
- Shop-made aligning dogs

6.5.1 Straight Butt Welding Clamps

Straight butt clamps (*Figure 26*) are chain-type clamps used to secure two straight pipes together end to end. They consist of straight sections of steel approximately 14 to 18 inches long and two lengths of chain, one near each end, attached to screw locks that can tighten the chain once it has been wrapped around the pipe.

Follow these steps to use a straight butt welding clamp:

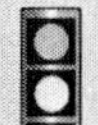

NOTE

For this exercise, assume that two previously prepared 4-inch steel pipes must be welded together end to end.

Step 1 Inspect the straight butt clamp for any obvious damage, such as broken or worn-out chain links, bent or broken clamp feet, warped clamp back, bent or broken chain hooks, damaged chain tighteners, and grease, rust, or excessive dirt.

Step 2 Repair or replace any damaged part. Clean and oil any part as needed.

Step 3 Set up two 6-foot lengths of 4-inch pipe to be welded into one 12-foot length.

Step 4 Position the selected clamp on top of, and evenly spaced over, the two sections of pipe so that the chains drop down the back side of the clamp as you face the pipes.

Step 5 Reach under the pipe, while holding the clamp in place, and grasp the chain.

The Right Fit

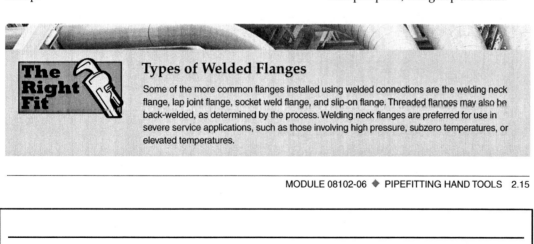

Types of Welded Flanges

Some of the more common flanges installed using welded connections are the welding neck flange, lap joint flange, socket weld flange, and slip-on flange. Threaded flanges may also be back-welded, as determined by the process. Welding neck flanges are preferred for use in severe service applications, such as those involving high pressure, subzero temperatures, or elevated temperatures.

MODULE 08102-06 ◆ PIPEFITTING HAND TOOLS 2.15

Emphasize the importance of careful handling of clamps and welded pipe.

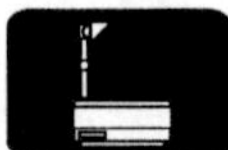

Show Transparency 11 (Figure 27). Describe flange welding clamps.

Explain how to use flange welding clamps.

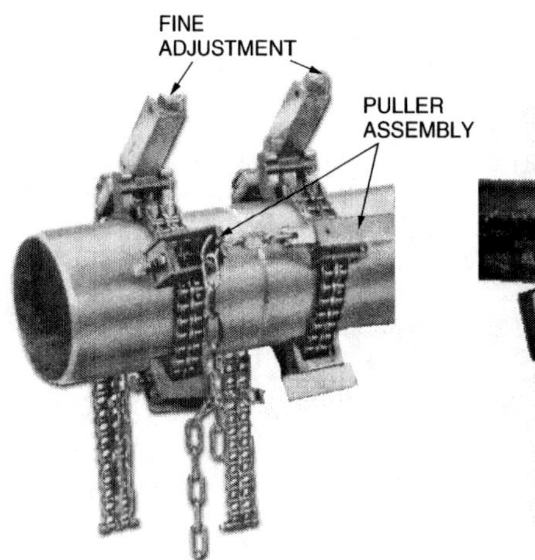

Figure 26 ◆ Straight butt welding clamps.

Step 6 Pull the chain around the pipe and up to the locking notch on the clamp.

Step 7 Secure the chain to the clamp.

Step 8 Rotate the clamp around the pipe to the bottom of the pipe.

Step 9 Turn the adjustable screw handle to tighten the chain. Tighten only until the chain is snug. Some readjustment may be needed later.

Step 10 Position the second (unclamped) section of pipe so that it is aligned with the first section of pipe and is ready to be welded.

Step 11 Wrap the chain around the second section of pipe.

Step 12 Attach the chain to the clamp.

Step 13 Tighten the second chain onto the second section of pipe.

Step 14 Adjust both chains until they are tight.

Step 15 Check the pipe sections again to see that they are still aligned with each other and ready for welding.

Step 16 After the welding has been completed, loosen the chains and remove the clamp.

WARNING!
Handle the pipe and the clamps with care. Welded objects remain hot for several minutes after the welding is completed.

Step 17 Clean the clamp if necessary, and inspect it for damage. If it is damaged, replace it.

Step 18 Store the straight butt welding clamp.

6.5.2 Flange Welding Clamps

Flange welding clamps (*Figure 27*) are chain-type clamps that are specifically designed to hold a pipe flange onto the end of a pipe for welding. The flange welding clamp is a straight section of steel approximately 12 inches long with a length of chain on one end and a screw clamp on the other. The clamp can be mounted on the top or bottom of the pipe and flange being welded.

Follow these steps to use a flange welding clamp:

NOTE
For this exercise, assume that a previously prepared 4-inch steel pipe will have a flange welded onto it.

Step 1 Inspect the flange welding clamp for any obvious damage, such as broken or worn-out chain links, bent or broken clamp feet, a warped clamp back, bent or broken chain hooks, damaged chain tighteners, a rusty or worn screw clamp, grease, rust, or excessive dirt.

Step 2 Repair or replace any damaged part. Clean and oil any parts as needed.

Instructor's Notes:

2.16 PIPEFITTING ◆ LEVEL ONE

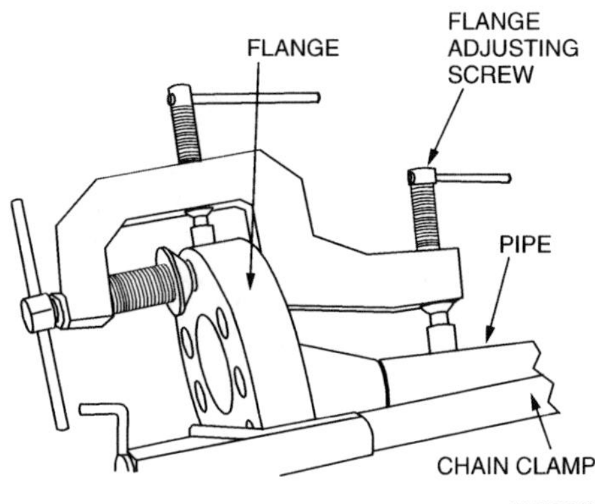

Figure 27 ◆ Flange welding clamp.

Step 3 Set up two adjustable pipe stands to support the pipe.

Step 4 Secure the section of pipe onto the stands, with the end to be welded extending at least 12 inches beyond one of the stands.

Step 5 Loosen the screw clamp on the end of the flange welding clamp.

Step 6 Position the edge of the larger part (lip) of the flange into the screw clamp opening.

Step 7 Align the flange to the pipe.

Step 8 Tighten the screw clamp onto the flange lip.

Step 9 Position the clamp and the flange so that the smaller opening of the flange is aligned with the end of the pipe where the weld will be.

Step 10 Insert a spacer between the flange and the butt end of the pipe.

Step 11 Hold the flange and the clamp in place, and wrap the clamp chain around the pipe.

Step 12 Attach the chain to the clamp.

Step 13 Turn the screw handle to tighten the chain.

Step 14 Check the flange and pipe joint again to see that they are still aligned with each other and ready for welding.

Step 15 After the welding has been completed, loosen the screw clamp from the flange lip.

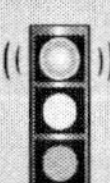

WARNING!
Handle the pipe and the clamps with care. Welded objects remain hot for several minutes after the welding is completed.

Step 16 Loosen the chain from the pipe.

Step 17 Remove the clamp from the flange and pipe.

Step 18 Clean the clamp if necessary, and inspect it for damage. If it is damaged, replace it.

Step 19 Store the flange welding clamp.

6.5.3 Other Chain-Type Welding Clamps

Other chain-type welding clamps include the T-joint welding clamp and the elbow welding clamp. The T-joint welding clamp is used to hold a pipe section perpendicular to another pipe to be welded. The elbow welding clamp is used to hold a 90-degree elbow to the end of a straight pipe to be welded. The procedures for using these are the same as for the other chain-type clamps. *Figure 28* shows T-joint and elbow welding clamps.

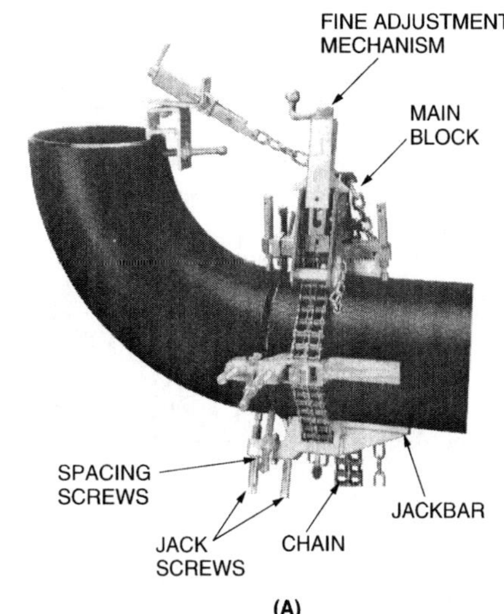

Figure 28 ◆ T-joint and elbow welding clamps.

Show trainees how to use flange welding clamps.

Emphasize the importance of careful handling of clamps and welded pipe.

Describe other chain-type welding clamps.

Have trainees practice using various types of clamps to secure pipe for welding.

Describe aligning dogs and explain how they are used.

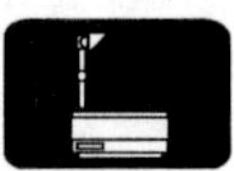

Show Transparency 12 (Figure 30). Describe Hi-Lo gauges and explain how they are used.

Describe a wraparound and explain how it is used.

Describe drift pins and explain how they are used.

6.5.4 Shop-Made Aligning Dogs

Many pipefitters fabricate their own aligning devices. These devices are known as aligning dogs in the pipefitting trade. *Figure 29* shows an example of an aligning dog.

The aligning dog shown in *Figure 29* is made from plate of the same material as the pipe being welded, to which a nut has been welded. The aligning devices are tack welded onto one section of the pipe so that the other section can be moved by tightening the bolt. Usually, two such dogs would be used to provide more control over the fit-up. Once the fit-up is complete, the aligning dogs are removed from the pipe. After the dog is removed from the pipe, the pipe must be ground to remove any welding deposits. Aligning dogs also require periodic grinding to remove welding deposits.

Welding codes in some areas prohibit anything being welded onto the pipe. Make sure you know the code before using any aids that must be welded to the pipe.

6.6.0 Hi-Lo Gauges

The primary purpose of a Hi-Lo gauge is to check for pipe joint misalignment (*Figure 30*). The name of the gauge comes from the relationship of the alignment of one pipe to another pipe. To check for internal misalignment, Hi-Lo gauges have two prongs, or alignment stops, that are pulled tightly against the inside diameter of the joint so that one stop is flush with each side of the joint. The variation between the two stops is read on a scale marked on the gauge. To measure misalignment using a Hi-Lo gauge, insert the prongs of the gauge into the joint gap. Pull up on the gauge until the prongs are snug against both inside surfaces, and read the misalignment on the end of the scale.

6.7.0 Wraparounds

A wraparound (*Figure 31*) is a piece of gasket material or leather belting that is used to draw a straight line around a pipe. A wraparound must have straight edges and must be long enough to wrap around the pipe one and a half times. When a wraparound is used correctly, the two overlapping edges should be aligned.

Figure 29 ◆ Aligning dog.

To use a wraparound, pull it tightly around the pipe until the edges overlap. Do not pull it tight enough to break it. Mark around the edge, using a scribe or soapstone.

6.8.0 Drift Pins

A drift pin is a round piece of steel that is tapered on both ends. Pipefitters use drift pins to align bolt holes when connecting flanges. Drift pins come in various sizes to fit different size bolt holes. *Figure 32* shows a common drift pin.

Follow these steps to use a drift pin:

Step 1 Select a pin of the proper size for the holes being aligned. The proper size drift pin will fit all the way through the bolt holes being aligned and will still be large enough and strong enough to pull the mating pieces together.

Step 2 Inspect the pin to make sure that it is not bent and that it does not have **burrs** that could cut your hands when using it.

Step 3 Insert the drift pin into one of the bolt holes of the mating flanges, and manipulate the pin to align the flanges.

Step 4 Repeat step 3 on the opposite side of the flanges.

Step 5 Insert bolts into the other bolt holes of the flanges, and hand tighten the bolts.

Step 6 Remove the drift pin from the flanges.

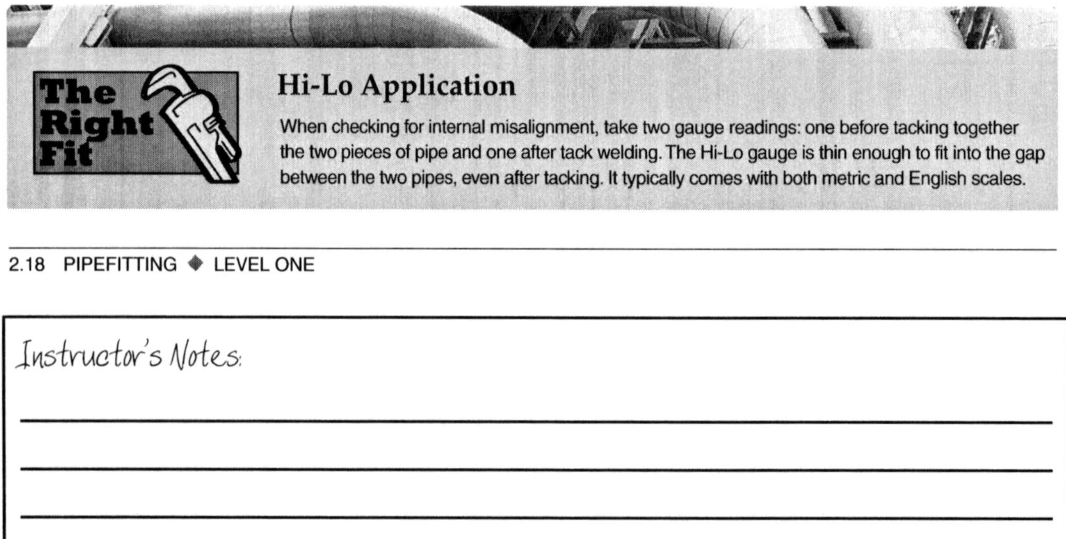

Hi-Lo Application

When checking for internal misalignment, take two gauge readings: one before tacking together the two pieces of pipe and one after tack welding. The Hi-Lo gauge is thin enough to fit into the gap between the two pipes, even after tacking. It typically comes with both metric and English scales.

Instructor's Notes:

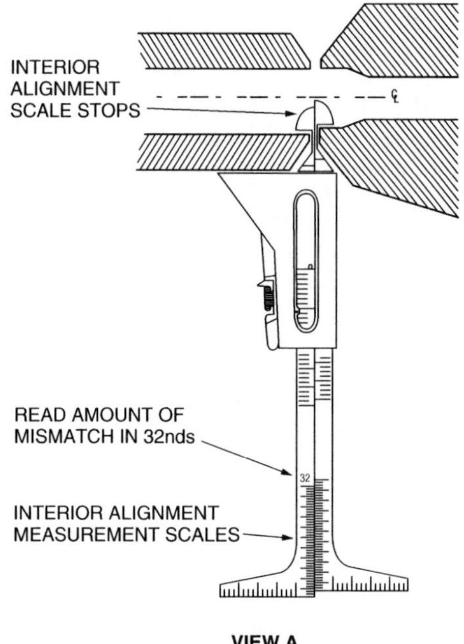

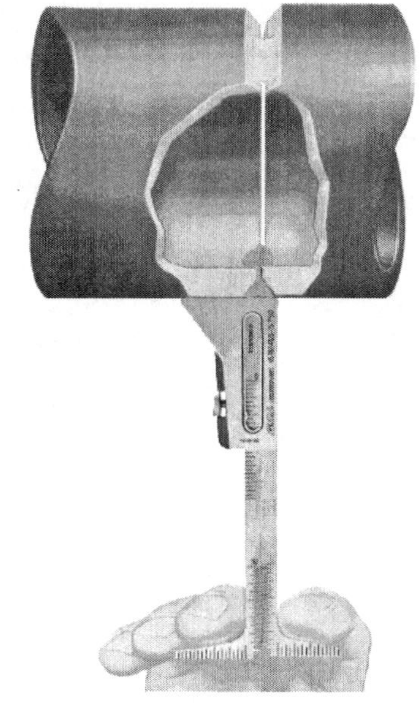

Figure 30 ◆ Checking internal misalignment using a Hi-Lo gauge.

Describe two-hole pins and explain how they are used.

Step 7 Clean the drift pin, using a rag and solvent if needed.

Step 8 Store the drift pin.

6.9.0 Two-Hole Pins

Two-hole pins are flange lineup pins used to orient a flange on the end of a pipe. These pins are available in a variety of sizes and are used to align the holes of a flange before it is welded. The pins are inserted into the top two holes on a flange. A level is then placed across them to enable you to align the flange. *Figure 33* shows two-hole pins used in conjunction with a flange lineup level.

Follow these steps to use two-hole pins:

Step 1 Select two pins of the proper size for the holes being aligned.

Step 2 Inspect the pins to make sure that they are not bent and that they do not have burrs that could cut your hands when using them.

Step 3 Unscrew the knurled flange nut from the flange pin if using the threaded pins and nuts.

Figure 31 ◆ Using a wraparound.

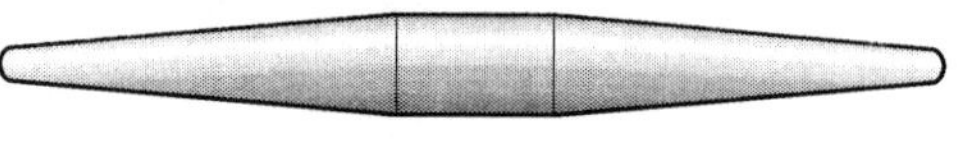

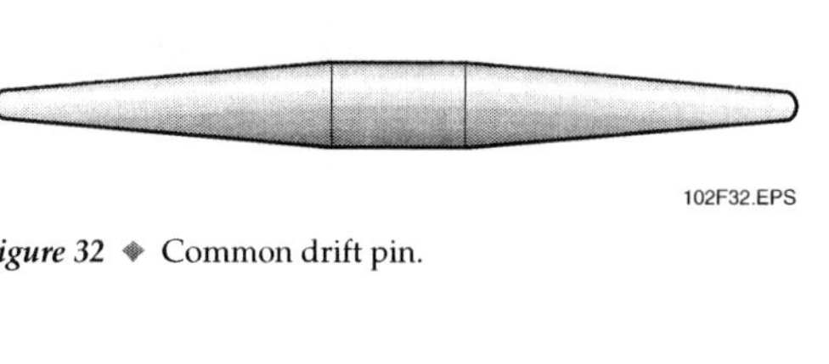

Figure 32 ◆ Common drift pin.

Describe flange spreaders and explain how they are used.

Show trainees how to use Hi-Lo gauges, wraparounds, drift pins, two-hole pins, and flange spreaders.

Have trainees practice using other pipe fabrication tools.

Have trainees review Sections 7.0.0–7.7.0.

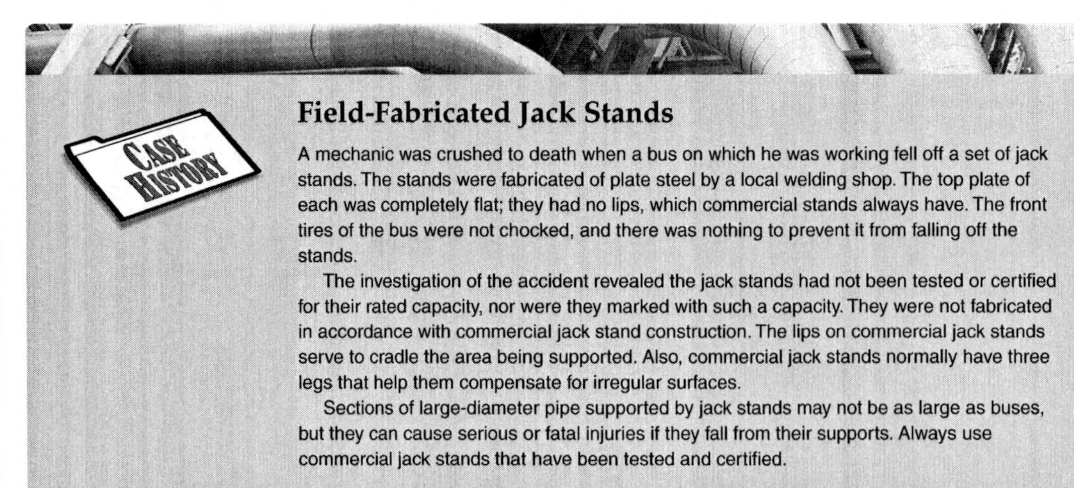

Field-Fabricated Jack Stands

A mechanic was crushed to death when a bus on which he was working fell off a set of jack stands. The stands were fabricated of plate steel by a local welding shop. The top plate of each was completely flat; they had no lips, which commercial stands always have. The front tires of the bus were not chocked, and there was nothing to prevent it from falling off the stands.

The investigation of the accident revealed the jack stands had not been tested or certified for their rated capacity, nor were they marked with such a capacity. They were not fabricated in accordance with commercial jack stand construction. The lips on commercial jack stands serve to cradle the area being supported. Also, commercial jack stands normally have three legs that help them compensate for irregular surfaces.

Sections of large-diameter pipe supported by jack stands may not be as large as buses, but they can cause serious or fatal injuries if they fall from their supports. Always use commercial jack stands that have been tested and certified.

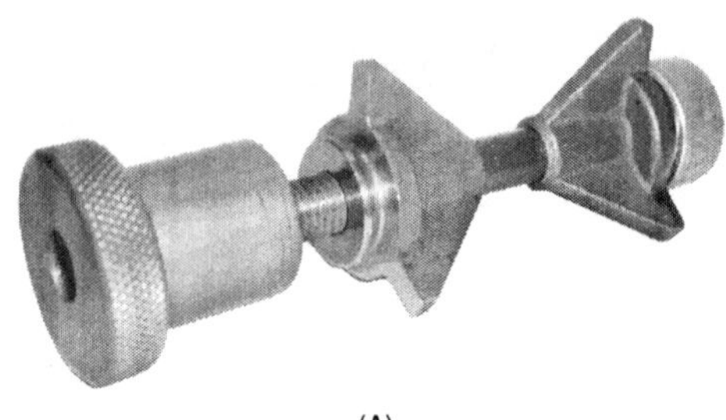

(A)

(B)

102F33.EPS

Figure 33 ◆ Two-hole flange lineup pins and level.

Step 4 Insert the flange pins into the two top bolt holes of the flange.

Step 5 Tighten the knurled flange nut onto the threaded end of the flange pin if using the threaded pins and nuts.

Step 6 Set a level over the two flange pins.

Step 7 Adjust the flange until it is level.

6.10.0 Flange Spreaders

Flange spreaders (*Figure 34*) are used to spread two mating flanges when gaskets need to be replaced without disassembling pipe lines. Flange spreaders exert tremendous pressure between the flanges smoothly and evenly, with no shock along the pipe line and without disturbing the alignment of the flanges. Flange spreaders must be used in pairs, one on each side of the flange, for even distribution of pressure.

102F34.EPS

Figure 34 ◆ Flange spreader.

Instructor's Notes:

2.20 PIPEFITTING ◆ LEVEL ONE

7.0.0 ◆ PIPE CUTTING TOOLS

Pipe cutting tools are used to properly cut and prepare pipe for joining and connection. A good connection depends on a clean, square cut and proper pipe reaming and threading. The following pipe cutting tools will be discussed:

- Hacksaws
- Soil pipe cutters
- Tube and pipe cutters
- Pipe reamers
- Hand pipe threaders

7.1.0 Hacksaws

A hacksaw (*Figure 35*) is a specialty tool designed for cutting metals, plastics, and other synthetics. It has a pistol grip handle and uses a variety of thin metal blades. Adjusting nuts on the frame secure the blade and allow you to vary the tightness of the blade. Blades are selected by the number of teeth per inch and by the material composition of the blade. Blades with finer and more numerous teeth per inch are used for cutting harder, stronger materials faster. *Table 1* shows which blades to use for various materials.

The blade should always be installed in the saw with the teeth pointing away from the handle because the forward stroke is the cutting stroke.

Table 1 Blades for Various Materials

Stock to Be Cut	Pitch of Blade (Teeth per Inch)	Explanation
Machine steel Cold rolled steel Structural steel	14	The coarse pitch makes the saw free and fast cutting.
Aluminum Babbitt Tool steel High-speed steel Cast iron	18	Recommended for general use.
Tubing Tin Brass Copper Channel iron Sheet metal (over 18 gauge)	24	Thin stock will tear and strip teeth on a blade of coarser pitch.
Small tubing Conduit Sheet metal (less than 18 gauge)	32	Recommended to avoid tearing.

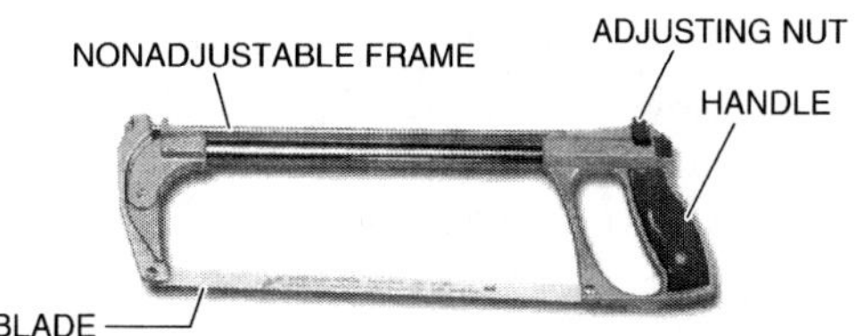

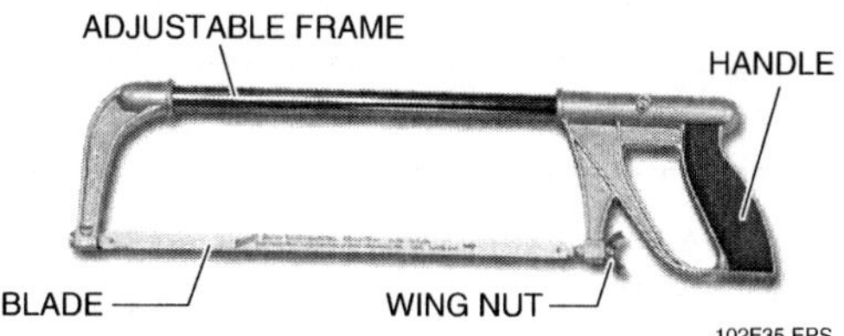

Figure 35 ◆ Hacksaws.

Because of their short length and brittleness, hacksaws require oiling to prevent rust and excessive wear.

Hacksaws are useful when an object cannot be taken to a power saw or the sawing job is small. Follow these steps to use a hacksaw:

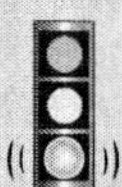

NOTE

For this exercise, assume that the blade needs to be replaced.

Step 1 Identify the kind of material to be cut.

Step 2 Select the hacksaw best suited for the job.

Step 3 Select the proper hacksaw blade for the metal to be cut. Certain types of metals require specific blades for cutting.

Step 4 Loosen the adjusting nut near the handle until the blade is loose on the holding posts.

Step 5 Remove the old hacksaw blade.

Step 6 Check the post, the saw frame, and the handle for excessive wear or damage.

Step 7 Install the new hacksaw blade onto the posts. Make sure the teeth of the new blade are pointed away from the handle.

Step 8 Tighten the adjusting nut until the blade is tight.

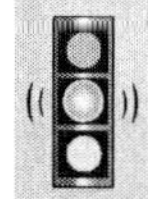

CAUTION

Do not overtighten the blade.

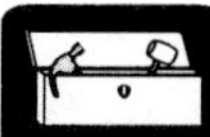

Ensure that you have everything required for teaching this session.

See the Teaching Tip for Section 7.0.0 at the end of this module.

Ask trainees to identify common pipe cutting tools.

Describe a hacksaw and explain how they are used.

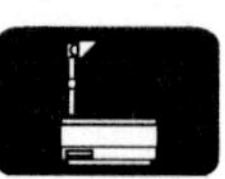

Show Transparency 13 (Table 1). Describe various hacksaw blades and explain what materials they are used for.

Explain how to use a hacksaw to cut pipe.

Show trainees how to install a blade on a hacksaw and use it to cut pipe.

Discuss the hazards posed by metal shavings from cut pipe.

Describe soil pipe cutters.

Step 9 Secure the object to be sawed in a vise.

Step 10 Measure and mark the area where the cutting will be performed.

Step 11 Position your body so that you are balanced and facing the work area.

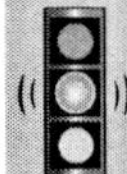

CAUTION

Always wear gloves when using a hacksaw.

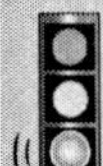

NOTE

When using a hacksaw, grasp the saw handle and face the cut area. If the saw handle is in your right hand, position your left foot forward and toward the work. Place your other hand on the top/front of the saw frame.

Step 12 Place the saw blade on the cutting line with most of the blade on the side of the mark away from you.

Step 13 Pull the saw back across the mark while applying a slight downward pressure. This will start the cut and allow for any needed correction. Keep the saw in line with your forearm as you begin with light, short strokes. Cutting takes place on the forward stroke.

Step 14 Press downward enough to make the teeth cut and push forward.

Step 15 Lift and pull the blade back to start another stroke. Do not lift the blade completely out of the groove. When sawing, keep your eyes on the sawing line rather than on the saw blade. Corrections may be made by a slight twist of the handle (and blade). Blow any cuttings off the line so that you can see.

Step 16 Continue sawing until the job is completed. When you near the end of the cut, saw slowly. Hold the waste piece in your other hand so that the material will not fall as you make the last stroke.

Step 17 Remove the saw from the work area, and carefully clean any buildup off the blade or the saw.

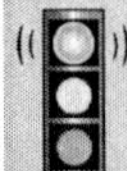

WARNING!

Be careful during cleanups. Freshly sawed objects and recently used blades may still be hot and can burn an unprotected hand. Also, metal shavings can puncture the skin.

Step 18 Oil the blade lightly to prevent rust.

Step 19 Clean the area.

Step 20 Store the hacksaw and any additional blades.

7.2.0 Soil Pipe Cutters

A soil pipe cutter (*Figure 36*) is used to cut cast iron soil pipe. It cannot be used to cut ductile iron pipe. It consists of a **ratchet** handle with a length of chain permanently attached to one side of the cutter. Each link of the chain contains a cutting wheel. The chain is wrapped around the pipe and locked into the side of the cutter. As the ratchet handle is pumped, the cutter tightens the chain around the pipe. As the

SOIL PIPE RATCHET CUTTER

SOIL PIPE SNAP CUTTER

102F36.EPS

Figure 36 ◆ Soil pipe cutters.

Instructor's Notes:

2.22 PIPEFITTING ◆ LEVEL ONE

chain tightens, the cutting wheels act as small chisels that penetrate the pipe until it is cut.

Follow these steps to use a soil pipe cutter:

Step 1 Select a length of cast iron soil pipe to be cut.

Step 2 Measure and mark the pipe to be cut.

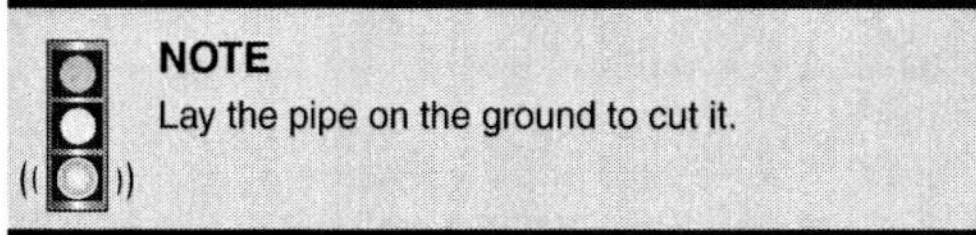

NOTE

Lay the pipe on the ground to cut it.

Step 3 Lift the pipe and place the chain underneath the cut mark.

Step 4 Place the cutter on top of the cut mark and pull the chain around the pipe.

Step 5 Attach the end of the chain in the open side of the cutter.

Step 6 Turn the chain tension knob clockwise to tighten the chain on the cut mark. The chain should be pulled as tight as possible.

Step 7 Pull out the lock knob and turn it so that the arrow points toward the handle.

Step 8 Push the handle down to start cutting the pipe.

Step 9 Continue to pump the handle until the pipe is broken. The ratchet will allow you to lift the handle without affecting the chain tension.

Step 10 Inspect the soil pipe cutter for any obvious damage.

Step 11 Store the cutter.

7.3.0 Tube and Pipe Cutters

Tube cutters (*Figure 37*) are used to cut thin-walled metal tubing made of copper, brass, aluminum, or steel. They are usually small, single-wheel cutters. Some models have rollers that are tightened to force the pipe against the cutting wheel, while other models have a sliding cutting wheel that is forced against the pipe. All tube cutters must be rotated completely around the tube to make the cut.

Pipe cutters (*Figure 38*) are larger and more durable than tube cutters, but they operate in the same manner. Manual pipe cutters are designed to cut pipe such as carbon steel, brass, copper, cast iron, stainless steel, and lead (up to 6 inches in diameter). There are several types of pipe cutters,

Explain how to use soil pipe cutters.

Show trainees how to use soil pipe cutters.

Describe various types of tube and pipe cutters.

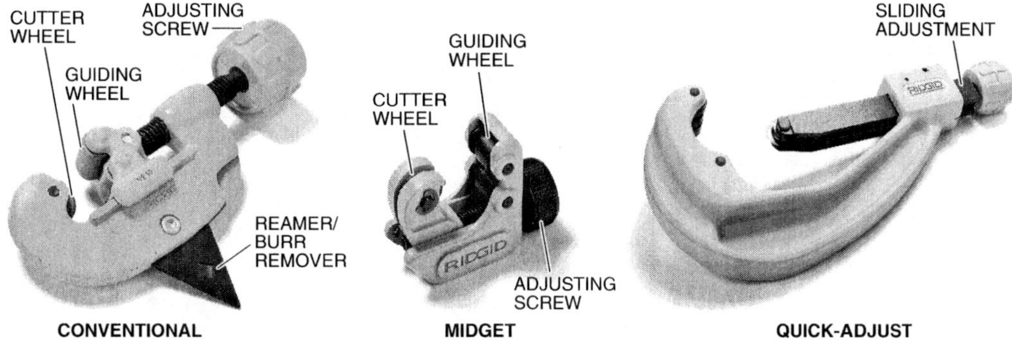

Figure 37 ◆ Types of tube cutters.

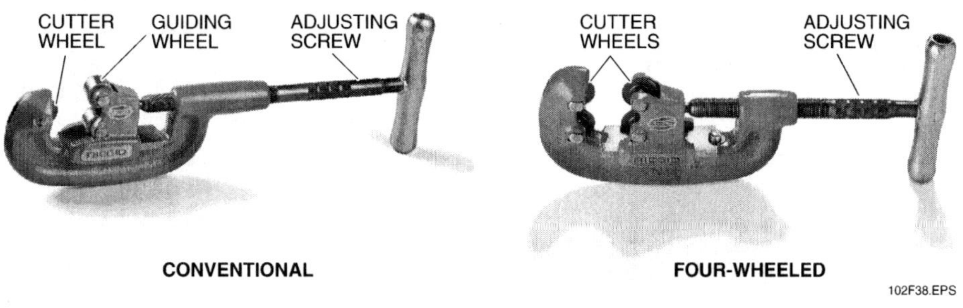

Figure 38 ◆ Pipe cutters.

MODULE 08102-06 ◆ PIPEFITTING HAND TOOLS 2.23

Explain how to use various types of tube and pipe cutters.

Emphasize the hazards posed by the sharp edges of newly cut pipe.

Show trainees how to how to use various types of tube and pipe cutters.

Explain that reamers are used to remove burrs after cuts.

with the one-wheel cutter being the most commonly used. The one-wheel cutter has one alloy cutting wheel and two guide rollers. The rollers help the wheel make a straight cut and prevent burrs from forming on the surface of the pipe. The pipe cutter's main advantage over the hacksaw is accuracy and smoothness of cut.

The single-wheel pipe cutter is rotated entirely around the pipe and can only be used where there is enough space. The four-wheel cutter can be used where you cannot rotate the cutter around the pipe more than one-third of a turn. Since the four-wheel cutter has no guide rollers to guide the cutter wheels, extra care must be taken to ensure perfect tracking.

After it is cut, the pipe must be reamed using a manual hand reamer. The outer edges of the cut should be filed smooth. Many tube cutters have a reamer attached to the body of the cutter. These can be used to clean the burrs inside the pipe after the cut has been made. Follow these steps to use a tube or pipe cutter:

Step 1 Select, measure, and mark the pipe to be cut.

Step 2 Identify the size and kind of material to be cut.

Step 3 Select a cutting tool to match the material being cut.

CAUTION

Pipe cutters may be too strong or heavy for cutting thin-walled tubing or plastic pipe. Use a tubing cutter in these cases.

Step 4 Inspect the selected cutter for obvious damage or excessive wear on the cutting edges and for grease, rust, and excessive dirt. If any damage is found, replace the cutter. If grease, rust, or excessive dirt is found, clean as necessary and lightly oil the moving parts.

Step 5 Secure the pipe in a vise, and install pipe supports if necessary. Secure the pipe near the cutting mark, but leave enough room to work the cutter.

Step 6 Put on safety glasses.

Step 7 Place a catch pan on the floor underneath the cutting mark to catch the cutting oil coming off the object.

Step 8 Place the cutter around the pipe at the cutting mark. Position the cutter so that the opening is facing up underneath the object being cut.

Step 9 Line up the cutting wheel with the cutting mark, and tighten the cutter handle until the cutting wheel is snug on the object.

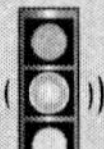

CAUTION

Do not overtighten the handle. This can break the cutting wheel or crush the object if it is plastic or light tubing.

Step 10 Rotate the cutter completely around the pipe. When using a four-wheel cutter in a confined space where you cannot rotate the cutter around the pipe, move the cutter back and forth on the pipe.

Step 11 Tighten the cutter handle one-quarter of a turn to apply more pressure.

Step 12 Make another complete circle around the pipe.

NOTE

Occasionally apply a small amount of cutting oil to help ensure a smooth and easy cut. The oil also helps to prevent wear on the cutter.

Step 13 Continue tightening the cutter handle and circling the pipe until the cut is complete. When the cut appears to be almost complete, hold the excess piece to keep it from dropping when the cut is completed.

Step 14 Remove the cutter.

Step 15 Wipe the cutting oil off the pipe with a towel.

WARNING!

The cut edge on the pipe is very sharp. Use caution when cleaning and handling to avoid cutting yourself.

Step 16 Clean the cutter and check it for any obvious damage.

Step 17 Close the cutter before storing it.

7.4.0 Manual Pipe Reamers

Hacksaws and cutter wheels leave internal burrs after cuts. Reamers are designed to remove these burrs and to make the inside edges of the cut

Instructor's Notes:

2.24 PIPEFITTING ◆ LEVEL ONE

smooth. Most manual reamers are of the ratchet type, consisting of a reamer cone, ratchet housing with handle, and handgrip. Reamer cones come in a variety of sizes and designs for specialty reaming jobs. A long-taper straight flute is used for smaller diameter pipe and rougher surfaces; a tapered spiral flute is designed for larger diameter, easier, and faster reaming. *Figure 39* shows ratchet-type manual pipe reamers.

Reamers should be kept lightly oiled and cleaned after each use. Always ream pipe before threading, because reaming will sometimes stretch the pipe. Follow these steps to use a ratchet-type pipe reamer:

Step 1 Identify the kind of reamer needed.

Step 2 Select the type and size of reamer needed to manually ream a given piece of pipe.

Step 3 Select a drive handle for the reamer according to the **die** head size.

Step 4 Inspect the selected reamer and handle for obvious damage or excessive wear on the cutting edges and for grease, rust, and excessive dirt. If any damage is found, replace the reamer. If grease, rust, or excessive dirt is found, clean as necessary.

Step 5 Connect the reamer and the drive handle.

Step 6 Locate the item to be reamed, and secure it with a vise. Secure the object near its end, but leave enough room to work the reamer.

Step 7 Position yourself in front of the work area, if possible, and balance your body as much as possible.

Step 8 Insert the tip of the reamer into the exposed end of the pipe. Guide it in with one hand and use the other hand for pressure on the rear of the reamer.

Step 9 Apply light forward pressure, and start rotating the reamer clockwise. The reamer should bite as soon as the proper pressure is applied.

Step 10 Rotate the reamer a couple of times; withdraw it while continuing to rotate it, and check the progress. Continue if needed.

CAUTION

The inside edge of the pipe should become very smooth. Do not over-ream. Stop when all burrs are gone.

Step 11 Remove the reamer.

Step 12 Clean the reamer with a shop cloth and check it for any obvious damage. If it is damaged, replace it.

Step 13 Disassemble the reamer and store all parts.

7.5.0 Hand Pipe and Bolt Threaders

Pipefitters often have to use hand pipe and bolt threaders. The ratchet-type hand pipe threader with removable, interchangeable die heads (*Figure 40*) is the most common. Die heads and dies are

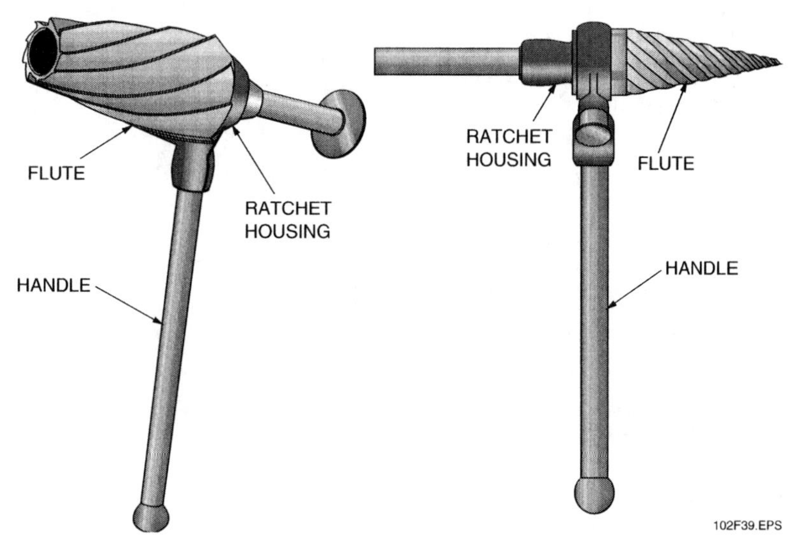

Figure 39 ◆ Ratchet-type manual pipe reamers.

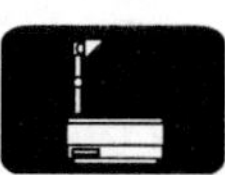

Show Transparency 14 (Figure 39). Describe a ratchet-type manual reamer.

Explain how to use a ratchet-type manual reamer.

Show trainees how to how to use a ratchet-type manual reamer.

Have trainees practice using various types of tools to cut pipe and clean it using a reamer.

Show Transparency 15 (Figure 40). Describe a ratchet-type hand pipe threader.

Discuss thread size and explain how to select the correct die for the threader.

Explain how to keep a pipe threader in good condition.

Explain how to replace a die head on a threader.

Show trainees how to how to replace a die head on a threader.

sized for specific diameters of pipe and are used to cut **male threads**. The size of the thread can be found on the face of the die heads. Special dies are available for threading brass, aluminum **conduit**, stainless steel, wrought iron, and cast iron, and for left-handed threads. These dies should not be used on any material other than what they were designed for, because the dies could be damaged or produce imperfect threads.

Most pipe threads are NPT threads, which designate National Pipe Threads. In some cases you will find metric threads, though. The most common bolt threads are NC and NF, which designate National Coarse and National Fine. Always check the engineering specifications before selecting threading dies.

7.5.1 Using and Caring for Pipe and Bolt Threaders

To cut threads correctly, the pipe threader must be kept in good condition. Follow these guidelines when using hand pipe threaders:

- Inspect the tool regularly for broken, worn, or damaged parts.
- Inspect dies regularly to make sure they are sharp and not chipped.
- Use a good quality cutting oil to lubricate the cutting edges and remove chips and curls.
- Use good techniques when cutting. Sloppy or careless work will not produce good threads.

7.5.2 Replacing Die Heads

Most handles will accept many different die heads for different jobs. Follow these steps to replace the die head on a pipe threader handle:

Step 1 Pull out the lock knob, and turn the knob so that the arrow points to the handle.

Step 2 Lift the die head straight out of the pipe threader handle.

Step 3 Insert a new die head into the pipe threader handle.

Step 4 Pull out and turn the lock knob so that the arrow is pointing away from the handle.

7.5.3 Replacing Dies

The die head holds the dies in place. When the dies become worn or damaged, they must be replaced. Follow these steps to replace dies:

Step 1 Remove the die head from the handle.

Step 2 Remove the four screws in the top of the die head using a screwdriver.

Step 3 Remove the cover plate from the top of the die head.

Step 4 Slide the four dies out of the die head.

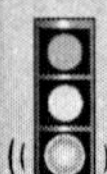

NOTE

Before replacing the dies, make sure the die head and the dies are clean, dry, and free of all metal chips. A small chip can raise a die and cause uneven pressure. This will tear the pipe thread and may break the die.

Step 5 Insert the new dies into the die head, making sure the number on each die matches the number on each die head slot.

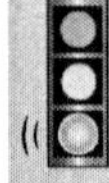

NOTE

It is very important to keep the die sets together. Do not mix dies of one set with dies from another set. Usually, each die is marked with a serial number and the four dies that are made together have the same number.

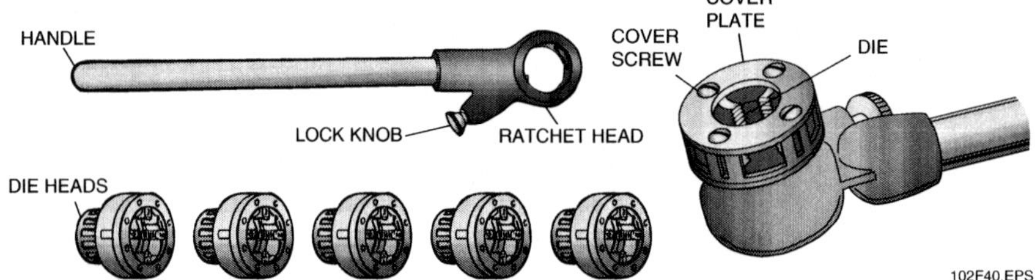

Figure 40 ◆ Hand threader.

Instructor's Notes:

2.26 PIPEFITTING ◆ LEVEL ONE

Step 6 Replace the cover plate onto the die head.

Step 7 Tighten the four screws into the die head.

Step 8 Replace the die head into the pipe threader handle.

7.5.4 Threading Pipe

To thread pipe using a hand threader and vise, proceed as follows:

Step 1 Identify the kind of material being threaded.

Step 2 Determine the diameter of the pipe.

Step 3 Select the die head best suited for the threading process.

Step 4 Insert the die head into the pipe threader handle.

Step 5 Inspect the die and die handle to ensure that all parts are clean and in good condition.

Step 6 Secure the pipe in a vise.

Step 7 Make sure that the end of the pipe is square, clean, and deburred.

Step 8 Position your body so that you are balanced, facing the work area, and within easy reach of the work area.

Step 9 Position the die flush on the end of the pipe. Make sure that the arrow on the lock knob points in the direction in which you are threading the pipe.

Step 10 Press the die against the end of the pipe, and turn the die handle clockwise to start the threading process.

Step 11 Turn the die handle two full rotations around the pipe. Add cutting oil as necessary to keep a thin coat on the threads while cutting.

Step 12 Stop the work, pull out the lock knob, and turn the knob so that the arrow points in the opposite direction.

Step 13 Back the dies off the pipe. This allows you to make sure the threads are properly started and allows the threads to be cleaned from the cutting die.

Step 14 Clean the cutting threads from the die if necessary.

Step 15 Pull out the lock knob, and turn it so that the arrow points in the direction in which you are threading.

Step 16 Rethread the die onto the pipe, and continue the process until the threading is complete. The threading is complete when one full thread extends past the back edge of the dies.

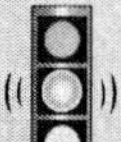

CAUTION

Do not thread the pipe too far. This can cause an improper fit between the fitting and the pipe, which will result in leaks.

Step 17 Back the die off the pipe.

Step 18 Inspect the new threads and use a rag to remove any debris from the threads.

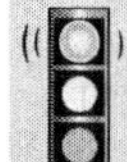

WARNING!

Freshly cut threads are very sharp. Use care in handling to avoid cuts.

7.5.5 Checking Threads

A **thread gauge** is used to check the threads on a bolt. *Figure 41* shows a thread gauge.

A thread gauge has many leaves. Each leaf measures 1 inch across and has a certain number of thread markers per inch as noted on the side of each leaf. This measures the number of threads per inch, also known as the pitch. The number of threads per inch depends on the diameter of the bolt. These are not commonly used on pipe threads.

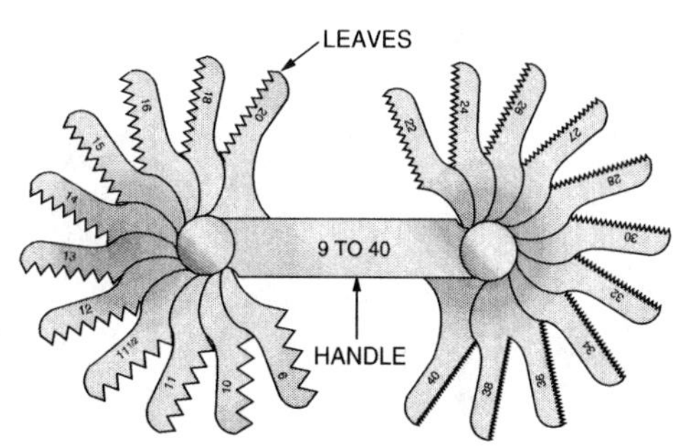

Figure 41 ◆ Thread gauge.

Explain how to thread pipe using a pipe threader.

Show trainees how to thread pipe using a pipe threader.

Show Transparency 16 (Figure 41). Describe a thread gauge and explain how they are used.

Have trainees practice using a pipe threader and checking the gauge.

MODULE 08102-06 ◆ PIPEFITTING HAND TOOLS 2.27

Describe a pipe extractor and explain how it is used.

Show trainees how to how use a pipe extractor.

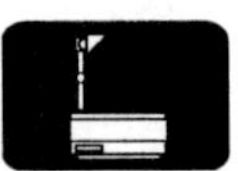

Show Transparency 17 (Figure 44). Describe a pipe tap and explain how it is used.

7.6.0 Pipe Extractors

Pipe extractors (*Figure 42*), also referred to as easy-outs, are used to remove broken threaded ends of pipe, plugs, and fittings. Each is plainly marked, showing the size pipe it should be used with and the size hole that must be drilled if fittings must be drilled to receive the extractor.

Follow these steps to use a pipe extractor:

Step 1 Identify the size of broken pipe or fitting that needs to be removed.

Step 2 Select the proper size pipe extractor to use with the broken pipe or fitting.

Step 3 Drive the cutting end into the broken pipe or fitting, using a hammer.

Step 4 Place a wrench on the square end of the extractor, and unscrew the broken fitting or pipe to remove it.

Step 5 Remove the broken pipe or fitting from the extractor.

Step 6 Inspect the extractor for any damage, and store it.

7.7.0 Pipe Taps

Taps are used to cut female threads of a pipe. *Figure 43* shows a pipe tap and a tap and die set.

The tap shown in *Figure 43* has the first three threads ground off. This helps start the tap in a hole and evens the cutting pressure over several threads on the tap.

Pipe taps are used to cut threads that are larger than the stated tap size. A ½-inch pipe tap cuts threads that are almost 11/16 inch in diameter. This is because on a ½-inch pipe, the **outside diameter** is 11/16 inch (*Figure 44*).

The pipe wall must be considered when drilling a hole for **female threads**. A pipe tap marked ½ inch will tap threads for a ½-inch pipe, but the tap is actually larger than ½ inch to allow for the pipe walls. Use a ½-inch pipe tap and a ½-inch pipe will fit.

The hole that is drilled before making the pipe threads must be the proper size. If the hole is too small, the tap will have to be forced in and may break. If the hole is too large, the tap will be loose and will not cut proper threads. *Table 2* shows what size drill to use for each size of tap. Notice that the tap drill is always larger than the tap size. Again, this is because the tap is actually larger than its listed size.

Follow these steps to cut female threads in pipe using a pipe tap:

102F42.EPS

Figure 42 ◆ Pipe extractors.

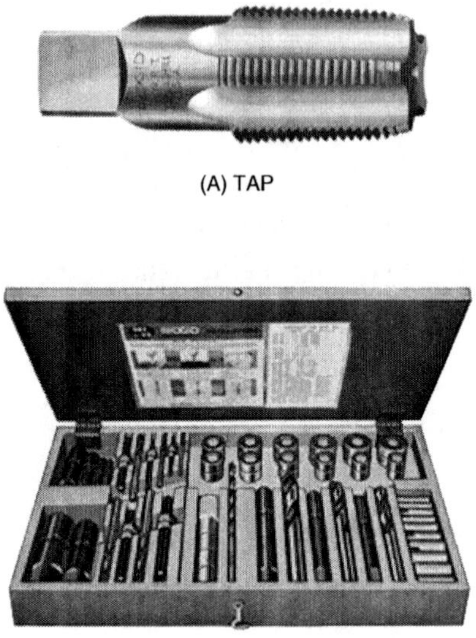

(A) TAP

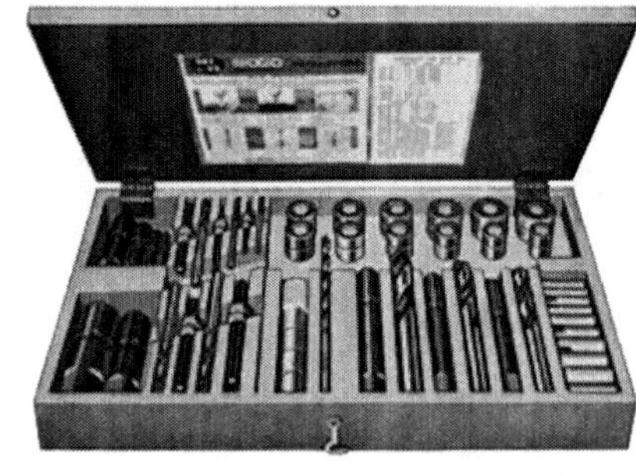

(B) TAP AND DIE SET

102F43.EPS

Figure 43 ◆ Pipe taps.

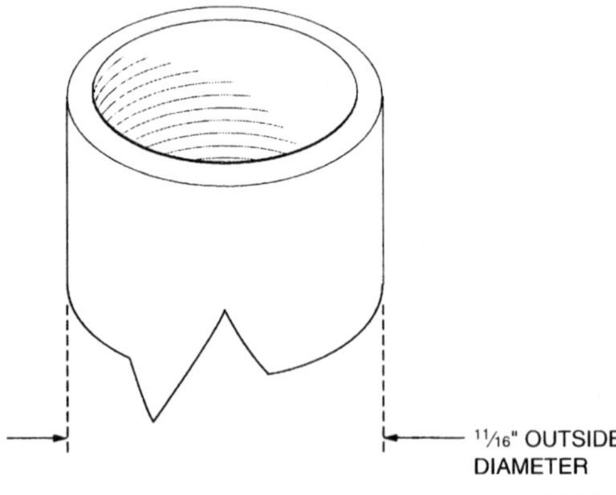

102F44.EPS

Figure 44 ◆ ½-inch diameter pipe.

Instructor's Notes:

2.28 PIPEFITTING ◆ LEVEL ONE

Step 1 Use a center punch to mark the spot where the hole is to be made.

Step 2 Refer to *Table 2* to determine the proper drill and tap to use.

Step 3 Obtain the drill and tap.

Step 4 Put on safety goggles and gloves.

Step 5 Drill the hole straight in toward the center of the pipe.

> **WARNING!**
> If the pipe has been in use, make sure the pipe is empty and contains no pressure. If the pipe has been used to move acids, alkalis, or chemicals, purge the pipe before drilling.

Step 6 Remove the drill, and store it in a safe place.

Step 7 Fasten the tap to the head, and insert the tap into the hole.

Step 8 Apply thread-cutting oil to the tap.

Step 9 Apply pressure, and turn the tap in one-half turn.

Step 10 Back off the tap one-quarter turn to break off chips and clear the tap.

Step 11 Repeat steps 8 and 9 until the required number of threads are cut.

> **CAUTION**
> Do not allow chips to clog the tap. If the tap will not turn and has a springy feeling, stop. Back off the tap one-quarter turn to break off chips, and use a piece of wire to clean the chips from the tap.

Table 2 Drill and Tap Sizes

Pipe Tap Size (Inches)	Drill Size (Inches)	Pipe Tap Size (Inches)	Drill Size (Inches)
1/8	11/32	2	2 3/16
1/4	7/16	2 1/2	2 5/8
3/8	19/32	3	3 1/4
1/2	23/32	3 1/2	3 3/4
3/4	15/16	4	4 1/4
1	1 5/32	4 1/2	4 3/4
1 1/4	1 1/2	5	5 5/16
1 1/2	1 23/32	6	6 3/8

8.0.0 ◆ BENDERS AND FLARING TOOLS

Benders are the tools used to manually bend pipe and tubing. Spring benders, lever compression tube benders, and manual benders (hickeys) are the most common manual bending tools. It is important to get the desired bend without flattening, kinking, or wrinkling the tube or pipe. Pipe and tube bending requires detailed calculations, but this exercise only covers use of the benders. For this exercise, simply bend a piece of pipe or tubing without making prior calculations.

8.1.0 Spring Benders

The simplest hand bending tool is the spring bender (*Figure 45*). The spring bender should fit tightly over the tube and should be centered over the area to be bent. Spring benders are used to bend soft copper, aluminum, and other soft metals. They are available in five sizes for pipe between 1/4 inch and 5/8 inch in diameter.

Follow these steps to bend tubing using a spring bender:

Step 1 Determine the size of the tubing to be bent.

Step 2 Select the proper size spring bender for the tubing. The spring must fit tightly over the tubing to ensure a good bend.

Step 3 Position the spring over the center of the area to be bent.

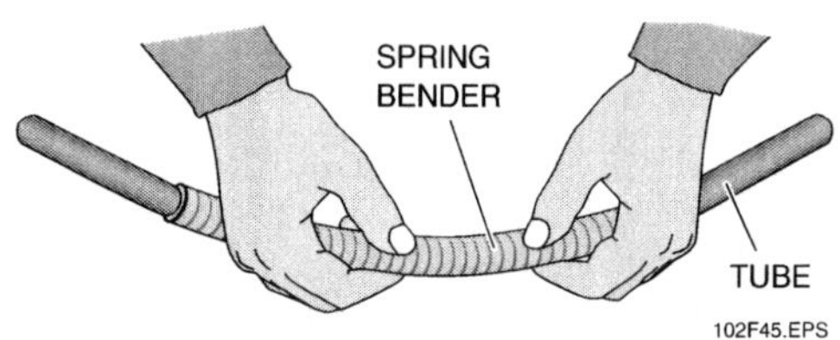

Figure 45 ◆ Tube-bending spring.

Show Transparency 18 (Table 2). Explain how to select the correct drill and tap size.

Emphasize the importance of checking that the pipe is empty before drilling.

Have trainees practice using a pipe extractor and a pipe tap.

Have trainees review Sections 8.0.0–8.4.0.

Ensure that you have everything required for teaching this session.

Identify the most commonly used pipe benders and explain how they are used.

Show Transparency 19 (Figure 45). Describe a spring bender and explain how it is used.

MODULE 08102-06 ◆ PIPEFITTING HAND TOOLS 2.29

Describe a lever compression tube bender and explain how it is used.

Describe a manual bender and explain how it is used.

Show trainees how to use various types of bending tools.

Describe a flare joint.

Step 4 Grasp the spring between your fingers and thumbs, and carefully bend the tubing until the proper radius and angle are formed.

Step 5 Slip the spring off the tube.

8.2.0 Lever Compression Tube Benders

Lever compression tube benders (*Figure 46*) are available in nine sizes for outside diameters between 3⁄16 inch and 7⁄8 inch. They can be used with soft and hard copper, brass, aluminum, steel, and stainless steel tubing. If the tubing is marked correctly prior to bending, bends are accurate to about 1⁄32 of an inch. A 90-degree bend in tubing adds approximately one tubing size diameter to the overall length of the tubing. For example, if you bend a 10-inch long, ½-inch diameter piece of tubing 90 degrees, the overall length of the tubing will be 10½ inches after bending.

Follow these steps to bend tubing using a lever compression bender:

Step 1 Hold the sheave block handle in your left hand, and open the slide bar to its widest position.

Step 2 Slide the tube into the bender and align it so that the start of the bend will be on the zero mark on the bender.

Step 3 Snap the clip over the tube to hold the tube in place.

Step 4 Turn the slide bar around until the full length of the groove and the slide bar are in contact with the tube.

Step 5 Pull the slide bar handle around in a smooth, continuous motion to bend the tube.

Step 6 Stop the bend at the required angle.

CAUTION

Be careful not to bend the tube past the required angle. It is difficult to straighten a tube without damaging it.

Step 7 Open the bender to its widest position.

Step 8 Raise the clip and remove the tube from the bender.

Step 9 Check the bender for the proper angle and for any signs of damage to the tube.

Step 10 Return the tools to their proper place.

8.3.0 Manual Benders

Manual benders are designed to bend electrical conduit, but are sometimes used by pipefitters to bend pipe. There are two types of manual benders, one for pipe and heavy-wall conduit and one for thin-wall conduit. *Figure 47* shows two types of manual benders.

Follow these steps to use a manual bender or hickey:

Step 1 Determine the size of the pipe to be bent.

Step 2 Select the proper size bender.

Step 3 Mark the pipe at the beginning of the bend.

Step 4 Lay the pipe on the floor.

Step 5 Slide the bender onto the pipe at the start of the bend.

Step 6 Grip the handle and pull the bender down toward the floor to make the bend.

NOTE

The thin-wall conduit bender will make a smooth bend with one long pull of the handle; it has a heel that you can press with your foot to assist in bending the pipe. The pipe and heavy-wall conduit bender requires you to make a small bend, slide the bender along the pipe a short distance, and make another bend until the proper bend is achieved.

Step 7 Stop the bend when the required bend angle is achieved.

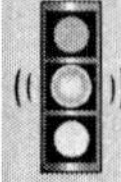

CAUTION

Be careful not to bend the pipe past the required angle. It is difficult to straighten pipe without damaging it.

Step 8 Remove the bender from the pipe.

Step 9 Check the pipe for cracking or kinking.

Step 10 Store the bender in its proper place.

8.4.0 Flaring Tools

Flare joints may be required in an installation in which a fire hazard exists and a torch for soldering or brazing is not permitted. A flare joint is made by measuring, cutting, and reaming the pipes and using flare fittings. This kind of joint is commonly

Instructor's Notes:

2.30 PIPEFITTING ◆ LEVEL ONE

Figure 46 ◆ Lever compression tube benders.

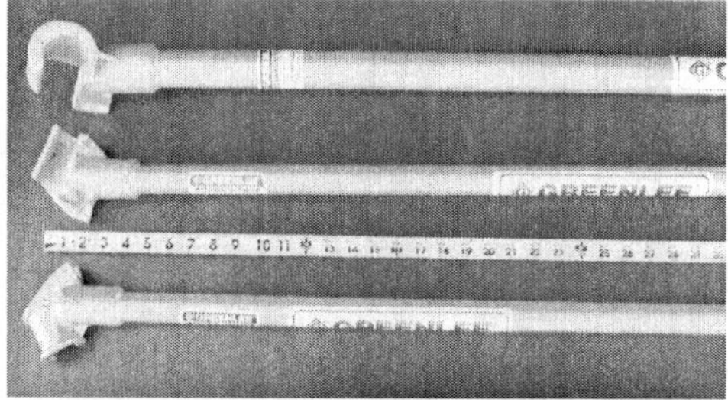

HICKEYS FOR PIPE AND HEAVY WALL CONDUIT

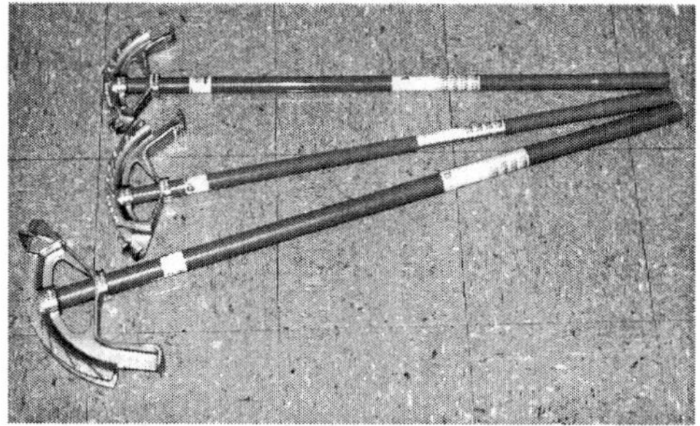

THIN WALL CONDUIT BENDERS

102F47.EPS

Figure 47 ◆ Manual benders.

Figure 48 ◆ Flaring tool.

used to join soft copper tubing with diameters from ¼ to 2 inches. Soft copper fittings should be leakproof and easily dismantled with the right tools.

Two kinds of flare fittings are popular: the single-thickness flare and the double-thickness flare. For both types, you use a special flaring tool (*Figure 48*) to expand the end of the tube outward into the shape of a cone, or flare. The single-thickness flare forms a 45-degree cone that fits against the face of a flare fitting (see *Figure 49*). In a single operation, the single-thickness flare is formed, and then the lip is folded back and compressed to make a double-thickness flare. The double-thickness flare (see *Figure 50*) is preferable with larger-size tubing. A single-thickness flare may be weak when used under excessive pressure or expansion. Double-thickness flare connections are easier to dismantle and reassemble without damage than single-thickness flare connections.

To make a flare joint, follow these steps:

Step 1 Measure, cut, and ream the tubing.

Step 2 Slip the flare nut over the tubing.

Step 3 Use the flaring tool to flare the tubing's ends until the fit is perfect.

Step 4 Slide the nuts onto each end of the flare tubing. Then gently bend or shape the tubing by hand over the male thread of each fitting.

Step 5 Use a smooth-jaw adjustable wrench to tighten until the fitting is snug.

Step 6 Test the joint for leaks.

Show Transparencies 20 and 21 (Figures 49 and 50). Describe single-thickness and double-thickness flares.

Explain how to make a flare joint.

Show trainees how to use flaring tools.

Have trainees practice fabricating pipe and checking the square and level. Note the proficiency of each trainee. This laboratory corresponds to Performance Task 5.

Have trainees practice identifying various pipefitting hand tools. Note the proficiency of each trainee. This laboratory corresponds to Performance Task 1.

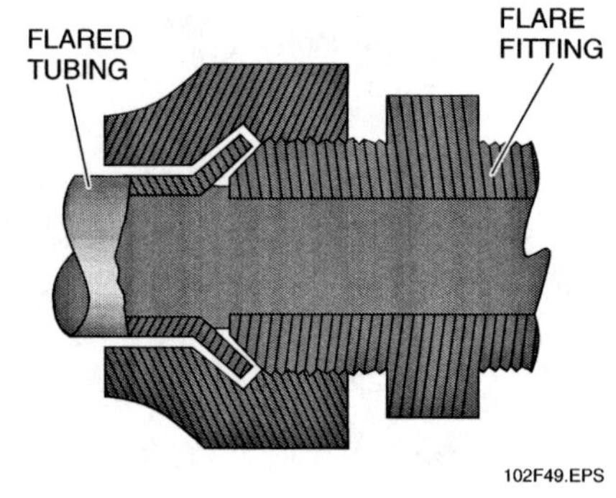

Figure 49 ◆ Single-thickness flare.

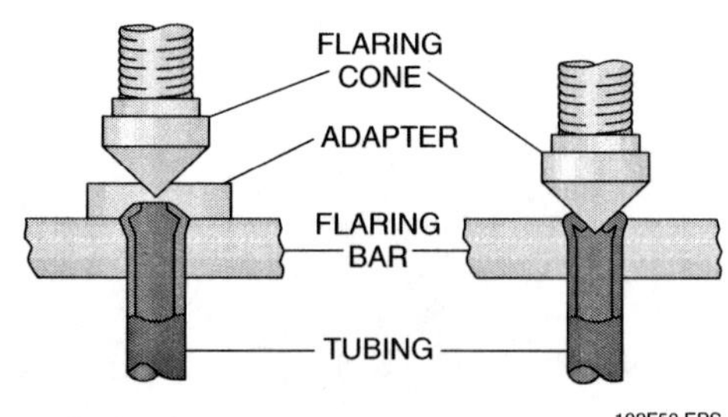

Figure 50 ◆ Double-thickness flare.

Instructor's Notes:

2.32 PIPEFITTING ◆ LEVEL ONE

Review Questions

1. A screwdriver may be used as a pry bar as long as it is sturdy enough for the application.
   a. True
   b. False

2. A three-legged pipe jack stand can be used for pipe up to _____ inches in diameter.
   a. 2
   b. 4
   c. 6
   d. 8

3. The height of a three-legged jack stand is typically adjusted by placing it on blocks.
   a. True
   b. False

4. The type of wrench that would be used on a chrome-plated pipe is the _____ wrench.
   a. strap
   b. chain
   c. Crescent®
   d. offset pipe

Figure 1

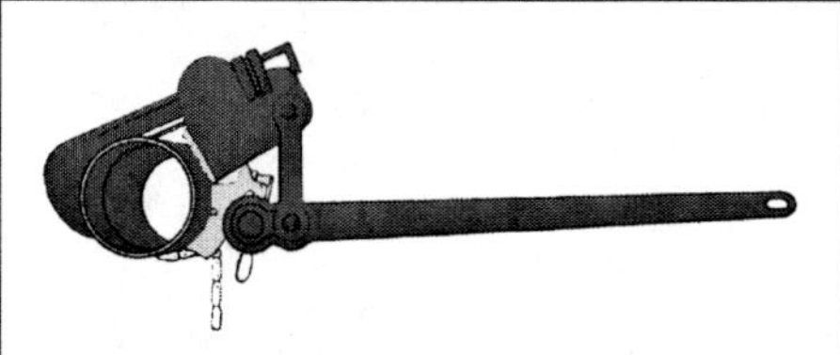

102RQ01.EPS

5. The tool shown in *Figure 1* is a(n) _____.
   a. pipe vise
   b. offset pipe wrench
   c. strap wrench
   d. compound-leverage pipe wrench

6. The term plumb relates to _____.
   a. horizontal alignment
   b. vertical alignment
   c. pipe height
   d. pipe location

7. A torpedo level is a good choice when accuracy is important.
   a. True
   b. False

8. A framing level can be used to measure _____.
   a. plumb only
   b. level only
   c. plumb or level

9. Which of the following is a task that *cannot* be performed with a framing square?
   a. Measuring the diameter of a pipe
   b. Aligning two sections of pipe
   c. Checking squareness of a pipe end
   d. Finding the center line of a pipe

Figure 2

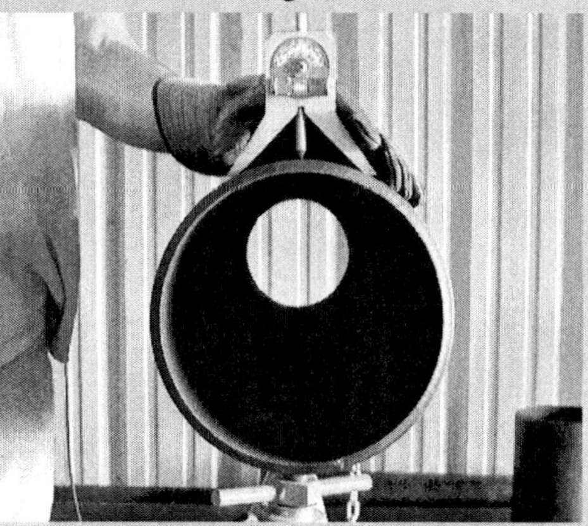

102RQ02.EPS

10. *Figure 2* shows a tool used to _____.
    a. tap a hole in a pipe
    b. locate the center of a pipe
    c. measure the outside diameter of a pipe
    d. bevel a pipe

11. A T-joint welding clamp is used to hold a pipe section perpendicular to another pipe.
    a. True
    b. False

Have the trainees complete the Review Questions, and go over the answers prior to administering the Module Examination.

MODULE 08102-06 ◆ PIPEFITTING HAND TOOLS 2.33

Review Questions

12. A hi-lo gauge is used to _____.
    a. check pressure in a piping system
    b. check for weld defects
    c. check pipe joint misalignment
    d. measure the inside diameter of a pipe

13. A wraparound is used to _____.
    a. align pipe joints for welding
    b. draw a straight line around a pipe
    c. measure the inside diameter of a pipe
    d. secure a pipe to a jack stand

14. Drift pins are used to _____.
    a. align bolt holes on pipe flanges
    b. secure pipes in jack stands
    c. keep pipe joints from moving while they are being welded
    d. check alignment of a pipe joint

15. A hacksaw blade used to cut copper tubing should have _____ teeth per inch.
    a. 14
    b. 18
    c. 24
    d. 32

Figure 3

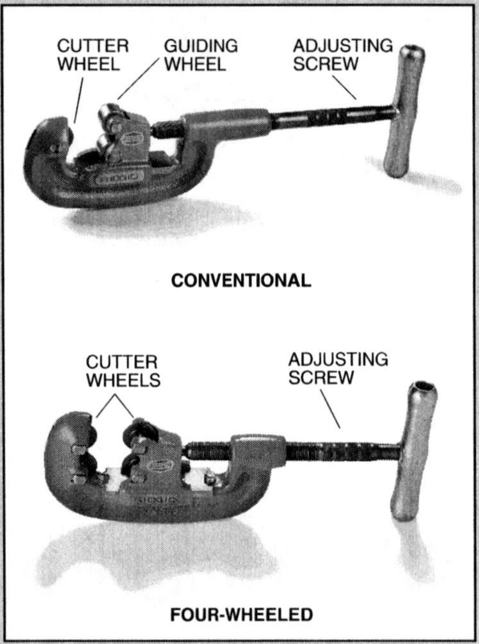

102RQ03.EPS

16. The tool shown in *Figure 3* is used to cut _____.
    a. cast iron soil pipe
    b. thin-wall copper tubing
    c. carbon steel pipe
    d. cement pipe

Figure 4

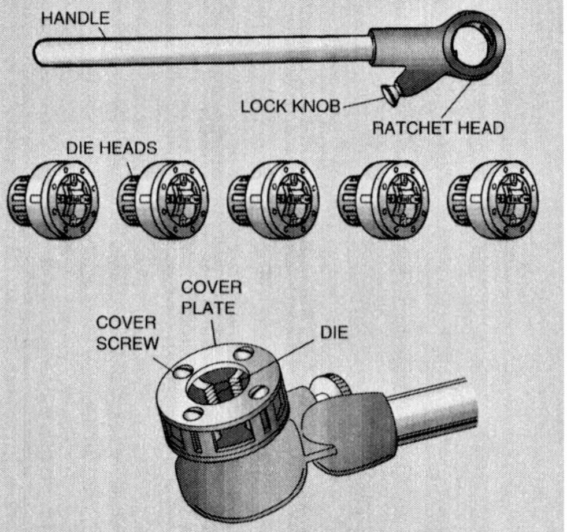

102RQ04.EPS

17. The tool shown in *Figure 4* is used to _____.
    a. thread pipe
    b. ream pipe
    c. bevel pipe ends
    d. position pipe for welding

18. Threading is complete when _____ full thread(s) extend(s) beyond the back edge of the die.
    a. one
    b. two
    c. three
    4. four

19. Pipe taps are used to _____.
    a. cut male threads of a pipe
    b. cut female threads of a pipe
    c. drill holes in pipe
    d. remove broken plugs and fittings

20. Lever compression tube benders can be used to bend pipe up to 1 inch in diameter.
    a. True
    b. False

Instructor's Notes:

2.34 PIPEFITTING ◆ LEVEL ONE

Summarize the major concepts presented in the module.

Learning the safe and proper use of pipefitting hand tools is an important step in your process toward becoming a successful pipefitter. You will use these tools every day of your professional life. Once you master the concepts, you will be able to perform your job more efficiently and more accurately. Use this module as a training resource to get you started, and as a reference for future use.

Notes

MODULE 08102-06 ◆ PIPEFITTING HAND TOOLS 2.35

Have the trainees complete the Trade Terms Quiz, and go over the answers prior to administering the Module Examination.

Administer the Module Examination. Record the results on Craft Training Report Form 200, and submit the results to the Training Program Sponsor.

Administer the Performance Test, and fill out Performance Profile Sheets for each trainee. If desired, trainee proficiency noted during laboratory sessions may be used to complete the Performance Test. Record the results on Craft Training Report Form 200, and submit the results to the Training Program Sponsor.

Trade Terms Quiz

Fill in the blank with the correct trade term that you learned from your study of this module.

1. To determine how many threads per inch are cut in a pipe, you would use a(n) _________.
2. Threads on the inside of a fitting are called _________.
3. Cutting pipe often leaves a ragged edge of metal called a(n) _________.
4. To cut male threads on a pipe, use a(n) _________.
5. A pipe end that has been forced open is called a(n) _________.
6. Before making the final weld, use _________ to hold the parts in place.
7. Threads on the outside of a pipe are called _________.
8. _________ is the outside width of a pipe.
9. A pipe-like raceway that contains electrical conductors is called a(n) _________.
10. A(n) _________ allows a tool to rotate in only one direction.
11. To change direction of a pipe or join two pipes, you would use a(n) _________.

Trade Terms

Burr
Conduit
Die
Female threads
Flare
Male threads
Outside diameter
Pipe fitting
Ratchet
Tack welds
Thread gauge

Instructor's Notes:

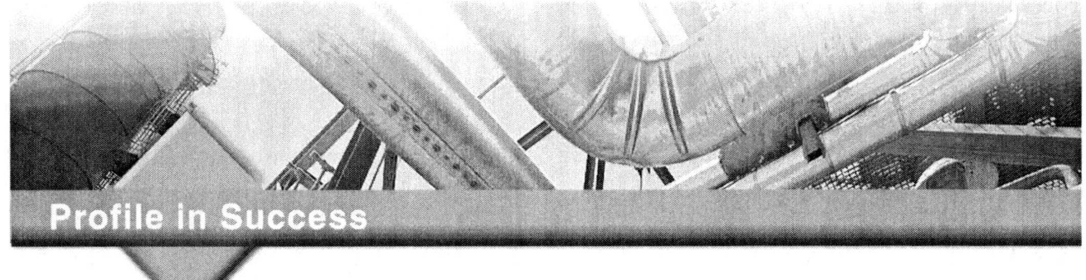

Glynn Allbritton

Kellogg Brown and Root
Scheduler and Planner

Glynn Allbritton has been working in the pipefitting field for 38 years. He started as a helper and has progressed in the trade through training and hard work. As an experienced Journeyman he wanted to help other pipefitters just starting out. He has worked with NCCER to help simplify the Pipefitting curriculum and as a Subject Matter Expert on the national committee.

How did you choose a career in the pipefitting field?
I started working in the pipefitting field in 1968 because it was one of the highest paying jobs available. I started as a helper and worked my way up through the ranks.

What types of training have you been through?
I have taken several courses offered by NCCER and other organizations. Over the years, I have taken Pipefitter I, II, and III; Supervision I, II, and III; OSHA Safety; and Business Management.

What kinds of work have you done in your career?
I started as a helper and worked my way up. Over the years I became a journeyman pipefitter and eventually a foreman. Now I am the scheduler and planner for pipe at Kellogg Brown and Root.

Tell us about your present job.
As a scheduler and planner I am responsible for planning jobs for the maintenance crew. I plan the assign the daily and weekly tasks for the crew. I am also responsible for ordering the parts and materials for the jobs.

What factors have contributed most to your success?
In a word, education. First learn all you can from your colleagues on the job. Don't be afraid to ask questions and learn from more experienced pipefitters and other trades. Second seek out training opportunities. There are always new developments in the field. Go back to school for more training and get the latest information. The more skills you have the more you can advance and become successful.

What advice would you give to those new to pipefitter field?
Get as much education as you can from technical schools, college, and on-the-job training. This gives you the fundamental skills to be a good pipefitter. Listen to more experienced pipefitters when they are talking about their job. You can learn a lot by listening and asking questions.

MODULE 08102-06 ◆ PIPEFITTING HAND TOOLS 2.37

Trade Terms Introduced in This Module

Burr: A sharp, ragged edge of metal usually caused by cutting pipe.

Conduit: A round raceway, similar to pipe, that contains conductors.

Die: A tool used to make male threads on a pipe or bolt.

Female threads: Threads on the inside of a fitting.

Flare: A pipe end that has been forced open to make a joint with a fitting.

Male threads: Threads on the outside of a pipe.

Outside diameter: A measurement of the outside width of a pipe.

Pipe fitting: A unit attached to a pipe and used to change the direction of fluid flow, connect a branch line to a main line, close off the end of a line, or join two pipes of the same size or of different sizes.

Ratchet: A device that allows a tool to rotate in only one direction.

Tack welds: Short welds used to hold parts in place until the final weld is made.

Thread gauge: A tool used to determine how many threads per inch are cut in a tap, die, bolt, nut, or pipe. Also called a pitch gauge.

Instructor's Notes:

Additional Resources

This module is intended to be a thorough resource for task training. The following reference work is suggested for further study. This is optional material for continued education rather than for task training.

Tools and Their Uses, Latest Edition. Naval Education and Training Program Development Center. Washington, DC: US Government Printing Office.

Figure Credits

Koike Aronson, Inc., 102F02 (table roller)

Sumner Mfg. Co., Inc., 102F02 (v-head jack, roller head jack, floor stand roller)

Reed Mfg. Co., 102F03, 102F06, 102F07, 102F09

Klein Tools, Inc., 102F08

Ridge Tool Company (RIDGID®), 102F05, 102F36–102F38, 102F42, 102F43A, 102F43B, 102F46, 102F48, 102RQ03

The Stanley Works, 102F12, 102F13, 102F15, 102F18

Pacific Laser Systems, Inc., 102F16

Zircon Corporation, 102F17

L.S. Starrett Co., 102F21

Topaz Publications, Inc., 102SA01, 102F35, 102F47

Mathey Dearman, 102F25, 102F26A, 102F28, 102F29, 102F31, 102F33, 102F34, 102RQ02

H&M Pipe Beveling Machine Co., 102F26B

G.A.L. Gage Co., 102F30

Zachary Construction Corporation, Title Page

MODULE 08102-06 ◆ PIPEFITTING HAND TOOLS 2.39

MODULE 08102-06 — TEACHING TIPS

The following are suggested activities or instructional methods to help you teach the material in this AIG.

General

When you call on someone to answer a question, the rest of the class relaxes or even tunes out because they expect that the question and answer will take place only between you and the trainee you called on. Instead, use this technique to involve more the trainees in answering questions and to keep them on their toes.

1. Ask the trainees to define a term or explain a concept.
2. After one trainee has answered, ask a trainee seated nearby if the answer is right. Then ask whether a trainee in the back of the room agrees.
3. Ask the trainees to explain why they think an answer is right or wrong.
4. Use the session to clear up incorrect ideas and encourage the trainees to learn from their mistakes.

Section 6.0.0 *Pipe Fabrication Tools*

Pipefitters must be able to use pipe fabrication tools. Trainees will need appropriate personal protective equipment, pencils, and paper. Set up several workstations and divide the class into small groups. Allow 20 to 30 minutes for this exercise.

1. Set up several workstations with a various pipe fabrication tools and pipe. Divide the trainees into small groups.
2. Have each group circulate among the tables and identify each type of tool and use it. If you wish, also have trainees write one fact about each tool.
3. At the last rotation, have one person in the group identify the tool at their station and explain how it is used.
4. Answer any questions they may have.

Section 7.0.0 *Pipe Cutting Tools*

Pipefitters must be able to cut different types of pipe using different pipe cutting equipment. Trainees will need appropriate personal protective equipment, pencils, and paper. Set up several workstations and divide the class into small groups. Allow 20 to 30 minutes for this exercise.

1. Set up several workstations with a various pipe cutting tools and pipe. Divide the trainees into small groups.
2. Have each group circulate among the tables and identify each type of tool and use it. If you wish, also have trainees write one fact about each tool.
3. At the last rotation, have one person in the group identify the tool at their station and explain how it is used.
4. Answer any questions they may have.

MODULE 08102-06 — ANSWERS TO REVIEW QUESTIONS

Answer	Section
1. b	2.0.0
2. c	3.2.0
3. b	3.2.0; Figure 2
4. a	4.3.0
5. d	4.1.0; Figure 4
6. b	5.0.0
7. b	5.1.0
8. c	5.2.0
9. a	6.1.0
10. b	6.4.0
11. b	6.5.3; Figure 28
12. c	6.6.0
13. b	6.7.0
14. a	6.8.0
15. c	7.1.0; Table 1
16. c	7.3.0
17. a	7.5.0
18. a	7.5.4
19. b	7.7.0
20. b	8.2.0

Answers to Trade Terms Quiz

1. Thread gauge
2. Female threads
3. Burr
4. Die
5. Flare
6. Tack welds
7. Male threads
8. Outside diameter
9. Conduit
10. Ratchet
11. Pipe fitting

CONTREN® LEARNING SERIES — USER UPDATE

NCCER makes every effort to keep these textbooks up-to-date and free of technical errors. We appreciate your help in this process. If you have an idea for improving this textbook, or if you find an error, a typographical mistake, or an inaccuracy in NCCER's Contren® textbooks, please write us, using this form or a photocopy. Be sure to include the exact module number, page number, a detailed description, and the correction, if applicable. Your input will be brought to the attention of the Technical Review Committee. Thank you for your assistance.

Instructors – If you found that additional materials were necessary in order to teach this module effectively, please let us know so that we may include them in the Equipment/Materials list in the Annotated Instructor's Guide.

Write: Product Development and Revision
National Center for Construction Education and Research
P.O. Box 141104, Gainesville, FL 32614-1104

Fax: 352-334-0932

E-mail: curriculum@nccer.org

Craft ______________ Module Name ______________

Copyright Date ______________ Module Number ______________ Page Number(s) ______________

Description ______________

(Optional) Correction ______________

(Optional) Your Name and Address ______________

Module 08103-06

Pipefitting Power Tools

NCCER STANDARDIZED CRAFT TRAINING PROGRAM

The National Center for Construction Education and Research (NCCER) provides a standardized national program of accredited craft training. Key features of the program include instructor certification, competency-based training, and performance testing. The program provides trainees, instructors, and companies with a standard form of recognition through the National Registry. The program is described in full in the *Guidelines for Accreditation*, published by NCCER. For more information on standardized craft training, contact NCCER by writing to P.O. Box 141104, Gainesville, FL 32614-1104; calling 352-334-0911; or e-mailing info@nccer.org. More information is available at www.nccer.org.

HOW TO USE THIS ANNOTATED INSTRUCTOR'S GUIDE

Each page presents two sections of information. The larger section displays each page exactly as it appears in the Trainee Module. The narrow column ties suggested trainee and instructor actions to each page and provides icons (detailed below) to call your attention to material, safety, audiovisual, or testing requirements. The bottom of each page includes space for your notes.

The **Audiovisual** icon indicates an appropriate time to show a transparency or other audiovisual aid.

The **Classroom** icon prompts you to define a term, stress a point, ask trainees to explain a concept, or give examples.

The **Demonstration** icon directs you to show trainees how to perform tasks.

The **Examination** icon tells you to administer the written module examination.

The **Homework** icon is placed where you may wish to assign reading for the next class, assign a project, or advise trainees to prepare for an examination.

The **Laboratory** icon is used when trainees are to practice performing tasks.

The **Materials** icon is a reminder for you to gather materials needed for classes, laboratories, and testing.

The **Performance Testing** icon tells you to administer a performance test or a portion thereof.

The **Safety** icon is used to emphasize safety issues. It is often keyed to *Caution* and *Warning!* statements in the Trainee Module.

The **Teaching Tip** icon indicated additional guidance is available, such as how to conduct an exercise, get the most educational value from a field trip, or encourage class participation. Teaching Tips may expand on a feature (*Think About It, Did You Know?*) or provide *Quick Quizzes* or similar exercises. You will be referred to the Teaching Tips section at the back of the module if there is additional material.

The **Combination** icon indicates that the laboratory listed corresponds with a performance task. If desired, you can note the proficiency of the trainees during the laboratory, and use it to satisfy performance testing requirements.

PREPARATION

Before teaching this module, you should review the Objectives, Performance Tasks, Materials and Equipment List, and Module Outline. Be sure to allow ample time to prepare your own training or lesson plan and gather all required materials and equipment.

MODULE OVERVIEW

This module identifies the hazards and explains general safety procedures that must be followed when using power tools, and explains specific guidelines for using electric and pneumatic power tools.

PREREQUISITES

Prior to training with this module, it is recommended that the trainee shall have successfully completed *Core Curriculum;* and *Pipefitting Level One*, Modules 08101-06 and 08102-06.

OBJECTIVES

Upon completion of this module, the trainee will be able to do the following:

1. State the safety procedures that must be followed when working with power tools.
2. Cut pipe using a portable band saw.
3. Identify and explain the uses of portable grinders.
4. Explain the proper and safe operation of machines used in pipe joint preparation:
   - Pipe threaders
   - Portable power drives
   - Pipe bevelers
5. Perform selected pipe joint preparation operations using power tools.

PERFORMANCE TASKS

Under the supervision of the instructor, the trainee should be able to do the following:

1. Cut pipe using a portable band saw (do not use threading machine).
2. Operate a portable grinder.
3. Replace dies in a threading machine.
4. Cut, ream, and thread pipe using a threading machine.
5. Cut and thread nipples using a nipple chuck.
6. Thread pipe using a portable power drive.
7. Identify several types of pipe bevelers.

MATERIALS AND EQUIPMENT LIST

Overhead projector and screen
Transparencies
Blank acetate sheets
Transparency pens
Whiteboard/chalkboard
Markers/chalk
Pencils and scratch paper
Appropriate personal protective equipment
Face shields
Gloves
Ground fault circuit interrupter
Abrasive saws
Portable band saws and accessories
Portable grinders and accessories
Assorted lengths of 1-, 1½-, and 2-inch pipe
Assorted lengths of 3-, 4-, and 6-inch pipe
Cut and beveled pipe
Soapstone
Band saw blades
Tripod chain vise
Wraparounds
Grinding wheels
Measuring tapes
Spanner wrenches

Geared threaders and accessories
Thread cutting oil
Nipple chucks
Ridgid 300 power drive
Ridgid 535 power drive
Pipe bevelers
Module Examinations*
Performance Profile Sheets*

*Located in the Test Booklet.

SAFETY CONSIDERATIONS

Ensure that the trainees are equipped with appropriate personal protective equipment and know how to use it properly. This module requires trainees to use power tools. Review basic power tool safety, electrical safety, and eye and hand protection.

ADDITIONAL RESOURCES

This module is intended to present thorough resources for task training. The following reference work is suggested for both instructors and motivated trainees interested in further study. This is optional material for continued education rather than for task training.

Tools and Their Uses, Latest Edition. Naval Education and Training Program Development Center. Washington, DC: U.S. Government Printing Office.

TEACHING TIME FOR THIS MODULE

An outline for use in developing your lesson plan is presented below. Note that each Roman numeral in the outline equates to one session of instruction. Each session has a suggested time period of 2½ hours. This includes 10 minutes at the beginning of each session for administrative tasks and one 10-minute break during the session. Approximately 15 hours are suggested to cover *Pipefitting Power Tools*. You will need to adjust the time required for hands-on activity and testing based on your class size and resources. Because laboratories often correspond to Performance Tasks, the proficiency of the trainees may be noted during these exercises for Performance Testing purposes.

Topic	Planned Time
Session I. Introduction, Safety, and Cutting	
A. Introduction	____________
B. Power Tool Safety	____________
C. Cutting Pipe Using Saws	____________
D. Laboratory – Trainees practice cutting pipe using a portable band saw. This laboratory corresponds to Performance Task 1.	____________
Session II. Portable Grinders	
A. Types of Portable Grinders	____________
B. Inspecting Grinders	____________
C. Operating Grinders	____________
D. Laboratory – Trainees practice operating a portable grinder. This laboratory corresponds to Performance Task 2.	____________

Session III. Threading Machines

A. Loading Pipe into a Threading Machine ______

B. Cutting and Reaming Pipe ______

C. Replacing Dies ______

D. Laboratory – Trainees practice replacing dies in a threading machine. This laboratory corresponds to Performance Task 3. ______

E. Threading Operations ______

F. Machine Maintenance ______

G. Laboratory – Trainees practice cutting, reaming, and threading pipe using a threading machine. This laboratory corresponds to Performance Task 4. ______

Session IV. Special Threading Applications

A. Cutting and Threading Nipples ______

B. Threading Pipe Using a Geared Threader ______

C. Laboratory – Trainees practice cutting and threading nipples using a nipple chuck. This laboratory corresponds to Performance Task 5. ______

Session V. Portable Power Drives and Power Bevelers

A. Portable Power Drives ______

B. Laboratory – Trainees practice threading pipe using a portable power drive. This laboratory corresponds to Performance Task 6. ______

C. Power Bevelers ______

D. Laboratory – Trainees practice identifying several pipe bevelers. This laboratory corresponds to Performance Task 7. ______

Session VI. Review, Module Examination, and Performance Testing

A. Review ______

B. Module Examination ______

1. Trainees must score 70 percent or higher to receive recognition from NCCER.
2. Record the testing results on Craft Training Report Form 200, and submit the results to the Training Program Sponsor.

C. Performance Testing ______

1. Trainees must perform each task to the satisfaction of the instructor to receive recognition from NCCER. If applicable, proficiency noted during laboratory exercises can be used to satisfy the Performance Testing requirements.
2. Record the testing results on Craft Training Report Form 200, and submit the results to the Training Program Sponsor.

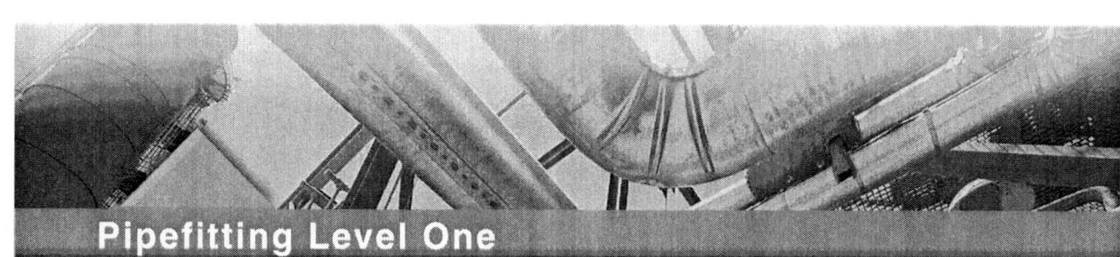

Assign reading of Module 08103-06.

08103-06

Pipefitting Power Tools

08103-06
Pipefitting Power Tools

Topics to be presented in this module include:

Overview

Pipefitters use power tools to work faster and more accurately. Power tools are used to cut, grind, thread, and shape all types of piping materials. Tools must be used correctly and safely. Power and pneumatic tools pose electrical and other hazards. Follow all operating instructions and safety precautions. Select the correct tool for the job and check that the tool is in good working order before use.

Band saws and abrasive saws are used to cut pipe. Grinders are used to prepare pipe for welding. Specialty tools, like threading machines and bevelers, should only be used for the specific jobs they were designed to perform. When used correctly, these tools can greatly increase a pipefitter's productivity.

Instructor's Notes:

Objectives

When you have completed this module, you will be able to do the following:

1. State the safety procedures that must be followed when working with power tools.
2. Cut pipe using a portable band saw.
3. Identify and explain the uses of portable grinders.
4. Explain the proper and safe operation of machines used in pipe joint preparation:
   - Pipe threaders
   - Portable power drives
   - Pipe bevelers
5. Perform selected pipe joint preparation operations using power tools.

Trade Terms

Assured equipment grounding conductor program
Bevel
Chamfer
Chuck
Die grinder
Fabrication
Ground fault circuit interrupter (GFCI)
Horsepower (hp)
Revolutions per minute (rpm)

Required Trainee Materials

1. Pencil and paper
2. Appropriate personal protective equipment

Prerequisites

Before you begin this module, it is recommended that you successfully complete *Core Curriculum;* and *Pipefitting Level One*, Modules 08101-06 and 08102-06.

This course map shows all of the modules in the first level of the *Pipefitting* curriculum. The suggested training order begins at the bottom and proceeds up. Skill levels increase as you advance on the course map. The local Training Program Sponsor may adjust the training order.

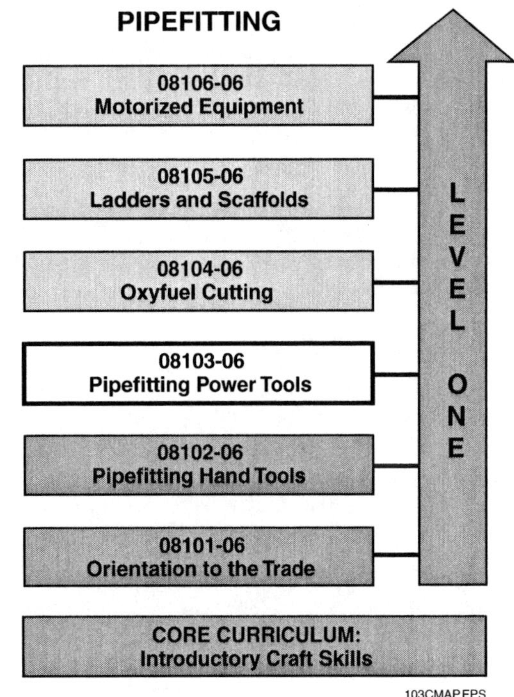

103CMAP.EPS

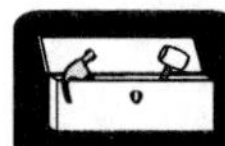

Ensure that you have everything required to teach the course. Check the Materials and Equipment list at the front of this module.

See the general Teaching Tip at the end of this module.

Show Transparency 1, Objectives, and Transparency 2, Performance Tasks. Review the goals of the module, and explain what will be expected of the trainee.

Explain that terms shown in bold are defined in the Glossary at the back of this module.

Review the modules covered in Level One and explain how this module fits in.

Describe the types of power tools used by pipefitters.

Explain the general safety rules that must be followed when using power tools.

Discuss electric power tool safety.

Ask a trainee to describe a ground fault circuit interrupter (GFCI). Explain how they are used on the job and in the classroom.

Point out the safety features in the classroom and laboratory, including circuit breakers, eye wash station, and first aid kit.

1.0.0 ◆ INTRODUCTION

The widespread use of power tools in the pipefitting trade has greatly increased the individual productivity of the pipefitter. Power tools are available for cutting, grinding, threading, and shaping all types of piping material. Pipefitters must know how to safely operate these power tools for the specific jobs for which the tools are designed. This module introduces the trainee to the different types of power tools, and the procedures for selecting, using, and maintaining these tools.

2.0.0 ◆ POWER TOOL SAFETY

If used improperly, power tools can cause severe injuries and even death. The following safety precautions must be followed when using power tools:

- Always keep the work area clean. Many accidents are caused by materials or tools carelessly spread about the work area.
- Store tools when they are not in use. Allowing power tools to lie around the work area when they are not in use increases the possibility that the tools may become damaged and unsafe for use.
- Use tools only for their intended purposes, and use the proper tool for each job. Know the capabilities of the tool, and work within those limits.
- Never force a tool beyond its capabilities. Forcing a tool to work beyond the limits of its design not only wears out the tool prematurely, but also places you in an unsafe situation.
- Always wear proper clothing. Do not wear loose-fitting clothing when working with or around power tools. There is always the danger that loose clothing, especially shirt tails, will get caught in the moving parts of the tool.
- Wear safety glasses when using power tools. Most power tools eject material, which could cause severe eye damage.
- Wear properly fitting gloves when threading pipe.
- Always maintain proper footing and balance when using a power tool. Do not overreach. Any fall can be dangerous, but a fall while using a power tool can be much worse.
- Respect your tools and always follow the recommended operating and maintenance procedures. A well-maintained tool is a safe tool.
- Disconnect the power source before performing maintenance on a tool or changing accessories.
- Allow only competent technicians to repair power tools.
- Do not alter or modify a power tool in any way.
- Inspect power tools before each use. Make sure that there is no damage to the tool or power cord and that all the safety guards are in place.
- Never carry a power tool by its power cord or an air tool by its air line.
- Avoid accidental starts. Make sure that the power switch is in the off position before plugging in the power cord.
- Be aware of the torque or kickback of the power tool being used. Maintain good balance and properly brace yourself when using power tools.

2.1.0 Electric Power Tool Safety

Electrical power tools are extremely dangerous and can cause severe injury and even death if misused. The following safety precautions apply to electric power tools:

- Do not use electric power tools in wet or damp locations. There is always a danger of electric shock, even if the tool is properly grounded.
- Respect power cords. Keep power cords out of situations where they can be frayed, cut, or damaged. Always disconnect a power cord by grasping the plug and pulling. Never yank on the cord to disconnect a power cord.
- Ensure that the current available is the same current for which the tool is designed. This will be indicated on the identification plate attached to the tool.
- All power cords must be tied off overhead, at least 7 feet off the floor, to prevent tripping hazards.
- Do not repair a frayed or damaged power cord. It must be replaced by a licensed electrician.
- Never remove the grounding pin from a three-wire power cord or extension cord.

> **NOTE**
> In some industries, such as mining and shipyards, trigger locks must be removed from power tools.

2.1.1 Ground Fault Circuit Interrupters

A **ground fault circuit interrupter (GFCI)** is a fast-acting circuit breaker that senses small imbalances in the circuit caused by current leakage to ground. A GFCI continually matches the amount of current going to an electrical device against the amount of current returning from the device. Whenever the two values differ by more than 5 milliamps, the GFCI interrupts the electric power within 1⁄40 of a second. *Figure 1* shows an extension cord with a GFCI.

Instructor's Notes:

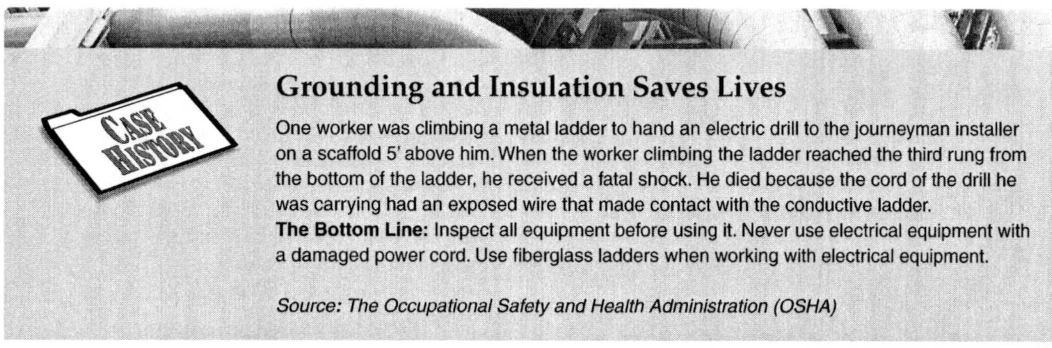

Grounding and Insulation Saves Lives

One worker was climbing a metal ladder to hand an electric drill to the journeyman installer on a scaffold 5' above him. When the worker climbing the ladder reached the third rung from the bottom of the ladder, he received a fatal shock. He died because the cord of the drill he was carrying had an exposed wire that made contact with the conductive ladder.

The Bottom Line: Inspect all equipment before using it. Never use electrical equipment with a damaged power cord. Use fiberglass ladders when working with electrical equipment.

Source: The Occupational Safety and Health Administration (OSHA)

Explain what kind of protection is provided by a GFCI. Discuss how water affects a GFCI.

Discuss OSHA requirements for ensuring proper grounding systems for temporary wiring.

103F01.EPS

Figure 1 ◆ Extension cord with a GFCI.

A GFCI will not protect you from line-to-line contact hazards such as holding either two hot wires or a hot and a neutral wire in each hand. It does provide protection against the most common form of electrical shock, which is a ground fault. It also provides protection from fires, overheating, and wiring insulation deterioration.

GFCIs can be used successfully to reduce electrical hazards on construction sites. Tripping of GFCIs—interruption of circuit flow—is sometimes caused by wet connectors and tools. Limit the amount of water that tools and connectors come into contact with by using watertight or sealable connectors. Having more GFCIs or shorter circuits can prevent tripping caused by cumulative leakage from several tools or from extremely long circuits.

Provide a ground fault circuit interrupter for the trainees to examine.

2.1.2 GFCI Requirements

OSHA allows two methods of ensuring a proper grounding system for temporary wiring. The first is to provide GFCIs for all 15A and 20A circuits. All GFCIs must be tested every three months. The second method is known as an **assured equipment grounding conductor program**. This is a monthly inspection of all electric tools and extension cords, which are not protected by GFCI, to ensure that the ground wire is intact and the equipment is safe to operate. A written program

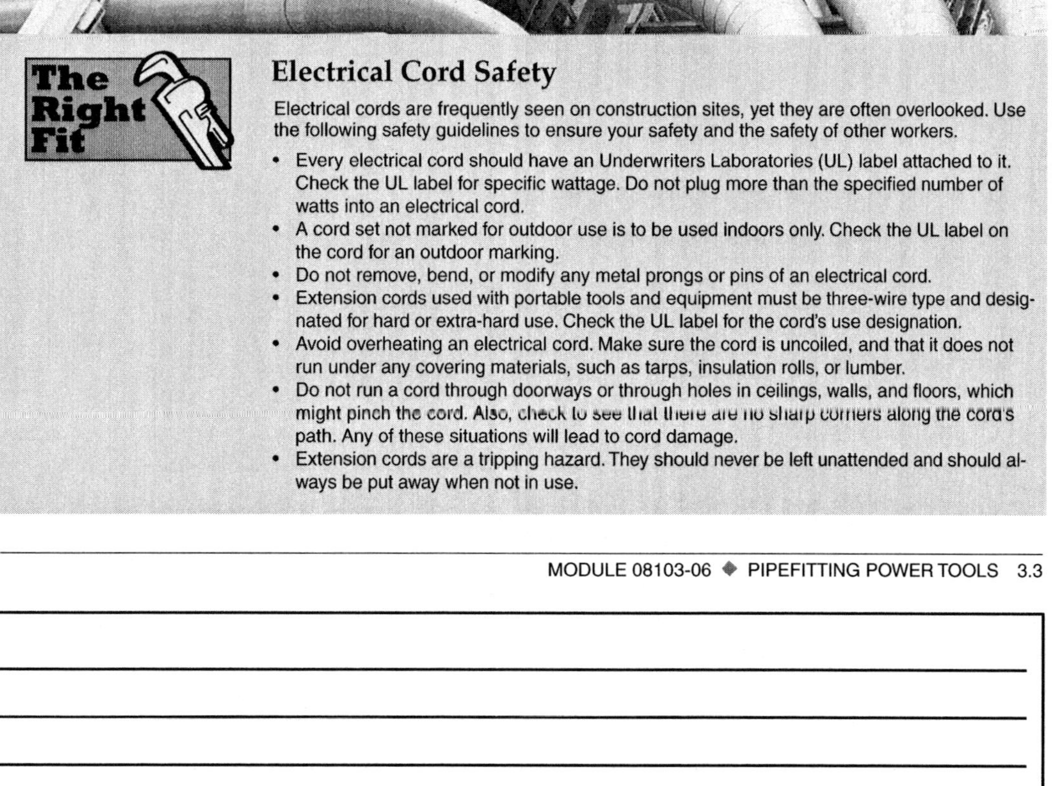

Electrical Cord Safety

Electrical cords are frequently seen on construction sites, yet they are often overlooked. Use the following safety guidelines to ensure your safety and the safety of other workers.

- Every electrical cord should have an Underwriters Laboratories (UL) label attached to it. Check the UL label for specific wattage. Do not plug more than the specified number of watts into an electrical cord.
- A cord set not marked for outdoor use is to be used indoors only. Check the UL label on the cord for an outdoor marking.
- Do not remove, bend, or modify any metal prongs or pins of an electrical cord.
- Extension cords used with portable tools and equipment must be three-wire type and designated for hard or extra-hard use. Check the UL label for the cord's use designation.
- Avoid overheating an electrical cord. Make sure the cord is uncoiled, and that it does not run under any covering materials, such as tarps, insulation rolls, or lumber.
- Do not run a cord through doorways or through holes in ceilings, walls, and floors, which might pinch the cord. Also, check to see that there are no sharp corners along the cord's path. Any of these situations will lead to cord damage.
- Extension cords are a tripping hazard. They should never be left unattended and should always be put away when not in use.

MODULE 08103-06 ◆ PIPEFITTING POWER TOOLS 3.3

Discuss pneumatic power tool safety.

Describe how pipefitters use portable band saws to cut pipe.

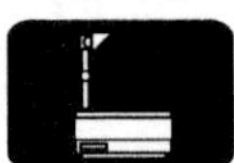

Show Transparency 3 (Table 1). Discuss how to select the proper band saw blade.

Explain how to replace a portable band saw blade.

Show trainees how to replace a portable band saw blade.

Emphasize the importance of ensuring that the machine is unplugged before changing the blade. Warn trainees of the hazards in removing a band saw blade.

must be in place to establish the frequency and method of logging the inspections. Some companies use a color code system to indicate when the equipment has been inspected. The current edition of the *National Electrical Code®* has more stringent requirements for temporary wiring.

2.2.0 Pneumatic Power Tool Safety

Pneumatic power tools can also be dangerous and can cause severe personal injury if misused. The following safety precautions apply to pneumatic power tools:

- All air hoses must be properly drained or blown clear before the hose is attached to the tool.
- Hose pressure must be checked to make sure that it is applicable to the tool.
- Ensure that the oiler reservoir is at the proper level before using the tool.
- Never direct the air flow from an air nozzle toward yourself or anyone else.
- Secure all hose connections to avoid accidental disconnection.
- Do not turn the air on until you trace the hose and ensure that the end is securely attached to an air tool.
- Do not crimp the hose to turn off the air to the tool or to change a tool.
- Inspect the air hose for damage before using the tool.
- When connecting into an air hose, make sure that it is an air hose and not an inert gas line, such as nitrogen or argon.
- Do not attempt to repair a damaged air hose. Always replace it.
- Make sure that the air hose is rated for the air pressure with which it will be used. The air hose should have a pressure rating printed on the hose.
- Air hoses must be tied off at least 7 feet above the ground or floor to prevent tripping hazards.

3.0.0 ◆ CUTTING PIPE USING SAWS

Pipefitters use portable band saws to cut carbon steel pipe, galvanized pipe, and pipe made from other materials. The larger diameter pipes, usually 4 inches and larger, require a special procedure to ensure that the pipe is cut square. The following sections explain the selection of proper blades for different types of materials and cutting pipe square using the portable band saw.

3.1.0 Selecting Band Saw Blades

Band saw blades are selected depending on the type and wall thickness of the material to be cut. Blades are classified by the material of the blade and the number of teeth per inch. A coarse blade has fewer teeth per inch and should be used on softer materials. A fine blade has more teeth per inch and should be used on harder materials. *Table 1* shows a blade selection chart and recommended usages.

3.2.0 Replacing Portable Band Saw Blades

The portable band saw blade must be replaced if it becomes worn and dull or if you are cutting a material that the blade in the saw was not designed to cut. Follow these steps to replace a band saw blade:

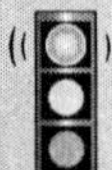

WARNING!
Make sure that the portable band saw is unplugged before replacing the blade.

Step 1 Rotate the band adjust knob 180 degrees to release the tension on the blade. *Figure 2* shows portable band saw components.

Step 2 Turn the saw upside down on your work table.

Step 3 Remove the blade from the blade pulleys underneath the saw.

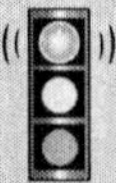

WARNING!
Be careful when removing the blade, because the blade has a tendency to spring out when you take it off the blade pulleys.

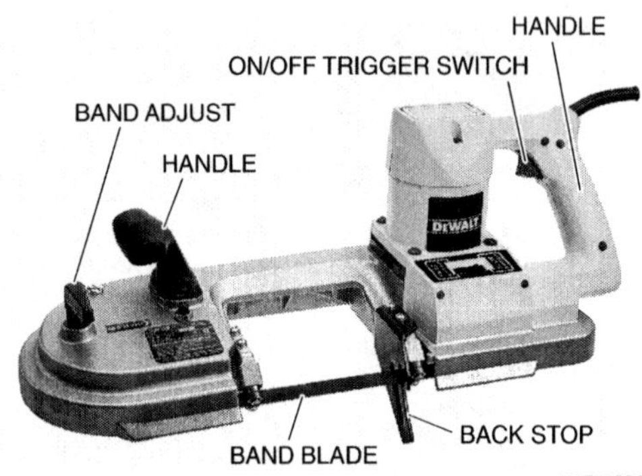

Figure 2 ◆ Portable band saw components.

Instructor's Notes:

Table 1 Blade Selection Chart and Recommended Usages

Band Saw Speed		Use Single-Speed Band Saw or Higher Speeds on Two-Speed and Variable-Speed Band Saws									Use Lower Speeds on Two-Speed or Variable-Speed Band Saws							
Material to be Cut		**Aluminum, Brass, Copper, Bronze, Mild Steel**					**Angle Iron, Cast Iron, Galvanized Pipe**				**Stainless Steel**				**Fiberglass, Asbestos, Plastics**			
Wall or Material Thickness		3/32 to 1/8	1/8 to 1/4	5/32 to 1/2	3/16 to 3/4	11/32 & over	3/32 to 1/8	1/8 to 1/4	5/32 to 1/2	3/16 & over	3/32 to 1/8	1/8 to 1/4	5/32 to 1/2	3/16 & over	3/32 to 1/8	1/8 to 1/4	5/32 to 1/2	3/16 & over
	Teeth per inch																	
Carbon Steel Blades	6					X												
	8					X												
	10				X					X								
	14			X					X									
	18		X					X										
	24	X					X											
Alloy Steel Blades	10				X					X								X
	14			X					X								X	
	18		X					X								X		
	24	X					X								X			
High-Speed Steel Blades	10				X					X				X				X
	14			X					X				X				X	
	18		X					X				X				X		
	24										X							

103T01.EPS

Step 4 Slip a new blade around the blade pulleys.

Step 5 Slip the new blade into the blade guide and the back stop.

Step 6 Turn the saw over, and rotate the band adjust knob to put tension on the blade.

3.3.0 Abrasive Saws

Abrasive saws use a special wheel that can slice through either metal or masonry. The most common types of abrasive saws are the demolition saw and the chop saw (*Figure 3*). The main difference between the two is that the demolition saw is not mounted on a base.

3.3.1 Demolition Saw

Demolition saws (*Figure 3A*) run on either electricity or gasoline. These saws are effective for cutting through most materials found at a construction site. The demolition saw is often used to cut ductile iron underground pipe. Although all saws are potentially hazardous, the demolition saw is a particularly dangerous tool that requires the full attention and concentration of the operator.

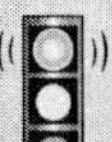

WARNING!

When sawing, be sure to use the appropriate personal protective equipment for your eyes, ears, and hands. If you have long hair, be sure to tie it back or cover it properly.

WARNING!

Gas-powered demolition saws must be handled with caution. Using a gas-powered saw where there is little ventilation can cause carbon monoxide poisoning. Carbon monoxide gas is hard to detect, but it is deadly. Use fans to circulate air, and have a trained person monitor air quality.

Source: Centers for Disease Control and Prevention website. www.cdc.gov/elcosh/docs/d0100/d000021/d000021.html, reviewed March 5, 2004.

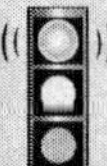

WARNING!

Because a demolition saw is handheld, there is a tendency to use it in positions that can easily compromise your balance and footing. Be very careful not to use the saw in any awkward positions.

Explain that abrasive saws can cut through both metal and masonry.

Describe a demolition saw.

Review personal protective equipment needed when operating saws.

Discuss the hazards posed from carbon monoxide from gas-powered saws.

Emphasize the importance of maintaining your balance when operating a saw.

Provide a band saw and an abrasive saw to the trainees for their examination.

Choosing the Proper Saw Blade

There are several different types of saw blades. Steel blades are generally the least expensive, but they dull faster. If you do not use your power tools frequently, these blades might be good enough. If you use your power saws a lot, however, carbide-tipped blades are a better choice. These blades are more expensive than steel ones, but they also last longer.

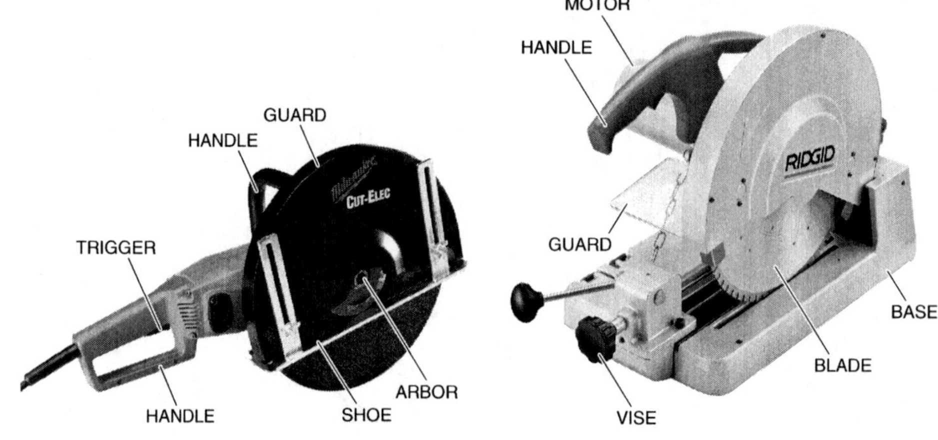

Figure 3 ◆ Abrasive saws.

Pneumatic Saw

This pneumatic reciprocating saw can be used to cut pipe and other materials. It uses hacksaw blades or modified reciprocating saw blades. It is ideal for use in situations where there is no electricity available to operate electrical power tools.

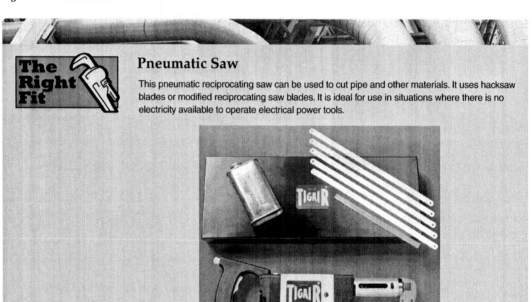

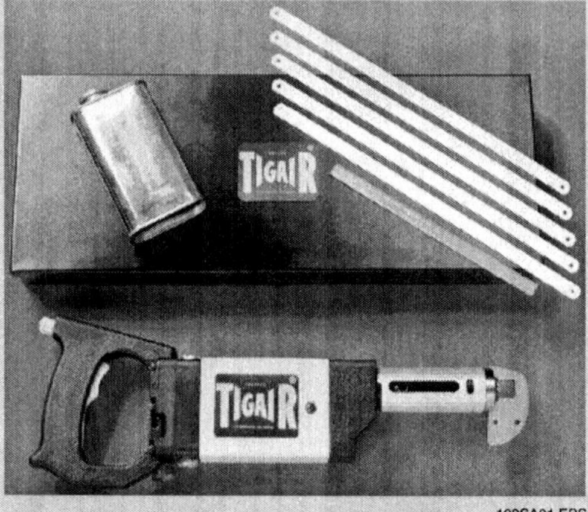

Instructor's Notes:

3.6 PIPEFITTING ◆ LEVEL ONE

Storing Gasoline

OSHA *29 CFR 1910.106* regulates the amount of flammable or combustible liquids that may be located outside of or inside of a gas storage cabinet or in any one fire area of a building.

Gas storage cabinets must be designed to limit the internal temperature to 325°F. Metal cabinets should be constructed of 18-gauge sheet metal and should be double walled. For added safety, gas storage cabinets should be self-closing. Cabinets must be labeled in conspicuous lettering with the statement "Flammable—Keep Fire Away."

Source: Occupational Safety and Health Administration website. "Regulations (Standards–*29 CFR*), *Flammable and Combustible Liquids–1910.106*." www.osha.gov/pls/oshaweb/owadisp.show_document?p_id=9752&p_table=STANDARDS, reviewed March 30, 2004.

Describe a chop saw and its uses.

3.3.2 Chop Saw

The chop saw is a versatile and accurate tool. It is a lightweight circular saw mounted on a spring-loaded arm that pivots and is supported by a metal base. This is a good saw to use to get exact square or angled cuts. It uses a blade (generally abrasive) to cut pipe, channel, tubing, conduit, or other light-gauge materials. The most common blade is an abrasive metal blade. Blades are available for other materials such as PVC or wood. A vise on the base of the chop saw holds the material securely and pivots to allow miter cuts. Some can be pivoted past 45 degrees in either direction.

Chop saws are sized according to the diameter of the largest abrasive blade they accept. Two common sizes are 12 inches and 14 inches. Each blade has a maximum safe speed. Never exceed that speed. Typically, the maximum safe speeds are 5,000 **revolutions per minute (rpm)** for 12-inch blades and 4,350 rpm for 14-inch blades.

Review the guidelines for using an abrasive saw.

3.3.3 Abrasive Saw Safety

Abrasive saws, such as demolition and chop saws, require extreme care during use. To avoid injuring yourself, follow these guidelines:

- Wear PPE to protect your eyes, ears, and hands.
- Use abrasive wheels that are rated for higher rpms than the tool can produce. This way, you will never exceed the maximum safe speed for the wheel you are using. The blades are matched to a safety rating based on their rpms.
- Be sure the wheel is secured on the arbor before you start using a demolition saw. Point the wheel of a demolition saw away from yourself and others before you start the saw.
- Use two hands when operating a demolition saw.
- Secure the materials you are cutting in a vise. The jaws of the vise hold the pipe, tube, or other material firmly and prevent it from turning while you cut.
- Be sure the guard is in place and the adjustable shoe is secured before using a demolition saw.
- Keep the saw and the blades clean.
- Inspect the saw each time you use it. Never operate a damaged saw. Ask your supervisor if you have a question about the condition of a saw.
- Inspect the blade before you use the saw. If you see damage, throw the blade away.

Explain how to cut pipe with a portable band saw.

Show trainees how to cut pipe using a portable band saw.

3.4.0 Cutting Pipe with a Portable Band Saw

When cutting pipe 4 inches in diameter or larger, it is important not to cut straight through the pipe because the blade may bend and may not produce a straight cut. The procedure for cutting pipe using a portable band saw consists of cutting around the pipe instead of straight through it. Follow these steps to cut pipe using a portable band saw:

Step 1 Select the pipe to be cut, and determine what kind of blade you need.

Step 2 Replace the blade in the portable band saw if necessary.

Step 3 Measure and mark the desired length on the pipe.

Step 4 Mark a straight line around the pipe at the cut mark using a wraparound.

Step 5 Start the band saw and cut the top of the pipe, keeping the blade on the cut line and the pipe snug against the back stop of the saw.

Step 6 Continue cutting on the cut line by rotating the saw over the top of the pipe.

Step 7 Stop cutting and remove the blade from the pipe when approximately ¾ of the blade between the front blade guide and back stop has cut into the pipe.

Step 8 Reposition the pipe in the vise so that the next section of uncut pipe is facing you.

Discuss how to avoid being injured from newly cut pipe.

MODULE 08103-06 ◆ PIPEFITTING POWER TOOLS 3.7

Have trainees practice cutting pipe using a portable band saw. Note the proficiency of each trainee. This laboratory corresponds to Performance Task 1.

Have trainees review Sections 4.0.0–4.3.0.

Ensure that you have everything required for teaching this session.

Describe the tasks pipefitters perform with a grinder.

Review the safety guidelines for using a portable grinder.

Describe different types of portable grinders. Discuss the need to inspect a grinder before using it.

See the Teaching Tip for Section 4.0.0 at the end of this module.

Step 9 Repeat steps 5 through 8 until the pipe is cut completely through.

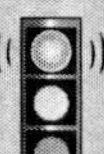

WARNING!

Avoid letting the pipe fall to the ground. It can fall on your foot or bounce back into your leg, causing injury. The freshly cut pipe end is extremely sharp and may be hot. Use care to avoid cutting yourself on the pipe end.

Step 10 Remove the pipe from the vise.

4.0.0 ◆ PORTABLE GRINDERS

Pipefitters use portable grinders for a variety of tasks, such as cleaning scale from pipe before welding, beveling pipe ends, and grinding away welding spatter. If not used properly, the portable grinder can cause severe injury. Follow these safety guidelines when using a portable grinder:

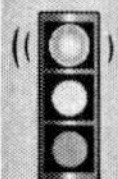

WARNING!

Compare the grinder wheel speed rating to the grinder operating speed. The wheel speed rating must meet or exceed the grinder operating speed.

- Inspect the grinding wheel visually before mounting it to the grinder. Do not mount a damaged wheel.
- Examine the grinder wheel guard for damage. It must be in place before starting the grinder.
- Inspect the power cord or air line for damage.
- Wear gloves and a safety shield to protect yourself from flying sparks and metal fragments when using a grinder.
- Run the grinder with the safety guard in place for a moment before grinding.

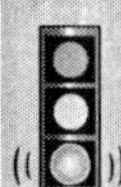

NOTE

A circular saw with an abrasive blade can be used to cut stainless steel pipe.

4.1.0 Types of Portable Grinders

Portable grinders can be either electrically or pneumatically powered. They are generally rated up to 5 **horsepower (hp)** and rotate up to 20,000 rpm at no load. Portable grinders are usually identified by the angle of the grinding wheel in relation to the motor housing. There are several types of portable grinders used for various purposes. *Figure 4* shows a right angle grinder and **die** grinder, along with a selection of attachments that are used with these grinders. There are also end grinders, which are similar to die grinders, but much larger.

Die grinders and small angle grinders work very well for small areas such as weld grooves and **bevel** edges (*Figure 5*). Grinding disks and wheels are made for specific types of metals. Be sure you have the correct disk or wheel for the type of metal being ground. Always use aluminum oxide disks for grinding aluminum or stainless steel. Do not use wheels that have been used on other metals when grinding stainless steel or aluminum. This will contaminate the surface of the aluminum or the stainless steel.

When using wire brush attachments on stainless steel or aluminum, brushes with stainless steel bristles must be used. Use only stainless steel brushes that have been used on stainless steel for brushing stainless steel. Use only stainless steel brushes that have been used on aluminum for brushing aluminum. Mild steel bristles will contaminate stainless steel or aluminum, causing weld defects.

When grinding, avoid grinding the base metal below the minimum allowable base metal thickness. If the base metal is ground below the minimum allowable thickness, the base metal will have to be replaced or, if allowable, built up with weld. Both of these are expensive alternatives.

WARNING!

If a grinding wheel or disk shatters while in use, the pieces of the wheel or disk become deadly projectiles as the weights of the pieces are multiplied by the centrifugal force and the revolutions per minute (rpm) of the grinding motor. The force with which a piece of wheel or disk is released is equivalent to that of an armor-piercing bullet. Never grind without full face protection. Avoid standing in the orbital path of the wheel or disk in case it disintegrates.

4.2.0 Inspecting Grinders Before Use

Whether the portable grinder is electric or pneumatic, a thorough inspection of the equipment is required before it is used. Follow these steps to inspect a grinder:

Step 1 Inspect the air inlet and the air line of pneumatic grinders to ensure there are no signs of damage that could cause a bad connection or loss of air.

Instructor's Notes:

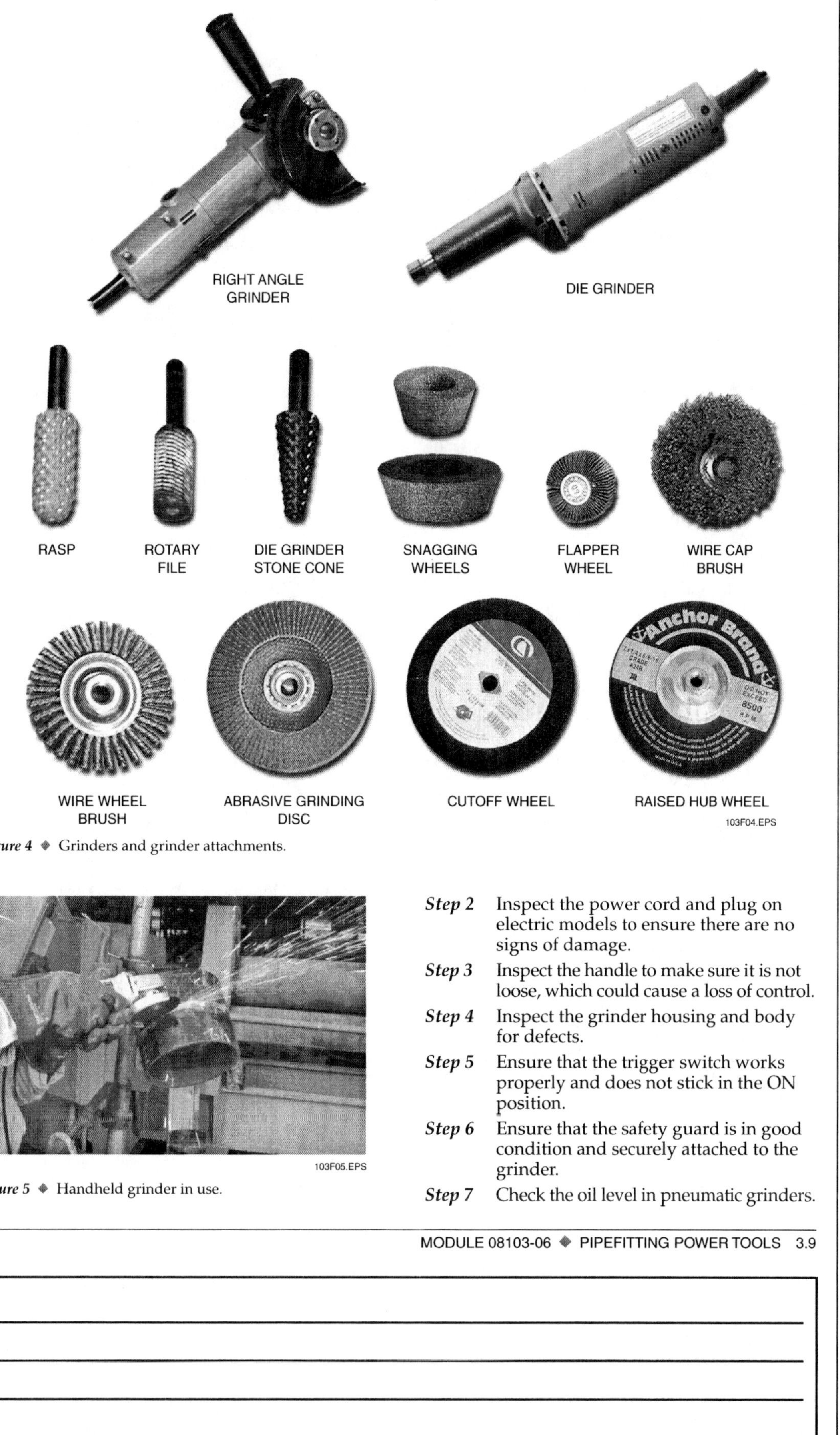

Figure 4 ◆ Grinders and grinder attachments.

Figure 5 ◆ Handheld grinder in use.

Step 2 Inspect the power cord and plug on electric models to ensure there are no signs of damage.

Step 3 Inspect the handle to make sure it is not loose, which could cause a loss of control.

Step 4 Inspect the grinder housing and body for defects.

Step 5 Ensure that the trigger switch works properly and does not stick in the ON position.

Step 6 Ensure that the safety guard is in good condition and securely attached to the grinder.

Step 7 Check the oil level in pneumatic grinders.

MODULE 08103-06 ◆ PIPEFITTING POWER TOOLS 3.9

Review the procedure for inspecting a grinder.

Provide a portable grinder and attachments for the trainees to examine.

Explain how to operate a portable grinder.

Review the safety precautions for using a portable grinder.

Show trainees how to operate a portable grinder.

Have trainees practice operating a portable grinder. Note the proficiency of each trainee. This laboratory corresponds to Performance Task 2.

Have trainees review Sections 5.0.0–5.5.0.

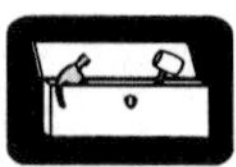

Ensure that you have everything required for teaching this session.

Describe a pipe threading machine.

Step 8 Ensure that the maximum rotating speed of the grinding wheel is higher than the maximum rotating speed of the grinder.

Step 9 Start the grinder and allow it to run for a moment while checking for visual abnormalities, excessive vibration, extreme temperature changes, or noisy operation.

Step 10 Inspect the work area to ensure the safety of yourself and others and to make sure that the heat and sparks generated by the grinder cannot start any fires.

4.3.0 Operating Grinders

When operating grinders, you must pay full attention to the grinder, the work being performed, and the flow of sparks and metal bits coming off the wheel. Each grinding accessory is designed to be used in only one way. When a flat grinding disc is being used, only put the flat surface of the disc against the work. If a cutoff wheel is being used, only use the edge of the wheel to do the cutting. Follow these steps to operate a grinder:

Step 1 Inspect the grinder.

Step 2 Put on personal protective equipment, including gloves, a face shield, ear plugs and safety shoes.

Step 3 Tuck in all loose clothing.

Step 4 Secure the object to be worked on in a vise or clamps to ensure that it does not move.

Step 5 Obtain any required hot work permits.

Step 6 Attach the grinder to the power source.

Step 7 Position yourself with good footing and balance, and establish a firm hold on the grinder to avoid kickback.

Step 8 Verbally warn bystanders before pulling trigger. Then, pull the trigger to start the grinder.

Step 9 Apply the grinder to the work, and direct the sparks to the ground whenever possible and away from any hazards in the area, such as combustible debris or acetylene tanks.

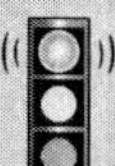

WARNING!

Special precautions must be taken to avoid kickback while prepping fittings and pipe. Remember that the wheel rotates clockwise, and you must position yourself and the wheel on the pipe end correctly to avoid the wheel grabbing the pipe end and kicking back toward you. Always grind with the direction of the bevel and position the wheel to the bevel, so that if it kicks off, it will kick off away from you. The wheel is properly positioned on the bevel when the torque pulls the wheel in the direction that the wheel would leave the pipe end if it grabbed. Reposition the wheel on the pipe end when moving from the top half to the bottom half of the pipe.

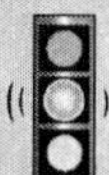

CAUTION

Protect nearby alloy metals, such as stainless steel tanks, from cross contamination of other metals being ground in the area.

Step 10 Apply proper force to the grinder on the grinding surface. If you are applying too much force on the grinder, you will hear the motor strain.

Step 11 Periodically stop grinding and inspect the work and the grinding wheel. Replace the wheel if necessary.

Step 12 Turn off and disconnect the grinder from the power source when grinding is complete.

Step 13 Inspect the grinder for any signs of damage.

Step 14 Return the grinder and any accessories to the storage area.

5.0.0 ◆ PIPE THREADING MACHINES

Pipe threading machines are multi-purpose power tools that center, hold, and rotate pipe, conduit, or bolt stock while cutting, reaming, and threading operations are performed (*Figure 6*). All pipe

Grinding Wheels

A grinding wheel must be replaced if it is worn or cracked. Cracks are not always visible, however. One way to check for cracks is to hold the wheel by the center hole and tap the edge with a screwdriver handle. If the wheel is cracked, you will hear a dull sound. If the wheel is good, you will hear a clear ring.

Instructor's Notes:

3.10 PIPEFITTING ◆ LEVEL ONE

threading machines are designed with a gear drive and **chucks**, which hold and turn the pipe, and a three-position power switch consisting of forward, off, and reverse drive positions. Depending on the size and complexity of the machine, pipe threading machines can be mounted on tripods, tables, or special stands (*Figure 7*).

You can use a variety of vises and wrenches to hold pipe while threading it. Vises have jaws that hold the pipe firmly and prevent it from turning. The teeth on vises may leave marks on the pipe, so you should use a vise only on pipe that will not be visible after installation. Use the standard yoke vise to hold pipe that is small in diameter. Use the chain vise to hold much larger pieces of pipe. Keep the chain on this vise oiled to prevent it from becoming stiff. Always make sure that you use the correct oil. A stiff chain will make the chain vise operate poorly. Refer to the manufacturer's specifications for the proper types of oil. Use pipe wrenches to grip and turn around stock.

5.1.0 Loading Pipe into a Threading Machine

On most threading machines, the pipe can be inserted from either end of the chuck. Some models, however, must be loaded from the rear. Either way, the operator must position the pipe so that the end to be threaded extends out away from the chuck. On most threading machines, there is a stop to help position the pipe properly. Follow these steps to load pipe into the threading machine:

Step 1 Measure and mark the length of pipe to be worked.

Step 2 Insert the pipe into the threading machine. If the pipe is long enough to be held by the centering device, insert the pipe through the front or rear of the machine. If the pipe is short, insert the pipe through the front of the machine. For long pipe, support the end of the pipe with an adjustable support.

Step 3 Make sure that the pipe is centered in the centering device.

Step 4 Tighten the centering device.

Step 5 Spin the handwheel counterclockwise to tighten the chuck jaws. This tightens the jaws on the pipe. A clockwise spin releases the jaws.

WARNING!

When you are working with a power threading machine, do not wear loose clothing, jewelry, or gloves that could easily be caught in the rotating parts of the machine. Be sure to wear appropriate PPE. Be sure to tie back long hair to prevent it from being entangled in the machine during operation. Use a piece of cloth or cardboard to absorb any oil that spills during the machine's operation, so the oil does not get on the bottom of your work shoes and you do not track the slippery oil around the job site.

103F06.EPS

Figure 6 ◆ Power threading machine.

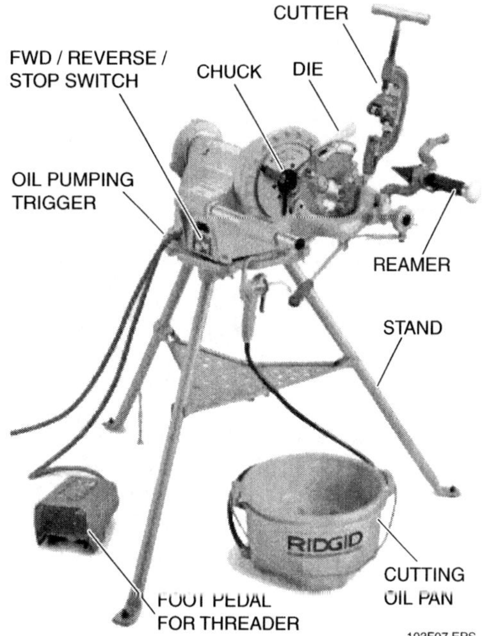

103F07.EPS

Figure 7 ◆ Cutting threads using a power threading machine.

Describe the various types of vises and wrenches used to hold pipe while threading it.

Explain how to load pipe onto a threading machine.

Review the safety precautions for using a threading machine.

MODULE 08103-06 ◆ PIPEFITTING POWER TOOLS 3.11

Review the steps for cutting and reaming pipe with a threading machine.

Show Transparencies 4 and 5 (Table 2 and Figure 8). Describe dies used with a pipe threading machine and explain how to select the correct die.

Review the steps for replacing dies in a threading machine.

5.2.0 Cutting and Reaming Pipe

Follow these steps to cut and ream pipe:

Step 1 Determine the length of pipe needed.

Step 2 Measure the pipe to determine how much must be cut off.

Step 3 Mark the pipe at the point where it is to be cut.

Step 4 Swing the cutter, threader, and reamer up and out of the way.

Step 5 Load the pipe into the threading machine.

Step 6 Place pipe stands as needed.

Step 7 Turn the centering device to center the pipe in the machine.

Step 8 Spin the chuck handwheel counter-clockwise to lock the pipe in place.

Step 9 Loosen the cutter until it fits over the pipe.

Step 10 Lower the cutter onto the pipe.

Step 11 Rotate the handwheel to move the cutter wheel to the cutting mark on the pipe.

Step 12 Rotate the cutter handle to tighten the cutter on the pipe.

Step 13 Direct the cutting oil to the cutting area.

Step 14 Turn the pipe machine switch to the FORWARD position.

Step 15 Step on the foot switch to start the machine and turn the pipe.

Step 16 Rotate the cutter handle to apply pressure to the cutter until the cut is complete.

Step 17 Release the foot switch.

Step 18 Raise the cutter to an out-of-the-way position.

CAUTION

Do not use a pipe cutter to cut pipe for a socket weld. Use a bandsaw. The pipe cutter will cause the pipe to flare. It may also interfere with the flow of material through the pipe.

Step 19 Lower the reamer into place.

Step 20 Press the latch on the reamer and slide the reamer bar toward the pipe until the latch catches.

Step 21 Step on the foot switch to start the machine.

Step 22 Rotate the handwheel to move the reamer into the end of the pipe.

Step 23 Ream the end of the pipe.

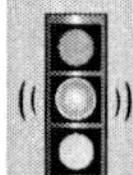

CAUTION

Do not overream the pipe. Only ream enough to remove the burrs.

Step 24 Rotate the handwheel to move the reamer out of the pipe.

Step 25 Release the foot switch.

Step 26 Slide the power switch to the OFF position.

Step 27 Raise the reamer out of the way.

Step 28 Loosen the chuck handwheel and the centering device.

Step 29 Remove the pipe from the machine.

5.3.0 Replacing Dies in a Threading Machine

Four sets of dies are used with pipe threading machines that thread pipe up to 2 inches in diameter and another set is used for pipe 2½ inches and larger. Dies are classified according to the number of threads per inch. *Table 2* shows proper die selection.

Once selected, the pipe thread dies must be installed and adjusted for different sizes of pipe. Pipe threading machines have universal die heads that adjust the dies for different sizes of pipe. *Figure 8* shows the die head details.

Each set of dies, except those used on ⅛-inch pipe, can be used for several sizes of pipe. To adjust the die head for different sizes of pipe, loosen the clamp lever and move the size bar until the line underneath the desired pipe size lines up with the index line. Tighten the clamp lever after the adjustment has been made. If oversized or undersized threads are required, set the index line in the direction of the over or under mark on the size bar.

The throw-out lever, which is marked OPEN and CLOSE, is a quick-release handle used to close the dies into the threading position and open the dies to retract the dies away from the pipe. After threading a pipe, the operator can lift this handle to the open position and quickly remove the dies from the pipe. Before cutting threads with the pipe threading machine, the proper dies must be installed in the die head. Follow these steps to replace the dies in a threading machine:

Step 1 Remove the die head from the machine and place it on a workbench with the numbers facing up.

Step 2 Open the throw-out lever.

Step 3 Loosen the clamp lever about three turns.

Step 4 Lift the tongue of the clamp lever washer up and out of the slot under the size bar.

Instructor's Notes:

3.12 PIPEFITTING ◆ LEVEL ONE

Table 2 Proper Die Selection

Nominal Pipe Size (inches)	Die Threads (per inch)	Thread Length (inches)	Number of Threads
⅛	27	⅜	10
¼	18	⅝	11
⅜	18	⅝	11
½	14	¾	10
¾	14	¾	10
1	11½	⅞	10
1¼	11½	1	11
1½	11½	1	11
2	11½	1	11
2½	8	1½	12
3	8	1½	12
3½	8	1⅝	13
4	8	1⅝	13
5	8	1¾	14
6	8	1¾	14
8	8	1⅞	15
10	8	2	16
12	8	2⅛	17

103T02.EPS

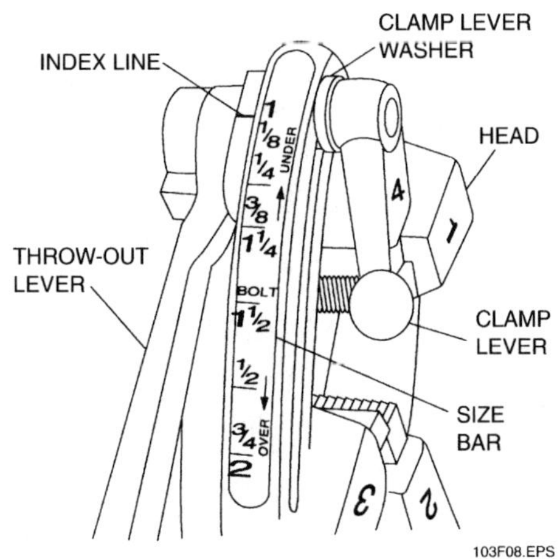

103F08.EPS

Figure 8 ◆ Die head details.

Step 5 Slide the throw-out lever all the way to the end of the slot.

Step 6 Remove the old dies from the die head.

Step 7 Select a matched set of replacement dies.

Step 8 Clean and inspect the new dies.

Step 9 Insert the new dies in the die head.

NOTE

The numbers 1 through 4 on the dies must match the numbers on the slots of the die head.

Step 10 Slide the throw-out lever back to where the tongue of the clamp lever washer drops in the slot under the size bar.

Step 11 Adjust the die head size bar until the index line is lined up with the proper size mark on the size bar.

Step 12 Tighten the clamp lever.

Step 13 Install the die head into the machine.

5.4.0 Performing Threading Operations

This section explains threading operations using a Ridgid® pipe threading machine. Before actually threading the pipe, the operator should check the thread cutting oil level and oil pump operation and prime the oil pump if necessary.

5.4.1 Checking Thread Cutting Oil Level

Before any threading operations are performed, it is extremely important to check the level of the special thread cutting oil. Follow these steps to check the thread cutting oil level:

Step 1 Slide the chip pan out from the base of the threading machine.

Step 2 Check to see if the oil is up to the fill-level line in the reservoir. Fill the reservoir with cutting oil as necessary.

Step 3 Lower the lubrication arm over the open reservoir. Oil cannot flow from the lubrication arm in its upright position.

Step 4 Turn the power selector switch to the FORWARD position.

Step 5 Step on the foot switch. Oil should flow from the lubrication arm. If it does not, it may be necessary to prime the pump.

Step 6 Turn the power selector switch to the OFF position to stop the threading machine.

Step 7 Raise the lubrication arm and slide the chip pan back into the base of the threading machine.

5.4.2 Priming the Oil Pump

If the oil does not flow from the lubrication arm, it may be necessary to prime the pump. Follow these steps to prime the oil pump:

Show trainees how to replace dies in a threading machine.

Have trainees practice replacing dies in a threading machine. Note the proficiency of each trainee. This laboratory corresponds to Performance Task 3.

Explain that the operator must check the oil and prime the pump if necessary before using a threading machine.

Review the steps for checking the cutting oil level on a threading machine.

Explain how to check if the pump needs priming.

Review the steps for priming the pump on a threading machine.

Show trainees how to check the oil and prime the pump on a threading machine.

Review the steps for threading pipe using a threading machine.

Review safety precautions for using a threading machine. Stress the importance of using gloves.

Emphasize the importance of maintaining a threading machine.

Describe procedures for maintaining a threading machine.

Step 1 Remove the button plug on the machine cover.

Step 2 Remove the primer screw through the opening.

Step 3 Fill the pump with cutting oil.

Step 4 Replace the primer screw and the button plug.

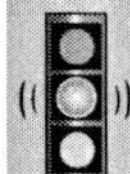

CAUTION

Failure to replace the primer screw and button plug causes the pump to drain itself immediately when the power is turned on.

Step 5 Turn the power on in the reverse direction, and check the flow of oil from the lubrication arm. Running the machine in reverse primes the pump more efficiently than running the machine forward.

5.4.3 Threading Pipe

After checking the cutting oil level and priming the pump, the operator is ready to thread the pipe. Follow these steps to thread pipe:

Step 1 Load the pipe into the threading machine.

Step 2 Place pipe stands under the pipe as needed.

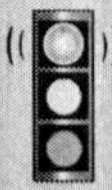

WARNING!

Do not wear loose clothing or jewelry. Keep hair pulled back while operating a threading machine. Wear a pair of properly fitting gloves to prevent cuts from sharp threads.

Step 3 Cut and ream the pipe to the required length.

Step 4 Make sure that the proper dies are in the die head.

Step 5 Loosen the clamp lever.

Step 6 Move the size bar to select the proper die setting for the size of pipe being threaded.

Step 7 Lock the clamp lever.

Step 8 Swing the die head down to the working position.

Step 9 Close the throw-out lever.

Step 10 Lower the lubrication arm and direct the oil supply onto the die.

Step 11 Turn the machine switch to the FORWARD position.

Step 12 Step on the foot switch to start the machine.

Step 13 Turn the carriage to bring the die against the end of the pipe.

Step 14 Apply light pressure on the handwheel to start the die.

Step 15 Release the handwheel once the dies have started to thread the pipe. The dies feed onto the pipe automatically as they follow the newly cut threads.

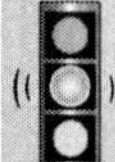

CAUTION

Make sure that the die is flooded with oil at all times while the die is cutting to prevent overheating the die and the pipe. Overheating can cause damage to the die and to the pipe threads.

Step 16 Open the throw-out lever as soon as two full threads extend from the back of the dies.

Step 17 Release the foot switch to stop the machine.

Step 18 Turn the carriage handwheel to back the die off the pipe.

Step 19 Swing the die and the oil spout up and out of the way.

Step 20 Screw a fitting that is the same size as the pipe you are threading onto the end of the pipe to check the threads.

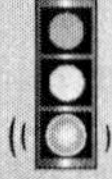

NOTE

If the threads are cut correctly, you should be able to screw a fitting three and a half revolutions by hand onto the new threads. If the threads are not deep enough, adjust the die using the clamp lever and repeat the threading operation.

Step 21 Turn the machine switch to the OFF position.

Step 22 Open the chuck and remove the pipe.

5.5.0 Threading Machine Maintenance

This section describes the procedures to keep the threading machine in proper working condition. By following a few simple guidelines, you can expect a long life of service from the threading machine:

Instructor's Notes:

- Never reverse the rotation of the machine while the machine is running. Always turn off the power and wait until the drive has stopped rotating before changing the direction of the drive mechanism.
- At the start of each day, clean the chuck jaws with a stiff brush to remove rust, scale, chips, pipe coating, or other foreign matter. Apply lubricating oil to the machine bedways, cutter rollers, and feed screw.
- Always use sharp dies, which produce smoother threads and require less motor power than dull dies.
- Refer to and follow the manufacturer's instructions for lubrication schedules for bearings and gears. The threading machine should be lubricated at least once for every 40 hours of running time.
- Ensure that there is always plenty of cutting oil in the machine. The cutting oil should be periodically cleaned to remove accumulated sludge, chips, and other foreign matter from the oil. Replace the cutting oil when it becomes dirty or contaminated.
- Remove the oil filter periodically and clean it using solvent. Blow the filter clean using compressed air. Do not operate the machine without the oil filter.
- Keep the chuck jaws in good condition, and replace the jaw inserts as necessary when they become worn.

6.0.0 ◆ SPECIAL THREADING APPLICATIONS

The special threading applications that can be performed with the threading machine include threading short nipples using a nipple chuck and threading pipe larger than 2 inches in diameter using a geared threader or mule.

6.1.0 Cutting and Threading Nipples

A nipple is a piece of pipe less than 12 inches long that is threaded on each end. Threaded pipe longer than 12 inches is considered cut pipe. The nipple chuck is a useful tool for holding nipples in the threading machine when threading. A nipple chuck with inserts and adapters will thread nipples from ⅛ inch to 2 inches in diameter. *Figure 9* shows a nipple chuck kit.

Follow these steps to cut and thread nipples:

Step 1 Load the pipe into the threading machine.

Step 2 Ream the end of the pipe.

Step 3 Thread the pipe.

103F09.EPS

Figure 9 ◆ Nipple chuck kit.

Step 4 Measure from the threaded end the desired length of the nipple and mark a line on the pipe at this point.

Step 5 Cut the pipe at the mark.

Step 6 Remove the pipe from the power drive chuck.

Step 7 Place the nipple chuck into the power drive chuck and tighten the jaws on the nipple chuck.

Step 8 Place the insert inside the end of the nipple chuck.

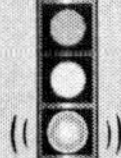

NOTE

Place the small end toward the outside if threading ⅛- to ¾-inch pipe. Place the large end toward the outside if threading 1-inch pipe. Do not use an insert with pipe that is 1¼ inch and larger.

Step 9 Select the proper size nipple chuck adapter depending on the size pipe you are threading.

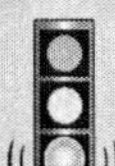

NOTE

If you are threading 2-inch pipe, you do not need a nipple chuck adapter.

Step 10 Screw the nipple chuck adapter into the nipple chuck.

Step 11 Tighten the nipple chuck adapter into the nipple chuck using a chuck wrench.

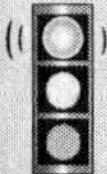

WARNING!

Do not start the threading machine with the chuck wrench over the nipple chuck adapter. The wrench could be thrown by the force of the machine.

Show trainees how to cut, ream, and thread pipe using a threading machine.

Have trainees practice cutting, reaming, and threading pipe using a threading machine. Note the proficiency of each trainee. This laboratory corresponds to Performance Task 4.

Have trainees review Sections 6.0.0–6.2.0.

Ensure that you have everything required for teaching this session.

Describe special applications that can be performed with a threading machine.

Describe a nipple and a nipple chuck kit.

Explain the procedure for cutting and threading nipples.

Explain the importance of removing the chuck wrench before starting the threading machine.

MODULE 08103-06 ◆ PIPEFITTING POWER TOOLS 3.15

Show trainees how to cut and thread nipples using a nipple chuck.

Have trainees practice cutting and threading nipples using a nipple chuck. Note the proficiency of each trainee. This laboratory corresponds to Performance Task 5.

Describe a geared threader or mule and explain how they are used.

Discuss the importance of securing a geared threader and other safety precautions necessary.

Have trainees review Sections 7.0.0–8.3.0.

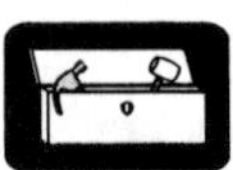

Ensure that you have everything required for teaching this session.

Describe a portable power drive and explain how they are used.

Review the steps for threading pipe using a portable power drive.

Step 12 Screw the nipple into the end of the nipple chuck.

Step 13 Tighten the nipple chuck release collar using the chuck wrench.

Step 14 Ream the end of the nipple.

Step 15 Thread the nipple.

Step 16 Insert the chuck wrench into the release collar and turn the release collar to loosen the nipple from the nipple chuck.

Step 17 Unscrew the nipple from the nipple chuck.

WARNING!

Use gloves when handling the threads on the freshly cut nipple. These threads are extremely sharp and can cut you.

Step 18 Place the chuck wrench over the nipple chuck adapter and unscrew the adapter from the nipple chuck.

Step 19 Wipe off all excess cutting oil from the nipple chuck adapter and the insert, and store them with the nipple chuck kit.

Step 20 Release the power drive chuck to remove the nipple chuck from the threading machine.

Step 21 Wipe off the nipple chuck and store it with the rest of the nipple chuck kit.

6.2.0 Threading Pipe Using a Geared Threader

A geared threader, also known as a mule, is used to thread pipe from 2½ to 6 inches in diameter (*Figure 10*). Two sets of dies are used with geared threaders. One set is used for pipe that is 2½ to 4 inches in diameter, and the other set is used for pipe that is 4 to 6 inches in diameter. Each die set contains five dies that are numbered just like the dies for the universal die head. The geared threader is connected to the pipe threading machine by a universal drive shaft.

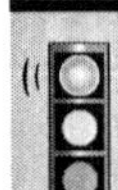

WARNING!

A geared threader produces an enormous amount of torque while threading. It can easily tip over the threading machine if the machine is not properly secured. If possible, bolt the threading machine to a steel table and secure the pipe in a vise that is bolted to a steel table. Follow the manufacturer's instructions when using this tool.

7.0.0 ◆ PORTABLE POWER DRIVES

A portable power drive (*Figure 11*) is a versatile handheld power unit that can be used for cutting threads, driving hoists and winches, and operating large valves. An electric motor provides the power to rotate a ring in the end of the tool. Threading dies and other accessories can be inserted into the circular ring, which has grooved spines to lock these accessories into place.

Threading die heads for pipe 2 inches and smaller either fit directly into the power drive or fit into a die head adapter and then into the power drive.

Usually, one hand can hold the power unit and control the power switch while oiling is done with the other hand. Follow these steps to thread pipe using a portable power drive:

Step 1 Secure the pipe to be threaded in a vise.

Step 2 Identify the size of the pipe to be threaded and insert the appropriate die into the portable power drive.

103F10.EPS

Figure 10 ◆ Geared threader.

103F11.EPS

Figure 11 ◆ Portable power drive.

Instructor's Notes:

3.16 PIPEFITTING ◆ LEVEL ONE

Step 3 Position the die head onto the end of the pipe with the face of the die stock facing away from the pipe (*Figure 12*). The drive unit should be positioned on the end of the pipe with the handle of the drive unit in a nearly vertical position above the pipe.

Step 4 Hold the handle of the drive unit with your right hand and center the die head onto the end of the pipe with your left hand.

Step 5 Pull the handle of the drive unit down sharply with your right hand while pushing the die stock against the end of the pipe with your left hand to make the dies bite into the end of the pipe.

Step 6 Continue to move the handle down to a position where you can straighten your elbow to provide a firm, strong grip on the drive unit. You should lower your right shoulder and balance your weight above the drive unit to resist the threading torque. If you are threading pipe larger than 1 inch, attach a torque arm to support the drive unit at this point. *Figure 13* shows the power drive support arm.

Step 7 Start the power drive and cut the thread.

CAUTION
Be sure to keep an adequate supply of cutting oil on the thread as it is being cut.

Step 8 Stop the power drive as soon as the thread has been cut to the required length.

Step 9 Turn the power switch to the reverse position and back the dies off the thread.

Step 10 Inspect the threads to make sure they have been cut properly.

8.0.0 ◆ POWER BEVELERS

Power bevelers are used to prepare pipe for butt weld **fabrication**. Bevelers place the correct angle, or bevel, on the end of a pipe before it can be welded.

There are many different ways to bevel the end of a piece of pipe to be butt welded. The joint can be beveled mechanically using grinders, nibblers, or cutters, or it can be beveled thermally using an oxyfuel cutting torch. Mechanical joint beveling is used most often on alloy steel, stainless steel, and nonferrous metal piping. It is often required for the materials that could be affected by the heat of the thermal process. Mechanical joint beveling is slower than thermal methods, such as oxyfuel cutting, but has the advantage of high precision with low heat input and the absence of oxides commonly left by the thermal methods. The method used to bevel the pipe generally depends on the base material type, ease of use, and code or procedure specifications. *Figure 14* shows mechanically beveled pipe ends.

Figure 12 ◆ Installing a die head.

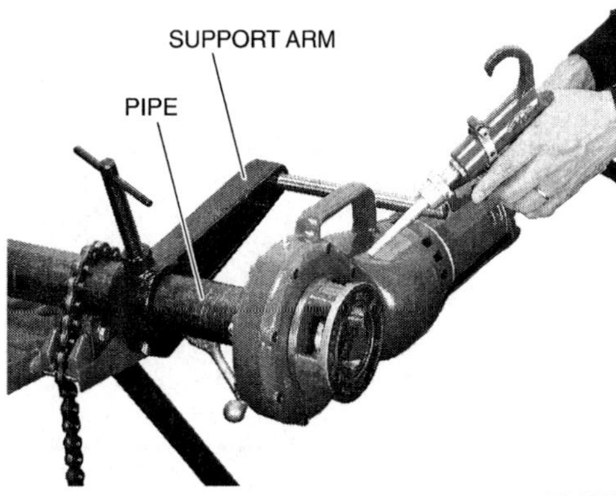

Figure 13 ◆ Power drive support arm.

8.1.0 Pipe Beveling Machines

Welded piping is found extensively in the utilities industry and even more in the petrochemical industry. Nearly every piece of pipe that is welded to a fitting, another piece of pipe, or any other connection requires that the edge be cut square and beveled according to specifications before welding. If a mechanical cutting and beveling process is used to accomplish this, it may be done using an electric or pneumatic beveling machine. Many of these are

Show trainees how to thread pipe using a portable power drive.

Have trainees practice threading pipe using a portable power drive. Note the proficiency of each trainee. This laboratory corresponds to Performance Task 6.

See the Teaching Tip for Section 8.0.0 at the end of this module.

Describe various types of power bevelers.

Explain that pipe must be cut and beveled before welding. Explain that powered beveling machines are often used by pipefitters to prepare pipe for welding.

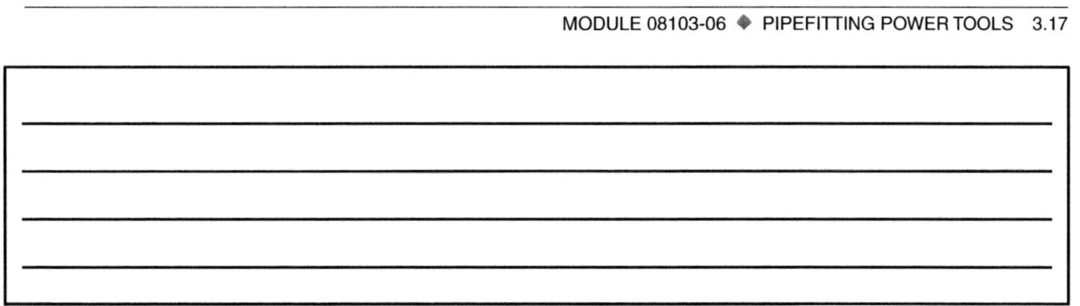

Describe a pipe beveling machine and explain how it is used.

Stress the importance of following the manufacturer's safety procedures when using bevelers and other grinding equipment.

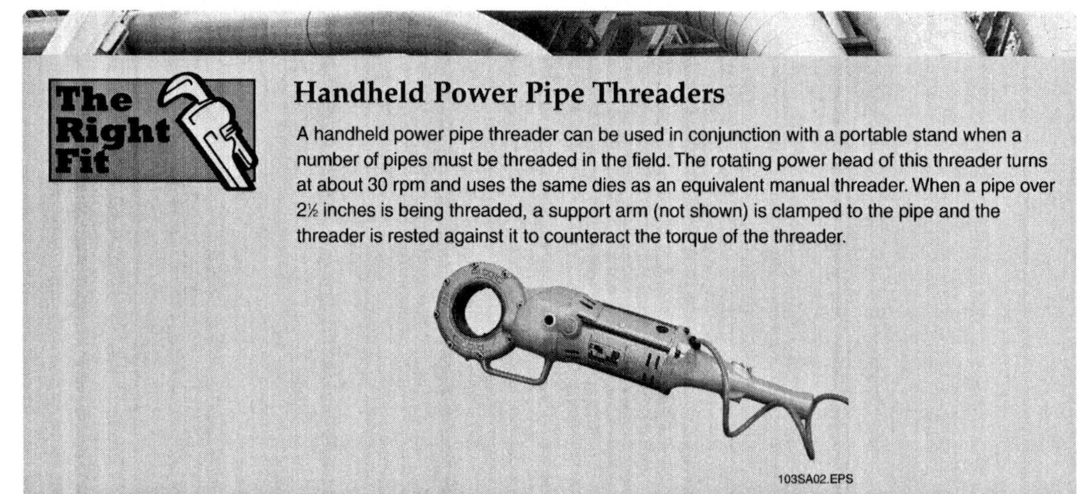

Handheld Power Pipe Threaders

A handheld power pipe threader can be used in conjunction with a portable stand when a number of pipes must be threaded in the field. The rotating power head of this threader turns at about 30 rpm and uses the same dies as an equivalent manual threader. When a pipe over 2½ inches is being threaded, a support arm (not shown) is clamped to the pipe and the threader is rested against it to counteract the torque of the threader.

103SA02.EPS

portable machines that operate much like a lathe. They have mandrels that hold various cutting tools, as well as numerous adjusting mechanisms that make it easy to set the bevel angle and depth. Various models are available to cut and bevel 2-inch to 60-inch pipe. When specifications call for tubing to be welded, machines similar to pipe beveling machines are available to face, square, and **chamfer** the ends of tubes in preparation for welding. Boiler tube ends are typically prepared using these smaller machines before welding the tubes in place.

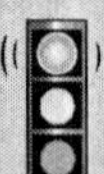

WARNING!

Always follow all manufacturers' safety procedures when using grinders. Failure to follow the manufacturer's safety recommendations could result in serious personal injury.

Special cutoff machines are also made to be mounted on the outside of the pipe. They can be mounted with a ring or a special chain with rollers. An electrically or pneumatically operated grinder with a cutoff blade is mounted and manually or electrically powered around the pipe to cut it off.

Figure 15 illustrates a pipe-beveling machine being properly applied to the end of a pipe to prepare it for welding.

103F14.EPS

Figure 14 ◆ Beveled pipe ends.

103F15.EPS

Figure 15 ◆ Portable beveling machine.

Instructor's Notes:

3.18 PIPEFITTING ◆ LEVEL ONE

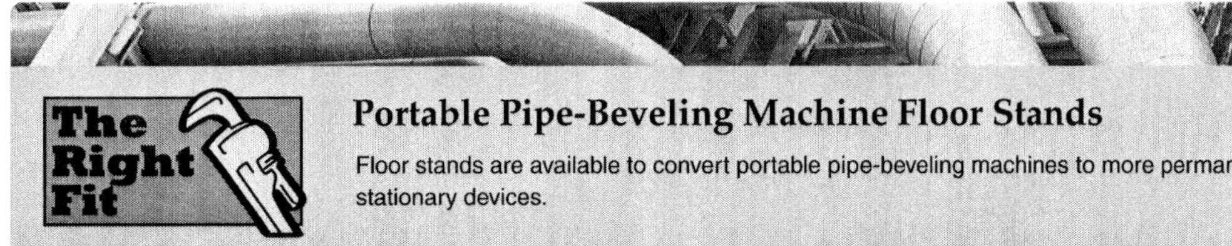

The Right Fit: Portable Pipe-Beveling Machine Floor Stands

Floor stands are available to convert portable pipe-beveling machines to more permanent stationary devices.

Describe nibblers and cutters and explain how they are used.

Stress the importance of following the manufacturer's safety procedures.

Show trainees how to identify several types of bevelers.

Have trainees practice identifying several types of bevelers. Note the proficiency of each trainee. This laboratory corresponds to Performance Task 7.

Explain thermal joint preparation.

8.2.0 Nibblers and Cutters

Nibblers prepare the edge of a plate or pipe with a reciprocal punch that cuts off a chip with each stroke. The bevel angle is set by adjusting the nibbler. Nibblers must have access to an edge to be used. *Figure 16* shows a nibbler.

Cutters use round cutting tools similar to mill cutting tools. The bevel angle is set by adjusting the cutter or changing the cutter blade. Cutters leave the surface much smoother than nibblers and can be used for cutoff operations. Another advantage of cutters is that they can easily be used to prepare J- or compound bevels by changing the shape of the cutter. Cutters made for pipe are sometimes called pipe-end-prep lathes.

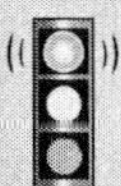

WARNING!

Always follow all manufacturers' safety procedures when using nibblers and cutters. Failure to follow the manufacturer's safety recommendations could result in serious personal injury.

8.3.0 Thermal Joint Preparation

Thermal joint preparation is done using the oxyfuel, plasma arc, or carbon arc cutting process. For beveling plate and pipe, it is best to use the oxyfuel or plasma arc cutting processes. The carbon arc cutting process is best for gouging seams, repairing cracks, and weld repairs.

The torch used for oxyfuel or plasma arc cutting can be handheld or mounted on a motorized carriage. The motorized carriage for plate cutting runs on flat tracks positioned on the surface of the plate to be cut. The carriage has on/off and forward/reverse switches and a speed adjustment. A handwheel on top of the carriage adjusts the torch holder transversely, and a handwheel at the torch holder adjusts the torch vertically. The bevel angle is set by pivoting the torch holder. The carriage can carry a special oxyfuel or plasma arc cutting torch designed to fit into the torch holder. *Figure 17* shows a motorized carriage for cutting and beveling plate.

Special equipment is used for cutting pipe. A steel ring or special chain with rollers is attached to the outside of the pipe. The torch is carried around the pipe on a torch holder that is powered by electricity, air, or a hand crank. The bevel angle is set by pivoting the torch holder, and the torch is adjusted vertically with a handwheel on the torch holder. An out-of-round attachment can be used to compensate for pipe that is out-of-round. For large-diameter pipe (54 inches or larger), special internal equipment is available. With this special equipment, the torch mechanism is mounted on the inside of the pipe. Special plasma or oxyfuel torches can be mounted in the torch holders of the external or internal pipe cutting equipment. *Figure 18* shows a pipe cutting and beveling tool.

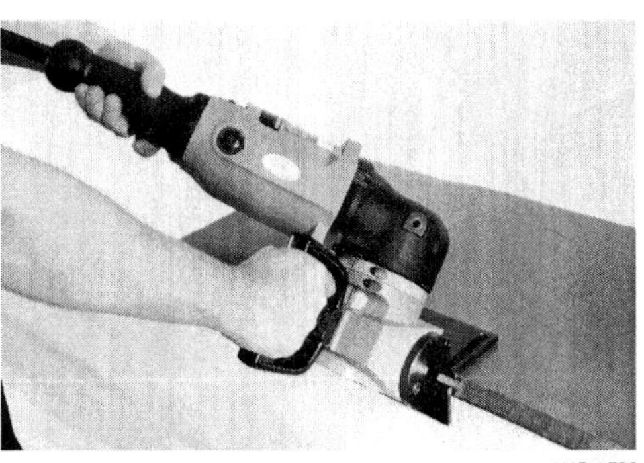

103F16.EPS

Figure 16 ◆ A nibbler.

MODULE 08103-06 ◆ PIPEFITTING POWER TOOLS 3.19

Thermal Joint Preparation Precautions

When preparing a pipe joint with the oxyfuel, plasma arc, or carbon arc cutting process, all dross (metal expelled during cutting) must be removed prior to welding. Any dross remaining on the joint during welding will cause weld defects. Joints prepared with the carbon arc cutting torch must be carefully inspected for carbon deposits. It is common for small carbon deposits to be left behind as the carbon electrode is consumed during the cutting or gouging operation. These carbon deposits will cause defects such as hard spots and cracking in the weld. Before welding, use a grinder to clean surfaces prepared with the carbon arc torch.

Automatic Pipe Bevelers

Automatic versions of pipe bevelers and cutters are an improvement over pattern cutters and other similar equipment. This is because there is no need to reset the preheat flame before each cut. Once an automatic pipe beveler is initially set up, all subsequent cuts can be made with the same settings. These systems also save time and cutting gases because all gases are extinguished immediately when the system is switched off.

103F17.EPS

Figure 17 ◆ Motorized carriage for cutting and beveling plate.

103F18.EPS

Figure 18 ◆ Pipe cutting and beveling tool.

Instructor's Notes:

3.20 PIPEFITTING ◆ LEVEL ONE

Have the trainees complete the Review Questions, and go over the answers prior to administering the Module Examination.

Review Questions

1. If a power cord has frayed insulation _____.
   a. it must be repaired with a special electrical tape
   b. fresh insulation must be melted into the opening
   c. it must be replaced by a licensed electrician
   d. it must be plugged into a GFCI outlet

2. Which of these statements regarding a GFCI is correct?
   a. It provides protection from all electrical shock hazards
   b. It does not provide protection from overheating
   c. It provides protection from line to line shocks
   d. It provides protection from ground faults caused by defective insulation

3. Which of the following materials would *not* be cut with a carbon steel blade?
   a. Aluminum
   b. Stainless steel
   c. Galvanized steel
   d. Copper

4. When grinding stainless steel, you should use a(n) _____ grinding wheel.
   a. stainless steel
   b. aluminum oxide
   c carbon steel
   d. mild steel

5. A _____ should be used when cleaning a bevel on a pipe end.
   a. flapper wheel
   b. buffing wheel
   c. wire brush
   d. grinding wheel

6. When small-diameter pipe is being threaded, it should be secured with _____.
   a. a standard yoke vise
   b. a chain vise
   c. duct tape
   d. locking pliers

7. When preparing a pipe for socket welding, you should cut the pipe with a bandsaw rather than a pipe cutter.
   a. True
   b. False

8. Pipe threading dies are classified according to _____.
   a. weight
   b. pipe size
   c. threads per inch
   d. pipe material

9. Each pipe diameter requires a different set of dies designed for that diameter.
   a. True
   b. False

10. To prime the cutting oil pump on a pipe threading machine, you should _____.
    a. run the machine forward
    b. run the machine in reverse
    c. simply fill the reservoir
    d. squirt oil on the die

11. If threads are cut correctly, you should be able to screw a fitting _____ turns onto a pipe by hand.
    a. 2
    b. 2½
    c. 3½
    d. 4½

12. A threading machine should be lubricated at least once for every _____ hours of running time.
    a. 8
    b. 40
    c. 80
    d. 100

13. A nipple is a piece of pipe that is less than _____ inches long and threaded on both ends.
    a. 2
    b. 6
    c. 8
    d. 12

MODULE 08103-06 ◆ PIPEFITTING POWER TOOLS 3.21

Review Questions

14. A nipple chuck is used to _____.
    a. cut threads on nipples
    b. hold the nipple in place during threading
    c. ream nipples
    d. measure nipples

15. A mule is another name for a _____.
    a. geared pipe threader
    b. nipple chuck
    c. pipe stand
    d. portable power drive

Figure 1

103RQ01.EPS

16. The device shown in *Figure 1* is a _____.
    a. geared pipe threader
    b. ratcheting pipe threader
    c. threading collar
    d. portable power drive

17. Which of the following tasks *cannot* be done with a portable power drive?
    a. Cutting threads
    b. Driving a winch
    c. Operating valves
    d. Cutting pipe

18. Pipe bevelers are used to _____.
    a. prepare pipe ends for threading
    b. remove burrs from pipe ends
    c. prepare pipe ends for welding
    d. cut stainless steel pipe

19. Which of the following thermal joint preparation processes should *not* be used for beveling plate and pipe?
    a. Oxyfuel
    b. Carbon arc
    c. Plasma arc
    d. Both oxyfuel and plasma arc

Figure 2

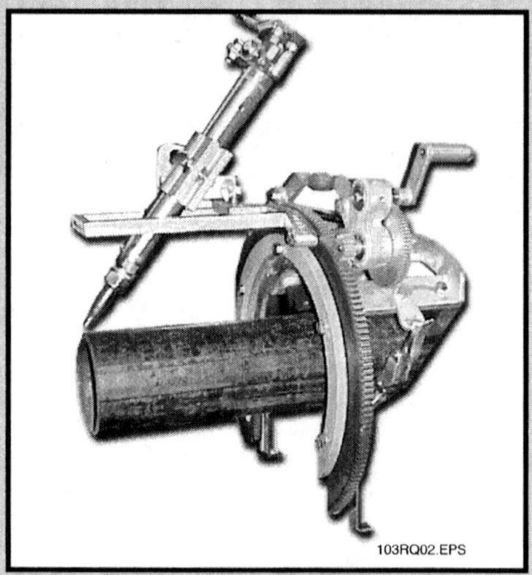

103RQ02.EPS

20. The device shown in *Figure 2* is used to _____.
    a. locate the center of a pipe
    b. measure the outside diameter of a pipe
    c. secure pipe ends for welding
    d. cut and bevel pipe

Instructor's Notes:

3.22 PIPEFITTING ◆ LEVEL ONE

Summary

Summarize the major concepts presented in the module.

Power tools are necessary items used by pipefitters, and the use of these tools has brought on many improvements in the pipefitting trade. The purpose of this module is to provide training on the safe use and application of these tools. Extra attention has been given to the portable grinders and pipe threading machines because, as helpers, these are the tools you will be involved with the most. You will learn more about the operations of the other power tools as you are trained in the fabrication and installation of the piping systems to which they apply. As in every aspect of your job, when using power tools you must remain safety conscious at all times. Ignoring safety precautions will most likely cause injuries.

Notes

MODULE 08103-06 ◆ PIPEFITTING POWER TOOLS 3.23

Have the trainees complete the Trade Terms Quiz, and go over the answers prior to administering the Module Examination.

Administer the Module Examination. Record the results on Craft Training Report Form 200, and submit the results to the Training Program Sponsor.

Administer the Performance Test, and fill out Performance Profile Sheets for each trainee. If desired, trainee proficiency noted during laboratory sessions may be used to complete the Performance Test. Record the results on Craft Training Report Form 200, and submit the results to the Training Program Sponsor.

Trade Terms Quiz

Fill in the blank with the correct trade term that you learned from your study of this module.

1. An angle cut or ground only on the edge of a piece of material is called a(n) _________.
2. A pipe or cutter is held in the machine by a(n) _________.
3. To cut male threads on a pipe, you would use a(n) _________.
4. The speed of revolutions of an object is measured in _________.
5. Power can be measured in _________, a unit equal to 745.7 watts.
6. An angle cut on the end of a piece of material is a(n) _________.
7. A detailed plan for required equipment inspections is called a(n) _________.
8. _________ is the act of putting components together.
9. Power tools should always be used with a(n) _________ to protect against electric shock.

Trade Terms

Assured equipment grounding conductor program
Bevel
Chamfer
Chuck
Die grinder
Fabrication
Ground fault circuit interrupter (GFCI)
Horsepower (hp)
Revolutions per minute (rpm)

Instructor's Notes:

____

____

____

3.24 PIPEFITTING ◆ LEVEL ONE

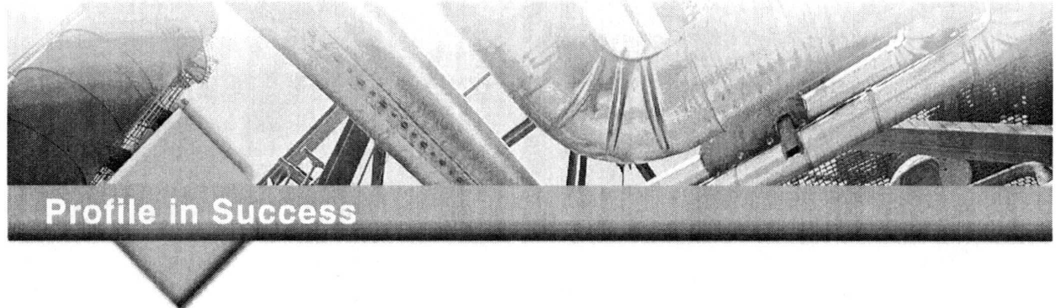

Myron Laurent

Alabama Department of Education
Career Technical Education Specialist
Alabama SkillsUSA, State Director

Myron Laurent became a pipefitter because it was a good paying job in southern Louisiana. While he was working as a pipefitter, several new industry groups required clients to document contract employee craft knowledge and skills. Myron took several classes with the local Associated Builders and Contractors (ABC) chapter and was asked to become an instructor. He discovered that he had a gift for teaching. Teaching gave him the confidence to continue his education. Over the years he earned an associate's, a bachelor's, and a master's degree. Now he promotes technical education throughout Alabama.

How did you choose a career in the pipefitting field?
I grew up near Baton Rouge, Louisiana and the petrochemical industry. A career as a pipefitter/welder offered job security and good money. At that time I thought no further than that.

What types of training have you been through?
The local ABC chapter is called the Pelican Chapter. I attended their classes in order to document my skills to meet new contractual requirements. While I was there, I was recruited to teach Pipefitting, Core, and Core Math to all crafts. I earned Master Trainer status and continued a long relationship with the chapter.

With the opportunity to teach came the realization that I was good at teaching and had much to offer all students. It was then that I decided to attempt college. I enrolled in a two-year Occupational Health and Safety program at Southeastern Louisiana University and continued teaching at ABC at night. With core courses out of the way, I continued my class work and earned a bachelors degree in Industrial Technology. In 2004 I earned a master's degree from LSU in Human Resource Education and Workforce Development.

I credit my returning to school to the experiences I had at the ABC Pelican Chapter training center. There we promoted competence. Competence developed confidence. Confidence enables learners to move forward. That is what education and technical training is all about.

What kinds of work have you done in your career?
In pipefitting I have primarily worked shutdowns, cutting, welding, and fitting on pipestills, cat crackers, and cokers.

Tell us about your present job.
Now I enjoy the opportunity to promote all areas of career/technical education across the state of Alabama. As the Director of Alabama SkillsUSA, I have the pleasure of working with some of the best teachers and nearly 10,000 students. These high school teachers try to give reason to reading and meaning to math by relating it to the trades.

We believe that craftsmen can be scholars and scholars craftsmen. I believe that anything that I can do to promote career/technical education will, in turn, benefit the construction industry that was so good to me.

What factors have contributed most to your success?
Generally, math skills are important to all crafts, especially pipefitting. Attention to detail is also important. Every job is an important job. Finally, Employability Skills, particularly dependability, responsibility, and good communication skills.

What advice would you give to those new to the pipefitting field?
Hone your listening and math skills and become the "go to guy" on your crew. Ask questions, learn from others, pay attention, and work safely. Lay-offs come last for the team player - the employee who gives his supervisor reasons to keep him until the end.

MODULE 08103-06 ◆ PIPEFITTING POWER TOOLS 3.25

Trade Terms Introduced in This Module

Assured equipment grounding conductor program: A detailed plan specifying an employer's required equipment inspections and tests and a schedule for conducting those inspections and tests.

Bevel: An angle cut or ground on the end of a piece of solid material.

Chamfer: An angle cut or ground only on the edge of a piece of material.

Chuck: The part of a machine that holds a piece of work tightly in the machine. A chuck is normally used only when the pipe or cutter will be rotated.

Die grinder: A tool used to make male threads on a pipe or a bolt.

Fabrication: The act of putting together component parts to form an assembly.

Ground fault circuit interrupter (GFCI): A fast-acting circuit breaker that senses small imbalances in the circuit caused by current leakage to ground and, in a fraction of a second, shuts off the electricity.

Horsepower (hp): A unit of power equal to 745.7 watts or 33,000 foot-pounds per minute.

Revolutions per minute (rpm): The number of complete revolutions an object will make in one minute.

Instructor's Notes:

3.26 PIPEFITTING ◆ LEVEL ONE

Resources & Acknowledgments

Additional Resources

This module is intended to be a thorough resource for task training. The following reference work is suggested for further study. This is optional material for continued education rather than for task training.

Tools and Their Uses, Latest Edition. Naval Education and Training Program and Development Center. Washington, DC: US Government Printing Offices.

Figure Credits

Coleman Cable, Inc., 103F01

Tigair®, 103SA01

DeWalt Industrial Tool Company, 103F02

Milwaukee Electric Tool Corporation, 103F03A

Ridge Tool Company (RIDGID®), 103F03B, 103F06, 103F07, 103F09-103F11, 103RQ01

Topaz Publications, Inc., 103F04, 103SA02, 103F14, 103F17

J.R. Yochum, 103F05

Reed Manufacturing Company, 103F12, 103F13

Mathey Dearman, 103F15

Heck Industries, Inc., 103F16

Magnatech Limited Partnership, 103F18, 103RQ02

Zachary Construction Corporation, Title Page

MODULE 08103-06 ◆ PIPEFITTING POWER TOOLS 3.27

MODULE 08103-06 — TEACHING TIPS

The following are suggested activities or instructional methods to help you teach the material in this AIG.

General

When you call on someone to answer a question, the rest of the class relaxes or even tunes out because they expect that the question and answer will take place only between you and the trainee you called on. Instead, use this technique to involve more the trainees in answering questions and to keep them on their toes.

1. Ask the trainees to define a term or explain a concept.
2. After one trainee has answered, ask a trainee seated nearby if the answer is right. Then ask whether a trainee in the back of the room agrees.
3. Ask the trainees to explain why they think an answer is right or wrong.
4. Use the session to clear up incorrect ideas and encourage the trainees to learn from their mistakes.

Section 4.0.0 *Grinders*

Pipefitters have must be able to grinders and accessories for pipe fabrication. Trainees will need appropriate personal protective equipment, pencils, and paper. This exercise familiarizes trainees with various types of grinders and their accessories. Allow 20 to 30 minutes for this exercise.

1. Set up several workstations with various grinders and accessories. Divide the trainees into small groups.
2. Have each group circulate among the tables and identify each tool or accessory. If you wish, also have trainees write one fact about each.
3. At the last rotation, have one person in the group identify the tool or accessory at their station and explain how it is used.
4. Answer any questions they may have.

Section 8.0.0 *Power Bevelers and Thermal Joint Preparation*

Pipefitters must be familiar with many different types of power bevelers. This exercise familiarizes trainees with the various types of power bevelers. Trainees will need pencils and paper. Arrange for a manufacturer's representative to give a presentation to the class on various types of power bevelers. Allow 30 to 60 minutes for this exercise.

1. Before the presentation, explain the different types of power bevelers available and their uses.
2. Ask the presenter to explain the functions of the various devices. Have trainees take notes and write down questions during the presentation.
3. Ask the presenter to answer any questions the trainees may have.

MODULE 08103-06 — ANSWERS TO REVIEW QUESTIONS

Answer	Section
1. c	2.1.0
2. d	2.1.1
3. b	3.1.0; Table 1
4. b	4.1.0
5. d	4.3.0
6. a	5.0.0
7. a	5.2.0
8. c	5.3.0
9. b	5.3.0
10. b	5.4.2
11. c	5.4.3
12. b	5.5.0
13. d	6.1.0
14. b	6.1.0
15. a	6.2.0
16. a	6.2.0
17. d	7.0.0
18. c	8.0.0
19. b	8.3.0
20. d	8.3.0

Answers to Trade Terms Quiz

1. Chamfer
2. Chuck
3. Die grinder
4. Revolutions per minute (rpm)
5. Horsepower (hp)
6. Bevel
7. Assured equipment grounding conductor program
8. Fabrication
9. Ground fault circuit interrupter (GFCI)

CONTREN® LEARNING SERIES — USER UPDATE

NCCER makes every effort to keep these textbooks up-to-date and free of technical errors. We appreciate your help in this process. If you have an idea for improving this textbook, or if you find an error, a typographical mistake, or an inaccuracy in NCCER's Contren® textbooks, please write us, using this form or a photocopy. Be sure to include the exact module number, page number, a detailed description, and the correction, if applicable. Your input will be brought to the attention of the Technical Review Committee. Thank you for your assistance.

Instructors – If you found that additional materials were necessary in order to teach this module effectively, please let us know so that we may include them in the Equipment/Materials list in the Annotated Instructor's Guide.

Write: Product Development and Revision
National Center for Construction Education and Research
P.O. Box 141104, Gainesville, FL 32614-1104

Fax: 352-334-0932

E-mail: curriculum@nccer.org

Craft Module Name

Copyright Date Module Number Page Number(s)

Description

(Optional) Correction

(Optional) Your Name and Address

Module 08104-06

Oxyfuel Cutting

NCCER STANDARDIZED CRAFT TRAINING PROGRAM

The National Center for Construction Education and Research (NCCER) provides a standardized national program of accredited craft training. Key features of the program include instructor certification, competency-based training, and performance testing. The program provides trainees, instructors, and companies with a standard form of recognition through the National Registry. The program is described in full in the *Guidelines for Accreditation*, published by NCCER. For more information on standardized craft training, contact NCCER by writing to P.O. Box 141104, Gainesville, FL 32614-1104; calling 352-334-0911; or e-mailing info@nccer.org. More information is available at www.nccer.org.

HOW TO USE THIS ANNOTATED INSTRUCTOR'S GUIDE

Each page presents two sections of information. The larger section displays each page exactly as it appears in the Trainee Module. The narrow column ties suggested trainee and instructor actions to each page and provides icons (detailed below) to call your attention to material, safety, audiovisual, or testing requirements. The bottom of each page includes space for your notes.

The **Audiovisual** icon indicates an appropriate time to show a transparency or other audiovisual aid.

The **Classroom** icon prompts you to define a term, stress a point, ask trainees to explain a concept, or give examples.

The **Demonstration** icon directs you to show trainees how to perform tasks.

The **Examination** icon tells you to administer the written module examination.

The **Homework** icon is placed where you may wish to assign reading for the next class, assign a project, or advise trainees to prepare for an examination.

The **Laboratory** icon is used when trainees are to practice performing tasks.

The **Materials** icon is a reminder for you to gather materials needed for classes, laboratories, and testing.

The **Performance Testing** icon tells you to administer a performance test or a portion thereof.

The **Safety** icon is used to emphasize safety issues. It is often keyed to *Caution* and *Warning!* statements in the Trainee Module.

The **Teaching Tip** icon indicated additional guidance is available, such as how to conduct an exercise, get the most educational value from a field trip, or encourage class participation. Teaching Tips may expand on a feature (*Think About It, Did You Know?*) or provide *Quick Quizzes* or similar exercises. You will be referred to the Teaching Tips section at the back of the module if there is additional material.

The **Combination** icon indicates that the laboratory listed corresponds with a performance task. If desired, you can note the proficiency of the trainees during the laboratory, and use it to satisfy performance testing requirements.

PREPARATION

Before teaching this module, you should review the Objectives, Performance Tasks, Materials and Equipment List, and Module Outline. Be sure to allow ample time to prepare your own training or lesson plan and gather all required materials and equipment.

MODULE OVERVIEW

This module explains the safety requirements for oxyfuel cutting. It identifies oxyfuel cutting equipment and setup requirements. It explains how to light, adjust, and shut down oxyfuel equipment. Trainees will perform cutting techniques that include straight line, piercing, bevels, and washing.

PREREQUISITES

Prior to training with this module, it is recommended that the trainee shall have successfully completed *Core Curriculum;* and *Pipefitting Level One*, Modules 08101-06 through 08103-06.

OBJECTIVES

Upon completion of this module, the trainee will be able to do the following:

1. Identify and explain the use of oxyfuel cutting equipment.
2. Set up oxyfuel equipment.
3. Light and adjust an oxyfuel torch.
4. Shut down oxyfuel cutting equipment.
5. Disassemble oxyfuel equipment.
6. Change empty cylinders.
7. Perform oxyfuel cutting:
    - Straight line and square shapes
    - Piercing and slot cutting
    - Bevels
    - Washing
8. Operate a motorized, portable oxyfuel gas cutting machine.

PERFORMANCE TASKS

Under the supervision of the instructor, the trainee should be able to do the following:

1. Set up oxyfuel equipment.
2. Light and adjust an oxyfuel cutting torch.
3. Shut down oxyfuel cutting equipment.
4. Disassemble oxyfuel equipment.
5. Change empty cylinders.
6. Perform straight line and square shape cutting.
7. Perform piercing and slot cutting.
8. Perform bevel cutting.
9. Perform washing.

MATERIALS AND EQUIPMENT LIST

Overhead projector and screen

Transparencies

Blank acetate sheets

Transparency pens

Whiteboard/chalkboard

Markers/chalk

Pencils and scratch paper

Appropriate personal protective equipment:
- Safety goggles
- Face shields
- Welding helmets
- Ear protection
- Welding cap
- Leather jacket
- Leather pants or chaps
- Gauntlet-type welding gloves
- Respirators

ANSI Z49.1-1999

OSHA 29 CFR 1910.146

MSDS for cutting products

Oxygen cylinder with cap

Fuel gas cylinder with cap

Regulators (oxygen and fuel gas)

Hose set

One-piece cutting torch

Combination cutting torch and torch tips

Assorted acetylene, liquefied fuel gas, and special-purpose cutting torch tips

Tip cleaners

Tip drills

Mechanical guide

Cylinder cart

Motorized oxyfuel track cutter

Framing squares

Combination squares with protractor head

Tape measure

Soapstone

Penknife

Pliers

Chipping hammer

Friction lighter

Vendor cutting tip chart

Wrenches (torch, hose, and regulator)

Steel plate
- Thin (16 to 10 gauge)
- Thick (¼ inch to 1 inch)

Television with VCR or DVD (optional)

Welding safety video (optional)

Module Examinations*

Performance Profile Sheets*

* Located in the Test Booklet.

SAFETY CONSIDERATIONS

Ensure that the trainees are equipped with appropriate personal protective equipment and know how to use it properly. This module requires that the trainees operate oxyfuel cutting equipment. Ensure that trainees are briefed on fire and shop safety policies prior to performing any work. Emphasize the special safety precautions associated with the use of cylinders and oxyfuel cutting equipment.

ADDITIONAL RESOURCES

This module is intended to present thorough resources for task training. The following reference works are suggested for both instructors and motivated trainees interested in further study. These are optional materials for continued education rather than for task training.

Safety in Welding, Cutting, and Allied Processes, ANSI Z49.1-99, 1999. Miami, FL: American Welding Society.

Welder's Handbook, Richard Finch, 1997. New York, NY: The Berkley Publishing Group, Inc.

TEACHING TIME FOR THIS MODULE

An outline for use in developing your lesson plan is presented below. Note that each Roman numeral in the outline equates to one session of instruction. Each session has a suggested time period of 2½ hours. This includes 10 minutes at the beginning of each session for administrative tasks and one 10-minute break during the session. Approximately 17½ hours are suggested to cover *Oxyfuel Cutting*. You will need to adjust the time required for hands-on activity and testing based on your class size and resources. Because laboratories often correspond to Performance Tasks, the proficiency of the trainees may be noted during these exercises for Performance Testing purposes.

Topic	Planned Time
Session I. Introduction, Safety, and Oxyfuel Cutting Equipment	
A. Introduction	________
B. Oxyfuel Cutting Safety	________
C. Oxyfuel Cutting Equipment	________
1. Cylinders, Regulators, and Hoses	________
2. Cutting Torch, Tips, and Tip Equipment	________
3. Friction Lighters	________
4. Cylinder Cart	________
5. Soapstone Markers	________
6. Specialized Equipment	________
Session II. Setting Up Oxyfuel Equipment	
A. Setting Up Oxyfuel Equipment	________
1. Cylinders	________
2. Hoses and Regulators	________
3. Torches and Tips	________
4. Purging and Testing	________
B. Laboratory – Trainees practice setting up oxyfuel equipment. This laboratory corresponds to Performance Task 1.	________
Session III. Torch Operations	
A. Controlling the Oxyfuel Torch Flame	________
B. Laboratory – Trainees practice lighting and adjusting an oxyfuel cutting torch. This laboratory corresponds to Performance Task 2.	________
C. Shutting Down Oxyfuel Equipment	________
D. Laboratory – Trainees practice shutting down an oxyfuel cutting outfit. This laboratory corresponds to Performance Task 3.	________
E. Disassembling Oxyfuel Equipment	________
F. Laboratory – Trainees practice disassembling an oxyfuel cutting outfit. This laboratory corresponds to Performance Task 4.	________
G. Changing Empty Cylinders	________
H. Laboratory – Trainees practice changing empty cylinders on an oxyfuel cutting outfit. This laboratory corresponds to Performance Task 5.	________

Sessions IV through VI. Performing Cutting Operations

A. Performing Cutting Procedures ____________

B. Portable Oxyfuel Cutting Machine Operation ____________

C. Laboratory – Trainees practice straight line and square shape cutting with an oxyfuel cutting torch. This laboratory corresponds to Performance Task 6. ____________

D. Laboratory – Trainees practice piercing and slot cutting with an oxyfuel cutting torch. This laboratory corresponds to Performance Task 7. ____________

E. Laboratory – Trainees practice bevel cutting with an oxyfuel cutting torch. This laboratory corresponds to Performance Task 8. ____________

F. Laboratory – Trainees practice washing with an oxyfuel cutting torch. This laboratory corresponds to Performance Task 9. ____________

Session VII. Review, Module Examination, and Performance Testing

A. Review ____________

B. Module Examination ____________

1. Trainees must score 70 percent or higher to receive recognition from NCCER.
2. Record the testing results on Craft Training Report Form 200, and submit the results to the Training Program Sponsor.

C. Performance Testing ____________

1. Trainees must perform each task to the satisfaction of the instructor to receive recognition from NCCER. If applicable, proficiency noted during laboratory exercises can be used to satisfy the Performance Testing requirements.
2. Record the testing results on Craft Training Report Form 200, and submit the results to the Training Program Sponsor.

PIPEFITTING

Pipefitters work on large, complex projects that help maintain a continuous supply of energy to keep our country moving forward.

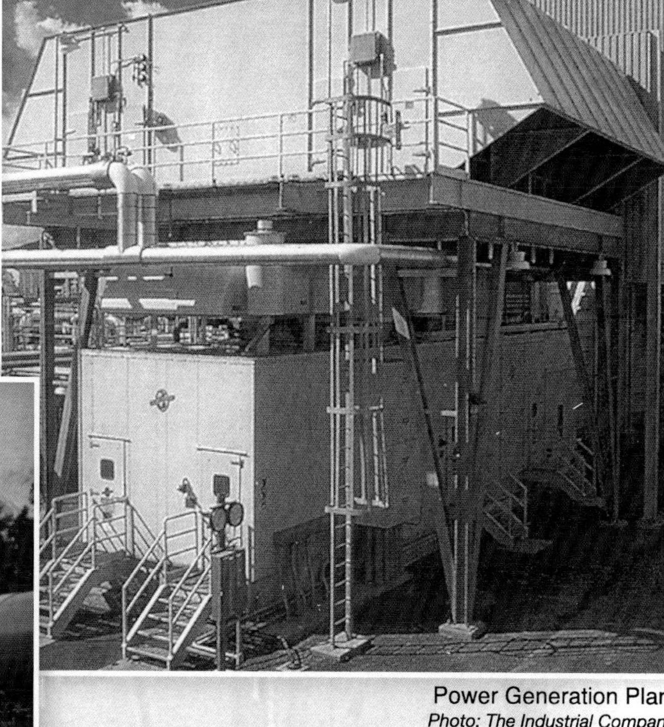

Power Generation Plant
Photo: The Industrial Company

The Alaska Pipeline
Photo: Courtesy of the Alyeska Pipeline Service Company

Oil Refinery at Night. *Photo: Holloway Houston Inc.*

Pipefitters learn to use a variety of specialized tools.

Power Pipe Cutter
Ridge Tool Co. (Ridgid®)

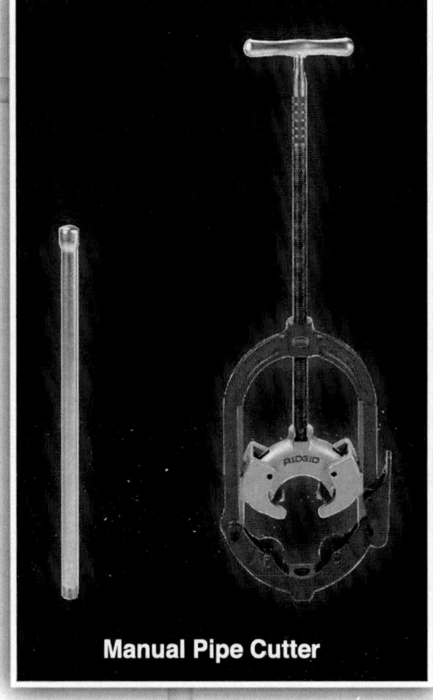

Manual Pipe Cutter

Ridge Tool Co. (Ridgid®)

Power Pipe Threader
Ridge Tool Co. (Ridgid®)

Pipe Groover
Ridge Tool Co. (Ridgid®)

Video Pipe Inspector
Ridge Tool Co. (Ridgid®)

Pipefitters work all over the world, under roof, under sky, and under sea.

Bolting on a valve. *Photo: Fluor Fernald*

Pipefitters laying concrete pipe. *Photo: Caterpillar, Inc.*

Welding a pipe. *Photo: Topaz Publications, Inc.*

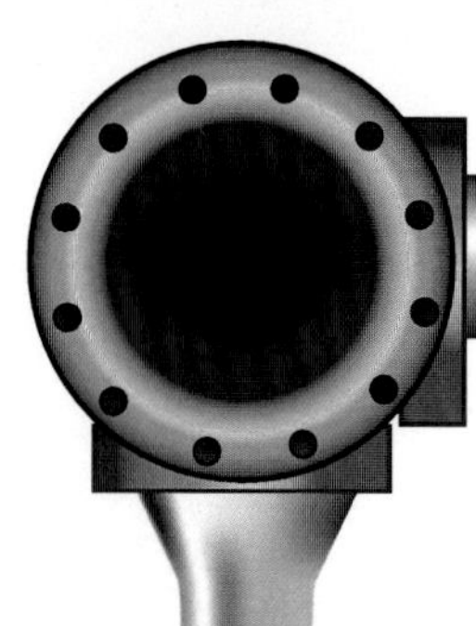

Some industrial processes require huge pieces of equipment.

A turbine-driven pump assembly. *Photo: Dresser-Rand Company*

Protective equipment allows pipefitters to work in potentially toxic environments. *Photo: Fluor Fernald*

Gate valves come in many sizes. *Photo: American Cast Iron Pipe Company*

Assign reading of Module 08104-06.

08104-06

Oxyfuel Cutting

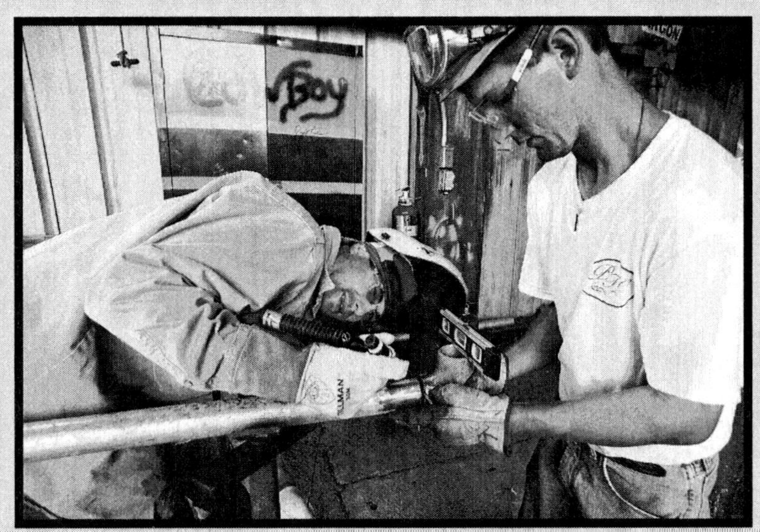

08104-06
Oxyfuel Cutting

Topics to be presented in this module include:

Overview

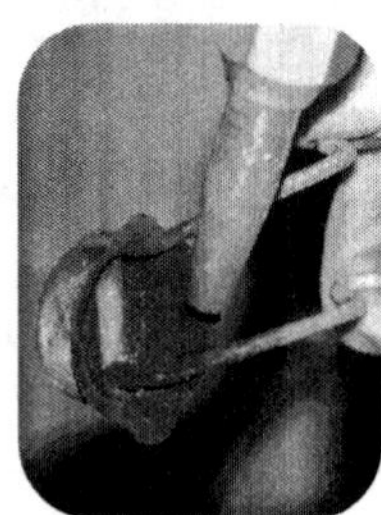

Oxyfuel cutting uses the flame and oxygen from a cutting torch to cut ferrous metals such as carbon steels. This process can be used to quickly cut, trim, and shape the hardest steel. It can be used to cut pipe, prepare joints for welding, clean metals, and disassemble structures.

The oxyfuel cutting process uses flammable gases under high pressure. Precautions must be taken to prevent fire and explosions. Wear leather protective clothing to protect your skin. Your eyes and ears must also be protected.

Specific procedures must be followed to set up, operate, and shut down oxyfuel equipment. Each piece of the equipment must be selected correctly, thoroughly inspected, and properly attached. With practice, oxyfuel cutting equipment will become a useful tool for the pipefitter.

Instructor's Notes:

Objectives

When you have completed this module, you will be able to do the following:

1. Identify and explain the use of oxyfuel cutting equipment.
2. Set up oxyfuel equipment.
3. Light and adjust an oxyfuel torch.
4. Shut down oxyfuel cutting equipment.
5. Disassemble oxyfuel equipment.
6. Change empty cylinders.
7. Perform oxyfuel cutting:
   - Straight line and square shapes
   - Piercing and slot cutting
   - Bevels
   - Washing
8. Operate a motorized, portable oxyfuel gas cutting machine.

Trade Terms

Backfire
Carburizing flame
Drag lines
Dross
Ferrous metals
Flashback
Kerf
Neutral flame
Oxidizing flame
Pierce
Slag
Soapstone

Required Trainee Materials

1. Pencil and paper
2. Appropriate personal protective equipment

Prerequisites

This course map shows all of the modules in the first level of the *Pipefitting* curriculum. The suggested training order begins at the bottom and proceeds up. Skill levels increase as you advance on the course map. The local Training Program Sponsor may adjust the training order.

Before you begin this module, it is recommended that you successfully complete *Core Curriculum; Pipefitting Level One*, Modules 08101-06 through 08103-06.

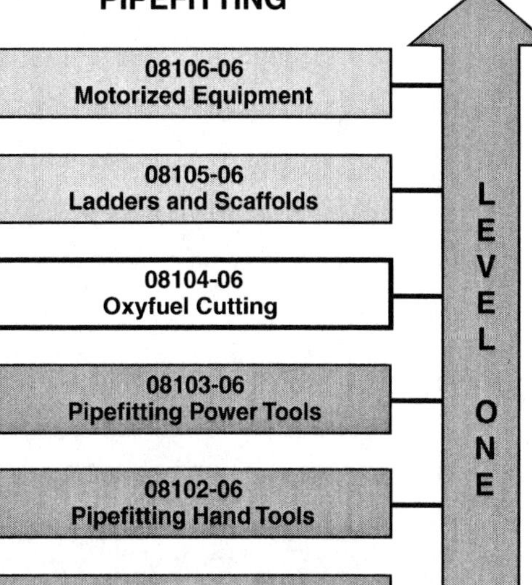

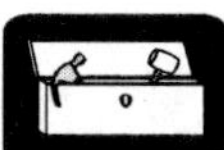

Ensure that you have everything required to teach the course. Check the Materials and Equipment list at the front of this module.

See the general Teaching Tip at the end of this module.

Explain that terms shown in bold are defined in the Glossary at the back of this module.

Show Transparency 1, Objectives, and Transparency 2, Performance Tasks. Review the goals of the module, and explain what will be expected of the trainee.

Explain that oxyfuel cutting (OFC) is a process that uses flame and oxygen to cut ferrous metals. Briefly describe the OFC process.

Describe the types of metals OFC is used on.

Explain that proper safety equipment and precautions must be taken when using OFC equipment.

Describe the protective clothing required for welding operations that produce sparks or spatter.

See the Teaching Tip for Section 2.0.0 at the end of this module.

1.0.0 ◆ INTRODUCTION

Oxyfuel cutting (OFC), also called flame cutting or burning, is a process that uses the flame and oxygen from a cutting torch to cut **ferrous metals**. The flame is produced by burning a fuel gas mixed with pure oxygen. The flame heats the metal to be cut to the kindling temperature (a cherry-red color); then a stream of high-pressure pure oxygen is directed from the torch at the metal's surface. This causes the metal to instantaneously oxidize or burn. The cutting process results in oxides that mix with molten iron and produce **dross**, which is blown from the cut by the jet of cutting oxygen. This oxidation process, which takes place during the cutting operation, is similar to a greatly sped-up rusting process. *Figure 1* shows oxyfuel cutting.

The oxyfuel cutting process is usually used only on ferrous metals such as straight carbon steels, which oxidize rapidly. This process can be used to quickly cut, trim, and shape ferrous metals, including the hardest steel.

Oxyfuel cutting can be used for certain metal alloys, such as stainless steel; however, the process requires higher preheat temperatures (white heat) and about 20 percent more oxygen for cutting. In addition, sacrificial steel plate or rod may have to be placed on top of the cut to help maintain the burning process. Other methods, such as carbon arc cutting, powder cutting, inert gas cutting, and plasma arc cutting, are much more practical for cutting steel alloys and nonferrous metals.

2.0.0 ◆ OXYFUEL CUTTING SAFETY

The proper safety equipment and precautions must be used when working with oxyfuel equipment because of the potential danger from the high-pressure flammable gases and high temperatures used. The following is a summary of safety procedures and practices that must be observed while cutting or welding. Keep in mind that this is just a summary. Above all, be sure to wear appropriate protective clothing and equipment when welding or cutting.

104F01.EPS

Figure 1 ◆ Oxyfuel cutting.

2.1.0 Protective Clothing and Equipment

- Always use safety goggles with a full face shield or a helmet. The goggles, face shield, or helmet lens must have the proper light-reducing tint for the type of welding or cutting to be performed. Never directly or indirectly view an electric arc without using a properly tinted lens (*Figure 2*).
- Wear proper protective leather and/or flame retardant clothing along with welding gloves that will protect you from flying sparks and molten metal, as well as heat.
- Wear 8-inch or taller high-top safety shoes or boots. Make sure that the tongue and lace area of the footwear will be covered by a pant leg. If the tongue and lace area is exposed or the footwear must be protected from burn marks, wear leather spats under the pants or chaps and over the top of the footwear.
- Wear a solid material (nonmesh) hat with a bill pointing to the rear or, if much overhead cutting or welding is required, a full leather hood with a welding face plate and the correctly tinted lens. If a hard hat is required, use a hard hat that allows the attachment of rear deflector material and a face shield.
- If a full leather hood is not worn, wear a face shield and snugly fitting welding goggles over safety glasses for gas welding or cutting. Either the face shield or the lenses of the welding goggles must be an approved shade 5 or 6 filter. Depending on the method used for electric arc cutting, wear safety goggles and a welding hood with the correctly tinted lens (shade 5 to 14).
- If a full leather hood is not worn, wear earmuffs, or at least earplugs, to protect your ear canals from sparks.

Instructor's Notes:

2.1.1 Ear Protection

Welding areas can be very noisy. In addition, if overhead work is being performed, hot sparks can cause burns to the ears and ear canals unless a leather hood is used. For maximum protection, earmuff-type hearing protectors (*Figure 3*) should also be used. They are available in varying degrees of protection from all noise, including low frequencies. Most earmuffs have adjustable headbands that can be worn over the head, behind the neck, or under the chin. To use earmuffs, adjust the tension on the headband and ear cushion pads to obtain the best possible seal. Check the earmuff shell for cracks and the ear cushion pads for tears before each use. Any damaged, cracked, or torn part must be repaired or replaced. As minimal protection, earplugs (*Figure 3*) can be used. Disposable earplugs are the most common form of hearing protection used in the industry. These devices usually have an outer layer of pliable foam and a core layer of acoustical fiber that filters out harmful noise yet allows you to hear normal conversation. To use disposable earplugs, simply roll each plug into a cone and insert the tapered end into the ear canal while pulling up on the upper portion of your ear. The earplugs will expand, filling the ear canal and creating a proper fit. Reusable earplugs that can be cleaned and worn repeatedly are also commonly used. These are typically cleaned with boiling water or alcohol. Plain cotton placed in the ear is not an acceptable protective device.

2.1.2 Eye, Face, and Head Protection

The heat and light produced by cutting or welding operations can damage the skin and eyes. Injury to the eyes may result in permanent loss of vision. Oxyfuel cutting and welding can cause eye fatigue and mild burns to the skin because of the infrared heat radiated by the process. Welding or cutting operations involving an electric arc of any kind produce UV radiation, which can cause severe burns to the eyes and exposed skin and permanent damage to the retina. A flash burn can harm unprotected eyes in just seconds. Welders should never view an electric arc directly or indirectly without wearing a properly tinted lens designed for electric arc use. If electric arc operations are occurring in the vicinity, safety goggles with a tinted lens (shades 3 to 5) and tinted side shields must be worn at all times.

Discuss the noise and spark hazards posed by welding. Describe ear muffs and ear plugs. Explain how they should be worn.

Discuss the heat and light hazards posed by OFC equipment. Describe different types of goggles and face shields used for OFC.

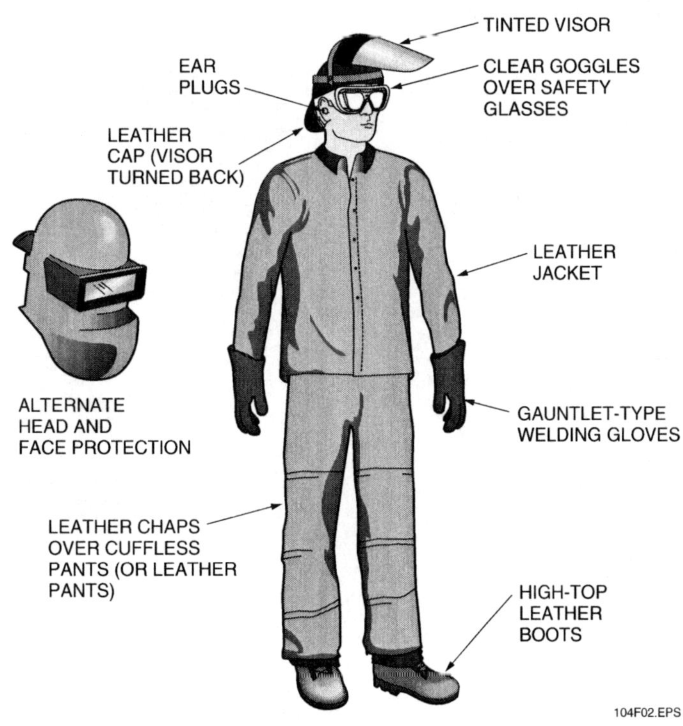

Figure 2 ◆ Typical personal protective equipment.

MODULE 08104-06 ◆ OXYFUEL CUTTING 4.3

Describe the welding helmets used for electric arc welding.

Provide different types of ear protection and eye, face, and head protection for the trainees to examine.

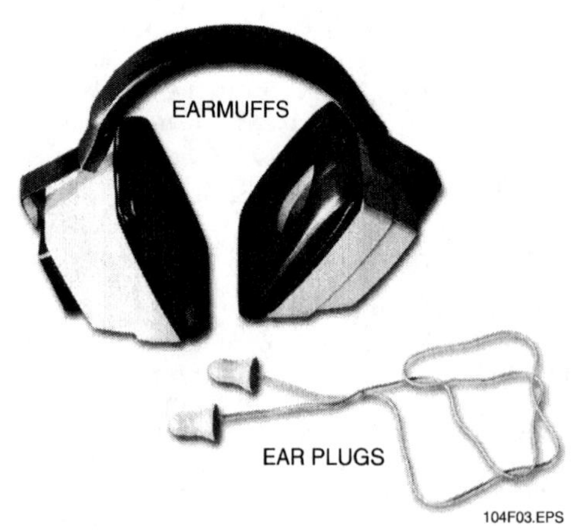

Figure 3 ◆ Typical ear protection.

Figure 4 ◆ Oxyfuel welding/cutting goggles and face shield combinations.

For oxyfuel welding and cutting, wear tinted welding goggles (shades 4 to 6) over safety glasses, and wear a clear face shield. Clear safety glasses and goggles with a tinted face shield can also be used. Most oxyfuel welders prefer the latter combination because, for clear vision, only the face shield has to be flipped up. For overhead oxyfuel operations, a leather hood may be used in place of the face shield to obtain protection from sparks and molten metal (*Figure 4*).

For electric arc operations, a leather hood or welding helmet with a properly tinted lens (shades 9 to 14) must be worn over safety goggles to provide proper protection. Many varieties of helmets are available; typical styles are shown in *Figure 5*. Some of the helmets are available with additional side-view lenses in a lighter tint so that welders can sense, by peripheral vision, any activities occurring beside them.

Most welding and cutting tasks require the use of safety goggles, chemical-resistant goggles, dust goggles, or face shields. Always check the material safety data sheet (MSDS) for the welding or cutting product being used to find out what type of eye protection is needed.

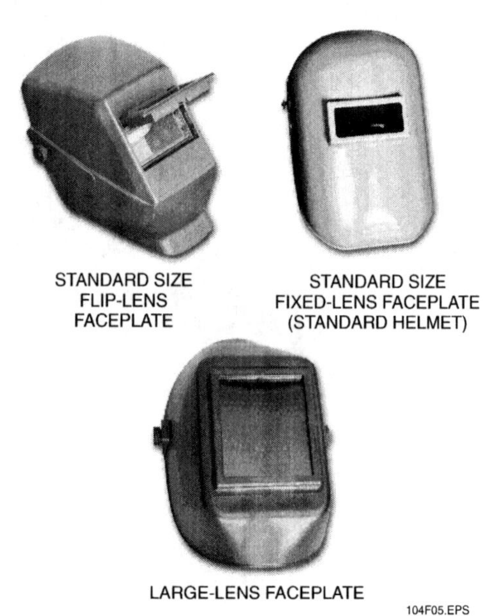

Figure 5 ◆ Typical electric arc welding helmets.

Instructor's Notes:

4.4 PIPEFITTING ◆ LEVEL ONE

2.2.0 Ventilation

Adequate mechanical ventilation must be provided to remove fumes that are produced by welding or cutting processes. *ANSI Z49.1-1999* on welding safety covers such ventilation procedures. The gases, dust, and fumes caused by welding or cutting operations can be hazardous if the appropriate safety precautions are not observed. The following general rules can be used to determine if there is adequate ventilation:

- The welding area must contain at least 10,000 cubic feet of air for each welder.
- There must be air circulation.
- Partitions, structural barriers, or equipment must not block air circulation.

Even when there is adequate ventilation, avoid inhaling welding or cutting fumes and smoke. The heated fumes and smoke generally rise straight up. Observe the column of smoke and position yourself to avoid it. A small fan may also be used to divert the smoke, but take care to keep the fan from blowing directly on the work area; the fumes and gases must be present at an electric arc in order to protect the molten metal from the air.

2.2.1 Fume Hazards

Welding or cutting processes create fumes. Fumes are solid particles consisting of the base metal, electrodes or welding wire, and any coatings applied to them. Most fumes are not considered dangerous as long as there is adequate ventilation. If ventilation is questionable, use air sampling to determine the need for corrective measures. Adequate ventilation can be a problem in tight or cramped working quarters. To ensure adequate room ventilation, local exhaust ventilation should be used to capture fumes *(Figure 6)*.

The exhaust hood should be kept four to six inches away from the source of the fumes. Welders should recognize that fumes of any type, regardless of their source, should not be inhaled. The best way to avoid problems is to provide adequate ventilation. If this is not possible, breathing protection must be used. Protective devices for use in poorly ventilated or confined spaces are shown in *Figures 7* and *8*.

If respirators are used, your employer should offer worker training on respirator fitting and usage. Medical screenings should also be given.

WARNING!

Fumes and gases can be dangerous. Overexposure can cause nausea, headaches, dizziness, metal fume fever, and severe toxic effects that can be fatal. Studies have demonstrated irritation to eyes, skin, and the respiratory system and even more severe complications. In confined spaces, fumes and gases may cause asphyxiation.

2.3.0 Respirators

Special metals require the use of respirators to provide protection from harmful fumes. Respirators are grouped into three main types based on how they work to protect the wearer from contaminants. The types are:

- Air-purifying respirators
- Supplied-air respirators (SARs)
- Self-contained breathing apparatus (SCBA)

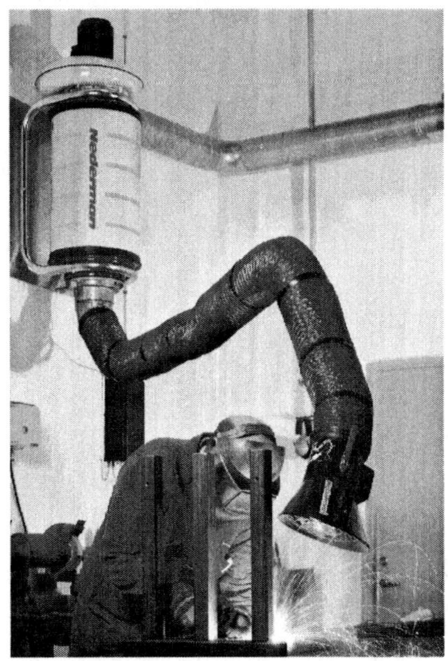

104F06.EPS

Figure 6 ◆ A flexible exhaust pickup.

Review the contents of *ANSI Z49.1-1999* regarding ventilation. Emphasize the harmful effects of exposure to fumes and gases.

Provide a copy of *ANSI Z49.1-1999* for the trainees to examine.

Explain that welding and cutting processes produce fumes. Discuss methods of controlling fume hazards, including ventilation, exhaust hoods, and respirators.

Discuss the symptoms of overexposure to fumes and gases.

Discuss the respiratory hazards of working with special metals. Describe different types of respirators.

MODULE 08104-06 ◆ OXYFUEL CUTTING 4.5

Describe an air-purifying respirator. Point out that air-purifying respirators provide the lowest level of protection. Identify and discuss the four classifications of these respirators.

Describe powered air-purifying respirators.

Provide an air-purifying respirator for the trainees to examine.

104F07.EPS

Figure 7 ◆ Typical respirator.

104F08.EPS

Figure 8 ◆ A belt-mounted respirator.

2.3.1 Air-Purifying Respirators

Air-purifying respirators provide the lowest level of protection. They are made for use only in atmospheres that have enough oxygen to sustain life (at least 19.5 percent). Air-purifying respirators use special filters and cartridges to remove specific gases, vapors, and particles from the air. The respirator cartridges contain charcoal, which absorbs certain toxic vapors and gases. When the wearer detects any taste or smell, the charcoal's absorption capacity has been reached and the cartridge can no longer remove the contaminant. The respirator filters remove particles such as dust, mists, and metal fumes by trapping them within the filter material. Filters should be changed when it becomes difficult to breathe. Depending on the contaminants, cartridges can be used alone or in combination with a filter/pre-filter and filter cover. Air-purifying respirators should be used for protection only against the types of contaminants listed on the filters and cartridges and on the National Institute for Occupational Safety and Health (NIOSH) approval label affixed to each respirator carton and replacement filter/cartridge carton. Respirator manufacturers typically classify air-purifying respirators into four groups:

- No maintenance
- Low maintenance
- Reusable
- Powered air-purifying respirators (PAPRs)

No-maintenance and low-maintenance respirators are typically used for residential or light commercial work that does not call for constant and heavy respirator use. No-maintenance respirators are typically half-mask respirators with permanently attached cartridges or filters. The entire respirator is discarded when the cartridges or filters are spent. Low-maintenance respirators generally are also half-mask respirators that use replaceable cartridges and filters. However, they are not designed for constant use.

Reusable respirators (*Figure 9*) are made in half-mask and full facepiece styles. These respirators require the replacement of cartridges, filters, and respirator parts. Their use also requires a complete respirator maintenance program. Air respirator maintenance is discussed later.

Powered air-purifying respirators (PAPRs) are made in half-mask, full facepiece, and hood styles. They use battery-operated blowers to pull outside air through the cartridges and filters attached to the respirator. The blower motors can be either mask- or belt-mounted. Depending on the cartridges used, they can filter particulates, dusts, fumes, and mists along with certain gases and vapors. PAPRs like the one shown in *Figure 10* have a belt-mounted, powered air-purifier unit connected to the mask by a breathing tube. Many models also have an audible and visual alarm that is activated when airflow falls below the required minimum level. This feature gives an immediate indication of a loaded filter or low battery charge condition. Units with the blower mounted in the mask do not use a belt-mounted powered air purifier connected to a breathing tube.

Instructor's Notes:

Describe a supplied-air respirator (SAR).

Know What You're Cutting or Welding

Cutting or welding operations involving materials, coatings, or electrodes containing cadmium, mercury, lead, zinc, chromium, and beryllium result in toxic fumes. For cutting or welding such materials, always use proper area ventilation and wear an approved full-face, supplied-air respirator (SAR) that uses breathing air from outside the work area. For occasional, very short-term exposure to fumes from zinc- or copper-coated materials, a high-efficiency particulate arresting (HEPA)-rated or metal-fume filter may be used on a standard respirator.

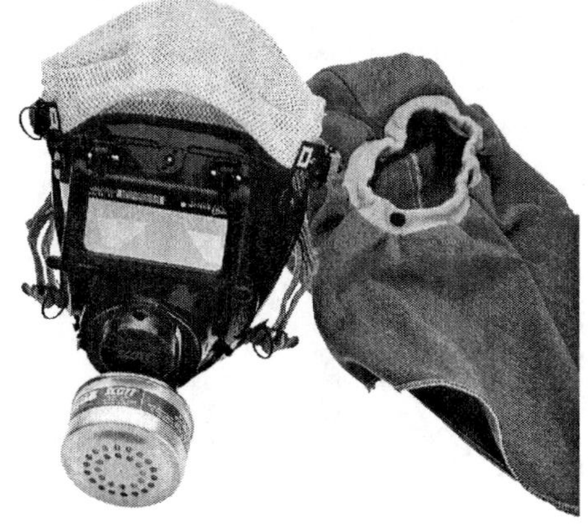

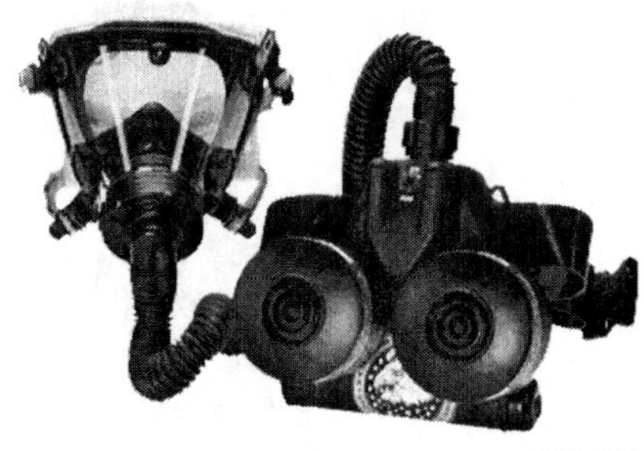

104F10.EPS

Figure 10 ◆ Typical powered air-purifying respirator (PAPR).

104F09.EPS

Figure 9 ◆ Reusable half-mask air-purifying respirator.

2.3.2 Supplied-Air Respirators

Supplied-air respirators (*Figure 11*) provide a supply of air for extended periods of time via a high-pressure hose that is connected to an external source of air, such as a compressor, compressed-air cylinder, or pump. They provide a higher level of protection in atmospheres where air-purifying respirators are not adequate. Supplied-air respirators are typically used in toxic atmospheres.

104F11.EPS

Figure 11 ◆ Supplied-air respirator.

MODULE 08104-06 ◆ OXYFUEL CUTTING 4.7

Discuss the conditions in which a SAR must be used.

Provide a full-face supplied air respirator for the trainees to examine.

Explain that a continuous-flow supplied-air respirator provides a constant stream of air to the user. Point out that the self-contained breathing apparatus (SCBA) provides the highest level of protection.

Respirator Inspection, Care, and Maintenance

A respirator must be clean and in good condition, and all of its parts must be in place for it to give you proper protection. Respirators must be cleaned every day. Failure to do so will limit their effectiveness and offer little or no protection. For example, suppose you wore the respirator yesterday and did not clean it. The bacteria from breathing into the respirator, plus the airborne contaminants that managed to enter the facepiece, will have made the inside of your respirator very unsanitary. Continued use may cause you more harm than good. Remember, only a clean and complete respirator will provide you with the necessary protection. Follow these guidelines:

- Inspect the condition of your respirator before and after each use.
- Do not wear a respirator if the facepiece is distorted or if it is worn and cracked. You will not be able to get a proper face seal.
- Do not wear a respirator if any part of it is missing. Replace worn straps or missing parts before using.
- Do not expose respirators to excessive heat or cold, chemicals, or sunlight.
- Clean and wash your respirator after every time you use it. Remove the cartridge and filter, hand wash the respirator using mild soap and a soft brush, and let it air dry overnight.
- Sanitize your respirator each week. Remove the cartridge and filter, then soak the respirator in a sanitizing solution for at least two minutes. Thoroughly rinse with warm water and let it air dry overnight.
- Store the clean and sanitized respirator in its resealable plastic bag. Do not store the respirator face down. This will cause distortion of the facepiece.

Some can be used in atmospheres that are immediately dangerous to life and health (IDLH) as long as they are equipped with an air cylinder for emergency escape. An atmosphere is considered IDLH if it poses an immediate hazard to life or produces immediate, irreversible, and debilitating effects on health. There are two types of supplied-air respirators: continuous-flow and pressure-demand.

The continuous-flow supplied-air respirator provides air to the user in a constant stream. One or two hoses are used to deliver the air from the air source to the facepiece. Unless the compressor or pump is especially designed to filter the air or a portable air-filtering system is used, the unit must be located where there is breathable air (grade D or better as described in *Compressed Gas Association [CGA] Commodity Specification G-7.1)*. Continuous-flow respirators are made with tight-fitting half-masks or full facepieces. They are also made with hoods. The flow of air to the user may be adjusted either at the air source (fixed flow) or on the unit's regulator (adjustable flow). Pressure-demand supplied-air respirators are similar to the continuous-flow type except that they supply air to the user's facepiece via a pressure-demand valve as the user inhales and fresh air is required. They typically have a two-position exhalation valve that allows the worker to switch between pressure-demand and negative-pressure modes to facilitate entry into, movement within, and exit from a work area.

2.3.3 Self-Contained Breathing Apparatus (SCBA)

SCBAs (*Figure 12*) provide the highest level of respiratory protection. They can be used in oxygen-deficient atmospheres (below 19.5 percent oxygen), in poorly ventilated or confined spaces, and in IDLH atmospheres. These respirators provide a supply of air for 30 to 60 minutes from a compressed-air cylinder worn on the user's back.

Instructor's Notes:

4.8 PIPEFITTING ◆ LEVEL ONE

104F12.EPS

Figure 12 ◆ Self-contained breathing apparatus.

Note that the emergency escape breathing apparatus (EEBA) is a smaller version of an SCBA cylinder. EEBA units are used for escape from hazardous environments and generally provide a five- to ten-minute supply of air.

2.3.4 Respiratory Program

Local and OSHA procedures must be followed when selecting the proper type of respirator for a particular job (*Figure 13*). A respirator must be properly selected (based on the contaminant present and its concentration level), properly fitted, and used in accordance with the manufacturer's instructions. It must be worn during all times of exposure. Regardless of the kind of respirator needed, OSHA regulations require employers to have a respirator protection program consisting of:

- Standard operating procedures for selection and use
- Employee training
- Regular cleaning and disinfecting
- Sanitary storage
- Regular inspection
- Annual fit testing
- Pulmonary function testing

As an employee, you are responsible for wearing respiratory protection when needed. When it comes to vapors or fumes, both can be eliminated in certain concentrations by the use of air purify ing devices as long as oxygen levels are acceptable. Examples of fumes are smoke billowing from a fire or the fumes generated when welding.

BEFORE USING A RESPIRATOR YOU MUST DETERMINE THE FOLLOWING:
1. THE TYPE OF CONTAMINANT(S) FOR WHICH THE RESPIRATOR IS BEING SELECTED
2. THE CONCENTRATION LEVEL OF THAT CONTAMINANT
3. WHETHER THE RESPIRATOR CAN BE PROPERLY FITTED ON THE WEARER'S FACE
ALL RESPIRATOR INSTRUCTIONS, WARNINGS, AND USE LIMITATIONS CONTAINED ON EACH PACKAGE MUST ALSO BE READ AND UNDERSTOOD BY THE WEARER BEFORE USE.

104F13.EPS

Figure 13 ◆ Use the right respirator for the job.

Always check the cartridge on your respirator to make sure it is the correct type to use for the air conditions and contaminants found on the job site.

When selecting a respirator to wear while working with specific materials, you must first determine the hazardous ingredients contained in the material and their exposure levels, then choose the proper respirator to protect yourself at these levels. Always read the product's MSDS. It identifies the hazardous ingredients and should list the type of respirator and cartridge recommended for use with the product.

Limitations that apply to all half-mask (air-purifying) respirators are as follows:

- These respirators do not completely eliminate exposure to contaminants, but they will reduce the level of exposure to below hazardous levels.
- These respirators do not supply oxygen and must not be used in areas where the oxygen level is below 19.5 percent.
- These respirators must not be used in areas where chemicals have poor warning signs, such as no taste or odor.

If your breathing becomes difficult, if you become dizzy or nauseated, if you smell or taste the chemical, or if you have other noticeable effects, leave the area immediately, return to a fresh air area, and seek any necessary assistance.

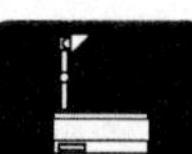

Show Transparency 3 (Figure 13). Discuss procedures that must be followed when selecting a respirator.

Discuss the OSHA requirements for respirator protection programs.

Explain how to select the correct respirator for working with specific materials.

Discuss limitations for half-mask (air-purifying) respirators.

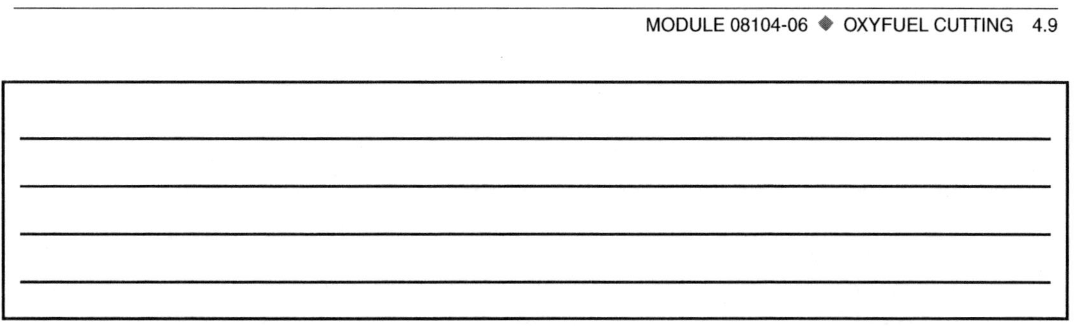

MODULE 08104-06 ◆ OXYFUEL CUTTING 4.9

Discuss the importance of proper fit for respirators. Review the procedures for a negative and positive fit test.

Explain the dangers of distorting the facepiece of a respirator.

Define confined spaces and review the provisions of *OSHA 29 CFR 1910.146.*

Show Transparency 4 (Table 1). Discuss the effects of an increase or decrease in oxygen levels.

Provide respirators to trainees and have them perform the positive and negative fit test.

Provide copies of *OSHA 29 CFR 1910.146* for the trainees to examine.

2.3.5 Positive and Negative Fit Checks

All respirators are useless unless properly fit-tested to each individual. To obtain the best protection from your respirator, you must perform positive and negative fit checks each time you wear it. These fit checks must be done until you have obtained a good face seal. To perform the positive fit check, do the following:

Step 1 Adjust the facepiece for the best fit, then adjust the head and neck straps to ensure good fit and comfort.

> **WARNING!**
> Do not overtighten the head and neck straps. Tighten them only enough to stop leakage. Overtightening can cause facepiece distortion and dangerous leaks.

Step 2 Block the exhalation valve with your hand or other material.

Step 3 Breathe out into the mask.

Step 4 Check for air leakage around the edges of the facepiece.

Step 5 If the facepiece puffs out slightly for a few seconds, a good face seal has been obtained.

To perform a negative fit check, do the following:

Step 1 Block the inhalation valve with your hand or other material.

Step 2 Attempt to inhale.

Step 3 Check for air leakage around the edges of the facepiece.

Step 4 If the facepiece caves in slightly for a few seconds, a good face seal has been obtained.

2.4.0 Confined Space Permits

A confined space refers to a relatively small or restricted space, such as a storage tank, boiler, or pressure vessel or small compartments, such as underground utility vaults, small rooms, or the unventilated corners of a room.

OSHA 29 CFR 1910.146 defines a confined space (*Figure 14*) as a space that:

- Is large enough and so configured that an employee can bodily enter and perform assigned work

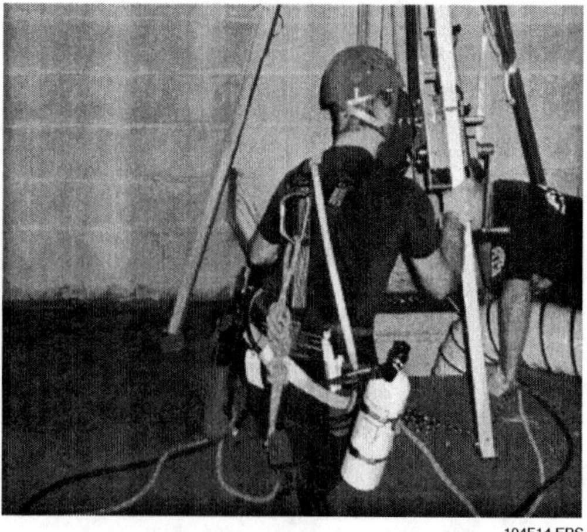

104F14.EPS

Figure 14 ◆ Worker entering a confined space with a restricted opening for entry and exit.

- Has a limited or restricted means of entry or exit; for example, tanks, vessels, silos, storage bins, hoppers, vaults, and pits
- Is not designed for continuous employee occupancy.

OSHA 29 CFR 1910.146 further defines a permit-required confined space as a space that:

- Contains or has the potential to contain a hazardous atmosphere
- Contains a material that has the potential for engulfing an entrant
- Has an internal configuration such that an entrant could be trapped or asphyxiated by inwardly converging walls or by a floor that slopes downward and tapers to a smaller cross section
- Contains any other recognized serious safety or health hazard

For safe working conditions, the oxygen level in a confined space atmosphere must range between 19.5 and 21.5 percent by volume, with 21 percent being considered the normal level. Oxygen concentrations below 19.5 percent by volume are considered deficient; those above 23.5 percent by volume are considered enriched (*Table 1*).

When welding or cutting is being performed in any confined space, the gas cylinders and welding machines are left on the outside. Before operations are started, the wheels for heavy portable equipment are securely blocked to prevent accidental movement. Where a welder must enter a confined space through a manhole or other opening, all means are provided for quickly removing

Instructor's Notes:

Table 1 Effects of an Increase or Decrease in Oxygen Levels

Oxygen Level	Effects
> 21.5 percent	Easy ignition of flammable material such as clothes
19.5 – 21.5 percent	Normal
17%	Deterioration of night vision, increased breathing volume, accelerated heartbeat
14 – 16 percent	Very poor muscular coordination, rapid fatigue, intermittent respiration
6 – 10 percent	Nausea, vomiting, inability to perform, unconsciousness
< 6 percent	Spasmodic breathing, convulsive movements, and death in minutes

the worker in case of emergency. When safety belts and lifelines are used for this purpose, they are attached to the welder's body so that he or she cannot be jammed in a small exit opening. An attendant with a pre-planned rescue procedure is stationed outside to observe the welder at all times and must be capable of putting the rescue operations into effect.

When welding or cutting operations are suspended for any substantial period of time, such as during lunch or overnight, all electrodes are removed from the holders, and the holders are carefully located so that accidental contact cannot occur. The welding machines are also disconnected from the power source.

In order to eliminate the possibility of gas escaping through leaks or improperly closed valves when gas welding or cutting, the gas and oxygen supply valves must be closed, the regulators released, the gas and oxygen lines bled, and the valves on the torch shut off when the equipment will not be used for a substantial period of time. Where practical, the torch and hose are also removed from the confined space. After welding operations are completed, the welder must mark the hot metal or provide some other means of warning other workers.

2.5.0 Area Safety

An important factor in area safety is good housekeeping. The work area should be picked up and swept clean. The floors and workbenches should be free of dirt, scrap metal, grease, oil, and anything that is not essential to accomplishing the given task. Collections of steel, welding electrode studs, wire, hoses, and cables are difficult to work around and easy to trip over. An electrode caddy can be used to hold the electrodes. Hooks can be made to hold hoses and cables, and scrap steel should be thrown into scrap bins.

The ideal welding shop should have bare concrete floors and bare metal walls and ceilings to reduce the possibility of fire. Never weld or cut over wood floors, as this increases the possibility of fire.It is important to keep flammable liquids as well as rags, wood scraps, piles of paper, and other combustibles out of the welding area.

If you must weld in an enclosed building, make every effort to eliminate anything that could trap a spark. Sparks can smolder for hours and then burst into flames. Regardless of where you're welding, be sure to have a fire extinguisher nearby. Also, keep a five-gallon bucket of water handy to cool off hot metal and quickly douse small fires. If a piece of hot metal must be left unattended, use soapstone to write the word HOT on it before leaving. This procedure can also be used to warn people of hot tables, vises, firebricks, and tools.

Never use a cutting torch inside your workshop unless a proper cutting area is available. Take whatever you're going to cut outside, away from flammables. Also be aware that welding sparks can ignite gasoline fumes in a confined space.

Whenever welding must be done outside a welding booth, use portable screens to protect other personnel from the arc or reflected glare (see *Figure 15*). The portable screen also prevents drafts of air from interfering with the stability of the arc.

The most common welding accident is burned hands and arms. Keep first-aid equipment nearby to treat burns in the work area. Eye injuries can also occur if you are careless. Post emergency phone numbers in a prominent location.

104F15.EPS

Figure 15 ◆ A typical welding screen.

Discuss safety precautions needed when working in a confined space.

Discuss how simple cleaning and housekeeping contribute greatly to area safety.

Describe an ideal welding area and precautions that must be taken in less than ideal conditions.

Review confined space safety procedures.

Review rules for keeping the work area safe.

Show Transparency 5 (Figure 16). Describe a hot work permit.

Review the basic safety procedures specific to oxyfuel gas welding.

Confined Space Precautions

Table 1 indicates the effect of increases and decreases in oxygen levels in a confined space. If too much oxygen is ventilated into a confined space, it can be absorbed by the welder's clothing and ignite. If too little oxygen is present, it can lead to the welder's death in minutes. For this reason, the following precautions apply:

- Make sure confined spaces are ventilated properly for cutting or welding purposes.
- Never use oxygen in confined spaces for ventilation purposes.

The following are some work-area reminders:

- Eliminate tripping hazards by coiling cables and keeping clamps and other tools off the floor.
- Don't get entangled in cables, loose wires, or clothing while you work. This allows you to move freely, especially should your clothing ignite or some other accident occur.
- Clean up oil, grease, or other agents that may ignite and splatter off surfaces while welding.
- Shut off the welder and disconnect the power plug before performing any service or maintenance.
- Keep the floor free of electrodes once you begin to weld. They could cause a slip or fall.
- Work in a dry area, booth, or other shielded area whenever possible.
- Make sure there are no open doors or windows through which sparks may travel to flammable materials.

2.6.0 Hot Work Permits and Fire Watches

A hot work permit (*Figure 16*) is an official authorization from the site manager to perform work that may pose a fire hazard. The permit includes information such as the time, location, and type of work being done. The hot work permit system promotes the development of standard fire safety guidelines. Permits also help managers keep records of who is working where and at what time. This information is essential in the event of an emergency or at other times when personnel need to be evacuated.

Oxyfuel Gas Welding Safety Precautions

- Always light the oxyfuel gas torch flame using an approved torch lighter to avoid burning your fingers.
- Never point the torch tip at anyone when lighting it or using it.
- Never point the torch at the cylinders, regulators, hoses, or anything else that may be damaged and cause a fire or explosion.
- Never lay a lighted torch down on the bench or work piece, and do not hang it up while it is lighted. If the torch is not in the welder's hands, it must be off.
- To prevent possible fires and explosions, check valves and flashback arrestors must be installed in all oxyfuel gas welding and cutting outfits.
- When cutting with oxyfuel gas equipment, clear the area of all combustible materials.
- Skin contact with liquid oxygen can cause frostbite. Be careful when handling liquid oxygen.
- Never use oxygen as a substitute for compressed air.
- All oxygen cylinders with leaky valves or safety fuse plugs and discs should be set aside and marked for the attention of the supplier. Do not tamper with or attempt to repair oxygen cylinder valves. Do not use a hammer or wrench to open the valves.

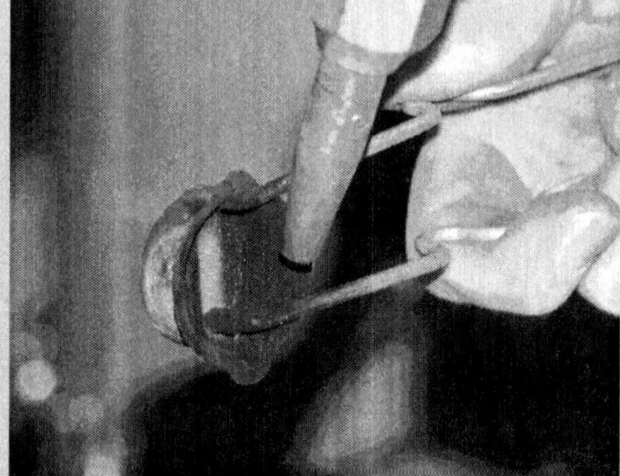

104SA01.EPS

Instructor's Notes:

4.12 PIPEFITTING ◆ LEVEL ONE

During a fire watch, a person other than the welder or cutting operator must constantly scan the work area for fires. Fire watch personnel must have ready access to fire extinguishers and alarms and know how to use them. Cutting operations must never be performed without a fire watch. Whenever oxyfuel cutting equipment is used, there is a great danger of fire. Hot work permits and fire watches are used to minimize this danger. Most sites require the use of hot work permits and fire watches. When they are violated, severe penalties are imposed.

Describe a fire watch.

HOT WORK PERMIT

FOR CUTTING, WELDING, OR SOLDERING WITH PORTABLE GAS OR ARC EQUIPMENT

Job Date __________ Start Time __________ Expiration __________ W/O# __________

Applicant Name ____________________ Company / Dep't. ____________ Phone ________

Supervisor ___________________________________ Phone ________________________

Location / Description of work____

____

IS FIRE WATCH REQUIRED? (ref. NFPA 51-B 3-3)

1.______(yes or no) Are combustible materials in building construction closer that 35 feet to the point of operation?

2.______(yes or no) Are combustibles more than 35 feet away but would be easily ignited by sparks?

3.______(yes or no) Are wall or floor openings within a 35 foot radius exposing combustible material in adjacent areas, including concealed spaces in floors or walls?

4.______(yes or no) Are combustible materials adjacent to the other side of metal partitions, walls, ceilings, or roofs which could be ignited by conduction or radiation?

5.______(yes or no) Does the work necessitate disabling a fire detection, suppression, or alarm system component?

YES to any of the above indicates that a qualified fire watch is required.

Fire Watcher Name(s) ______________________________________ Phone _________

NOTIFICATIONS

NOTIFY THE FOLLOWING GROUPS AT LEAST 72 HOURS PRIOR TO WORK AND 30 MINUTES AFTER WORK IS COMPLETED. Write in names of persons contacted.

*Facilities Management Service Desk (492-5522) ______________________________

**Facilities Management Fire Alarm Supervisor (492-0633) __________________________

*Facilities Management Fire Protection Group (FPG) (492-5681, 492-4042)________________

*Environmental Health and Safety Industrial Hygiene Group (492-6025)___________________

* Notify by phone or in person. If by phone, write down name of person and send them a completed copy of this permit.

**Notify in person.

SIGNATURES REQUIRED

University Project Manager _________________________ Date _________ Phone _______

I understand and will abide by the conditions described in this permit. I will implement the necessary precautions which are outlined on both sides of this permit form. Thirty minutes after each hot work session, I will reinspect work areas and adjacent areas to which spark and heat might have spread to verify that they are fire safe and contact Facilities Management Alarm Technicians to have any disabled fire protection systems reactivated.

________________________ , ________________ Date __________ Phone _________ Permit Applicant Company or Dep't

104F16.EPS

Figure 16 ◆ Hot work permit.

MODULE 08104-06 ◆ OXYFUEL CUTTING 4.13

Stress that a hot work permit must be obtained and a fire watch established before performing any heating, cutting, or welding.

Discuss the hazards posed when cutting containers.

Emphasize the importance of thoroughly emptying and cleaning containers before cutting them. Point out that heating containers can produce hazardous vapors.

Discuss hazardous vapors and emphasize the importance of proper ventilation.

Describe the procedure for cleaning or purging a container before cutting it.

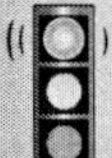

WARNING!
Never perform any type of heating, cutting, or welding until you have obtained a hot work permit and established a fire watch. If you are unsure of the procedure, check with your supervisor. Violation of hot work permit and fire watch procedures can result in serious injury or death.

2.7.0 Cutting Containers

Cutting and welding activities present unique hazards depending upon the material being cut or welded and the fuel used to power the equipment. All cutting and welding should be done in designated areas of the shop if possible. These areas should be made safe for welding and cutting operations with concrete floors, arc filter screens, protective drapes, curtains or blankets (*Figure 17*), and fire extinguishers. No combustibles should be stored nearby.

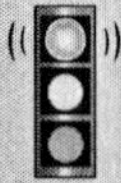

WARNING!
Welding or cutting must never be performed on drums, barrels, tanks, vessels, or other containers until they have been emptied and cleaned thoroughly, eliminating all flammable materials and all substances (such as detergents, solvents, greases, tars, or acids) that might produce flammable, toxic, or explosive vapors when heated.

Containers must be cleaned by steam cleaning, flushing with water, or washing with detergent until all traces of the material have been removed.

3,000°F INTERMITTENT, 1,500°F CONTINUOUS SILICON DIOXIDE CLOTH

104F17.EPS

Figure 17 ◆ A typical welding blanket.

WARNING!
Clean containers only in well-ventilated areas. Vapors can accumulate during cleaning, causing explosions or injury.

After cleaning the container (*Figure 18*), fill it with water or an inert gas such as argon or carbon dioxide (CO_2) for additional safety. Air, which contains oxygen, is displaced from inside the container by the water or inert gas. Without oxygen, combustion cannot take place. When using water, position the container to minimize the air space. When using an inert gas, provide a vent hole so the inert gas can purge the air.

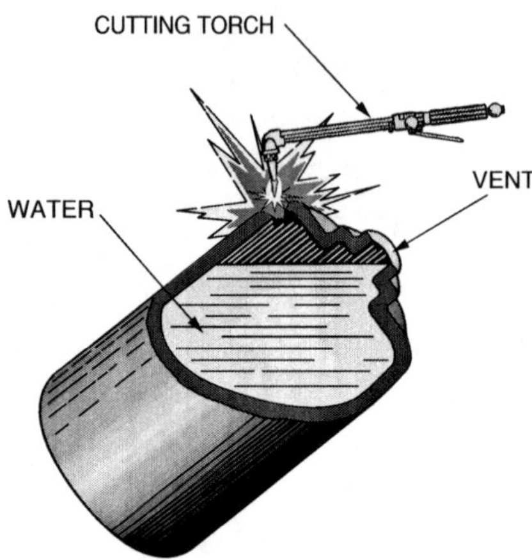

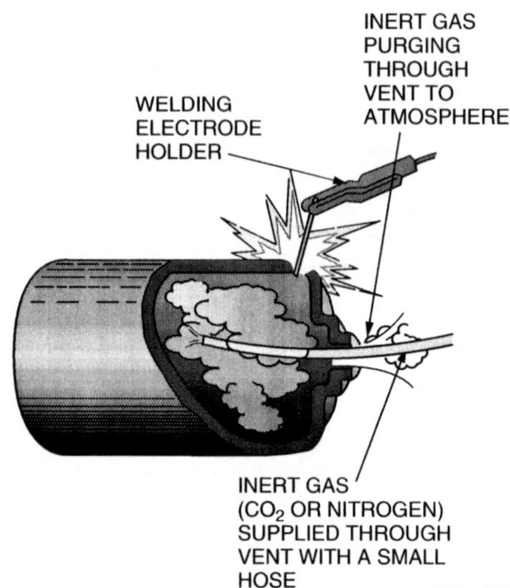

104F18.EPS

Figure 18 ◆ Purging containers of potential health hazards.

Instructor's Notes:

Emphasize that a container cannot be considered safe until it has been properly tested.

Identify the requirements for safe storage and handling of cylinders.

Identify the requirements of locating, securing, and identifying cylinder storage areas.

Discuss precautions and procedures specific to valve protection caps.

Explain how to handle various problems with cylinders.

WARNING!

Do not assume that a container that has held combustibles is clean and safe until proven so by proper tests. Do not weld in places where dust or other combustible particles are suspended in air or where explosive vapors are present. Removal of flammable materials from vessels/containers may be done by steaming or boiling.

Proper procedures for cutting or welding hazardous containers are described in the *American Welding Society (AWS) F4.1-1999, Recommended Safe Practices for the Preparation for Welding and Cutting of Containers and Piping*. You should also consult with the local fire marshal before welding or cutting such containers.

2.8.0 Cylinder Storage and Handling

Oxygen and fuel gas cylinders or other flammable materials must be stored separately. The storage areas must be separated by 20 feet or by a wall 5 feet high with at least a 30-minute burn rating. The purpose of the distance or wall is to keep the heat of a small fire from causing the oxygen cylinder safety valve to release. If the safety valve releases the oxygen, a small fire would become a raging inferno.

Inert gas cylinders may be stored separately or with either fuel cylinders or oxygen cylinders. Empty cylinders must be stored separately from full cylinders, although they may be stored in the same room or area. All cylinders must be stored vertically and have the protective caps screwed on firmly.

2.8.1 Securing Gas Cylinders

Cylinders must be secured with a chain or other device so that they cannot be knocked over accidentally. Even though they are more stable, cylinders attached to a manifold should be chained, as should cylinders stored in a special room used only for cylinder storage.

2.8.2 Storage Areas

Cylinder storage areas must be located away from halls, stairwells, and exits so that in case of an emergency they will not block an escape route. Storage areas should also be located away from heat, radiators, furnaces, and welding sparks. The location of storage areas should be where unauthorized people cannot tamper with the cylinders. A warning sign that reads *Danger—No Smoking, Matches, or Open Lights*, or similar wording, should be posted in the storage area.

2.8.3 Cylinders with Valve Protection Caps

Cylinders equipped with a valve protection cap must have the cap in place unless the cylinder is in use. The protection cap prevents the valve from being broken off if the cylinder is knocked over. If the valve of a full high-pressure cylinder (argon, oxygen, CO_2, or mixed gases) is broken off, the cylinder can fly around the shop like a missile if it has not been secured properly. Never lift a cylinder by the safety cap or valve. The valve can easily break off or be damaged. When moving cylinders, the valve protection cap must be replaced (*Figure 19*), especially if the cylinders are mounted on a truck or trailer. Cylinders must never be dropped or handled roughly.

104F19.EPS

Figure 19 ◆ Handle bottles with care.

2.8.4 General Precautions

Use warm water (not boiling) to loosen cylinders that are frozen to the ground. Any cylinder that leaks, has a bad valve, or has gas-damaged threads must be identified and reported to the supplier. Use a piece of soapstone to write the problem on the cylinder. If closing the cylinder valve cannot stop the leak, move the cylinder outdoors to a safe location, away from any source of

MODULE 08104-06 ◆ OXYFUEL CUTTING 4.15

Review the general precautions for working with cylinders.

Have trainees review Sections 3.0.0–3.12.3.

Ensure that you have everything required for teaching this session.

Refer to Figure 20 and identify the equipment used in oxyfuel cutting.

Discuss the properties of oxygen and the use of oxygen cylinders.

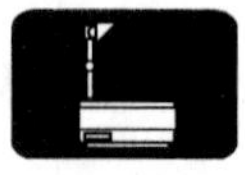

Show Transparency 6 (Figure 21). Explain how high-pressure oxygen cylinders are sized and marked.

ignition, and notify the supplier. Post a warning sign, then slowly release the pressure.

In its gaseous form, acetylene is extremely unstable and explodes easily. For this reason it must remain at pressures below 15 pounds per square inch (psi). If an acetylene cylinder is tipped over, stand it upright and wait at least 30 minutes before using it. If liquid acetone is withdrawn from a cylinder, it will gum up the safety check valves and regulators and decrease the stability of the acetylene stored in the cylinder. For this reason, acetylene must never be withdrawn at a per-hour rate that exceeds 10 percent of the volume of the cylinder(s) in use. Acetylene cylinders in use should be opened no more than one and one-half turns and, preferably, no more than three-fourths of a turn. Other precautions include:

- Use only compressed gas cylinders containing the correct gas for the process used and properly operating regulators designed for the gas and pressure used.
- Make sure all hoses, fittings, and other parts are suitable and maintained in good condition.
- Keep cylinders in the upright position and securely chained to an undercarriage or fixed support.
- Keep combustible cylinders in one area of the building for safety. Cylinders must be at a safe distance from arc welding or cutting operations and any other source of heat, sparks, or flame.
- Never allow the electrode, electrode holder, or any other electrically hot parts to come in contact with the cylinder.
- When opening a cylinder valve, keep your head and face clear of the valve outlet.
- Always use valve protection caps on cylinders when they are not in use or are being transported.

3.0.0 ◆ OXYFUEL CUTTING EQUIPMENT

The equipment used to perform oxyfuel cutting includes oxygen and fuel gas cylinders, oxygen and fuel gas regulators, hoses, and a cutting torch. A typical movable oxyfuel (oxyacetylene) cutting outfit is shown in *Figure 20*.

3.1.0 Oxygen

Oxygen (O_2) is a colorless, odorless, tasteless gas that supports combustion. Combined with burning material, pure oxygen causes a fire to flare and burn out of control. When mixed with fuel gases, oxygen produces the high-temperature flame required in order to flame cut metals.

Respect Explosive Mixtures

An accumulation of an oxygen and fuel-gas mixture can result in a dangerous explosion when ignited. A 2-inch balloon filled with an oxygen and acetylene mixture has the explosive power of an M80 firecracker (one-quarter stick of dynamite). The mixed gases from 100 feet of ¼-inch twin hose will fill a 9-inch balloon. A 9-inch balloon filled with mixed oxygen and acetylene gases has an explosive power that is 90 times more powerful than an M80 firecracker.

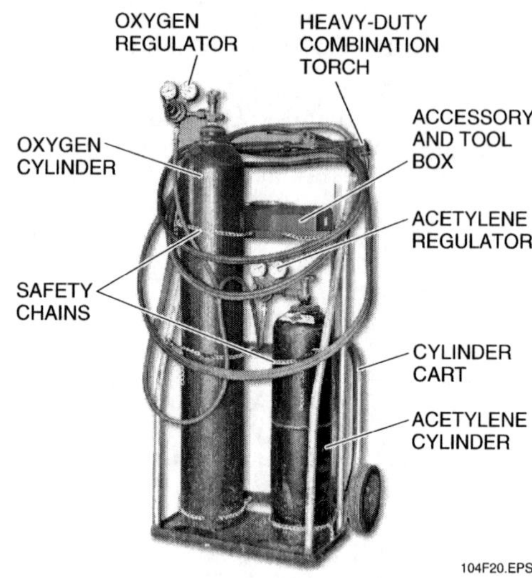

Figure 20 ◆ Typical oxyacetylene welding/cutting outfit.

3.1.1 Oxygen Cylinders

Oxygen is stored at more than 2,000 pounds per square inch (psi) in hollow steel cylinders. The cylinders come in a variety of sizes based on the cubic feet of oxygen they hold. The smallest standard cylinder holds about 85 cubic feet of oxygen, and the largest ultra-high-pressure cylinder holds about 485 cubic feet. The most common size oxygen cylinder used for welding and cutting operations is the 227 cubic foot cylinder. It is more than 4 feet tall and 9 inches in diameter. The shoulder of the oxygen cylinder has the name of the gas and/or the supplier stamped, labeled, or stenciled on it. *Figure 21* shows standard high-pressure oxygen cylinder markings and sizes. The cylinders must be tested every 10 years.

Instructor's Notes:

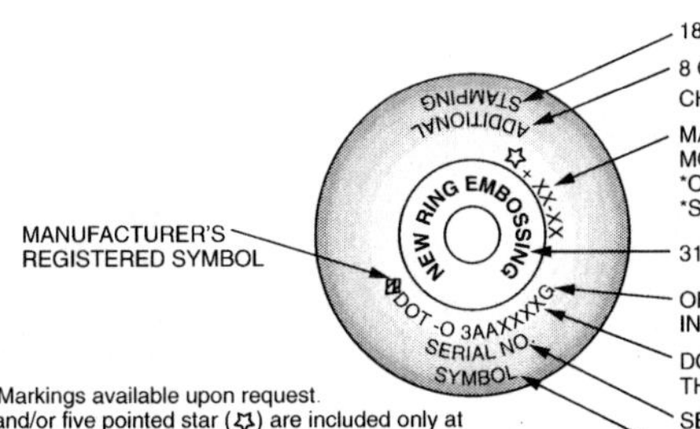

Transport Canada Markings available upon request.
*The plus sign (+) and/or five pointed star (☆) are included only at customer's request, and indicate compliance with applicable requirements of the Code of Federal Regulations, Title 49, Transportation.

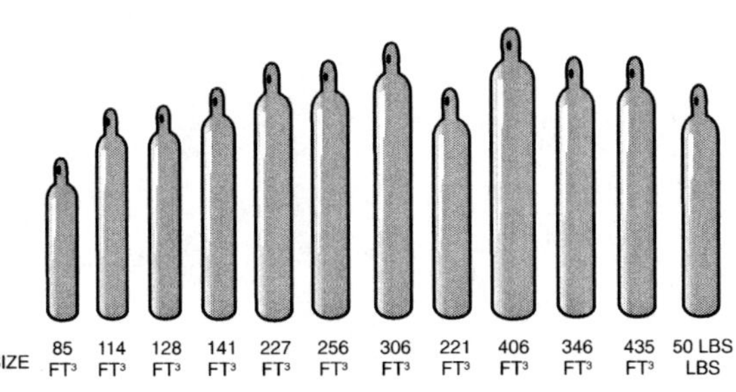

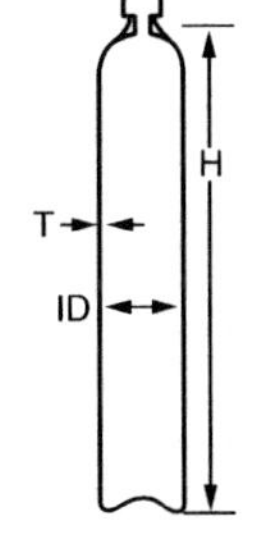

HIGH PRESSURE CYLINDER MARKINGS

DOT SPECIFICATIONS	O_2 CAPACITY (FT^3) AT RATED SERVICE PRESSURE	O_2 CAPACITY (FT^3) AT 10% OVERCHARGE	WATER CAPACITY (IN^3) MINIMUM	WATER CAPACITY (IN^3) MAXIMUM	NOMINAL DIMENSIONS (IN) AVG. INSIDE DIAMETER "ID"	NOMINAL DIMENSIONS (IN) HEIGHT "H"	NOMINAL DIMENSIONS (IN) MINIMUM WALL "T"	NOMINAL WEIGHT (LB)	PRESSURE (PSI) SERVICE	PRESSURE (PSI) TEST
STANDARD HIGH PRESSURE CYLINDERS[1]										
3AA2015	85	93	960	1040	6.625	32.50	0.144	48	2015	3360
3AA2015	114	125	1320	1355	6.625	43.00	0.144	61	2015	3360
3AA2265	128	140	1320	1355	6.625	43.00	0.162	62	2265	3775
3AA2015	141	155	1630	1690	7.000	46.00	0.150	70	2015	3360
3AA2015	227	250	2640	2710	8.625	51.00	0.184	116	2015	3360
3AA2265	256	281	2640	2710	8.625	51.00	0.208	117	2265	3775
3AA2400	306	336	2995	3060	8.813	55.00	0.226	140	2400	4000
3AA2400	405	444	3960	4040	10.060	56.00	0.258	181	2400	4000
ULTRALIGHT® HIGH PRESSURE CYLINDERS[1]										
E-9370-3280	365	NA	2640	2710	8.625	51.00	0.211	122	3280	4920
E-9370-3330	442	NA	3181	3220	8.813	57.50	0.219	147	3330	4995
ULTRA HIGH PRESSURE CYLINDERS[2]										
3AA3600	347[3]	374	2640	2690	8.500	51.00	0.336	170	3600	6000
3AA6000	434[3]	458	2285	2360	8.147	51.00	0.568	267	6000	10000
E-10869-4500	435[3]	NA	2750	2890	8.813	51.00	0.260	148	4500	6750
E-10869-4500	485[3]	NA	3058	3210	8.813	56.00	0.260	158	4500	6750

1. Regulators normally permit filling these cylinders with 10% overcharge, provided certain other requirements are met.
2. Under no circumstances are these cylinders to be filled to a pressure exceeding the marked service pressure at 70°F.
3. Nitrogen capacity at 70°F.

All cylinders normally furnished with 3/4" NGT internal threads, unless otherwise specified.
Nominal weights include neck ring but exclude valve and cap, add 2 lbs (.91 kg) for cap and 1 1/2 lb. (.8 kg) for valve.
Cap adds approximately 5 in. (127 mm) to height.
Cylinder capacities are approximately 5 in. (127 mm) to height.
Cylinder capacities are approximately at 70°F. (21°C).

104F21.EPS

Figure 21 ◆ High-pressure oxygen cylinder markings and sizes.

MODULE 08104-06 ◆ OXYFUEL CUTTING 4.17

Explain how oxygen cylinders are regulated by a bronze valve.

Emphasize the danger of removing protective caps from cylinders that are not secured.

Discuss the properties of acetylene and the use of acetylene cylinders.

Oxygen cylinders have bronze cylinder valves on top (*Figure 22*). Turning the cylinder valve handwheel controls the flow of oxygen out of the cylinder. A safety plug on the side of the cylinder valve allows oxygen in the cylinder to escape if the pressure in the cylinder rises too high. Oxygen cylinders are usually equipped with Compressed Gas Association (CGA) 540 valves for service up to 3,000 per square inch gauge (psig). Some cylinders are equipped with CGA 577 valves for up to 4,000 psig service or CGA 701 valves for up to 5,500 psig service.

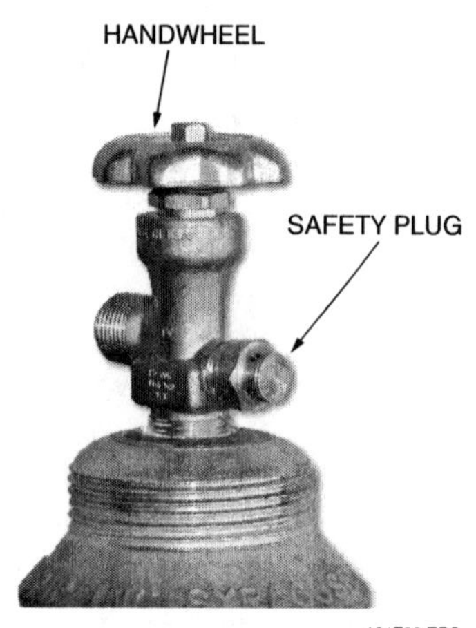

Figure 22 ◆ Oxygen cylinder valve.

Use care when handling oxygen cylinders because oxygen is stored at such high pressures. When it is not in use, always cover the cylinder valve with the protective steel safety cap (*Figure 23*).

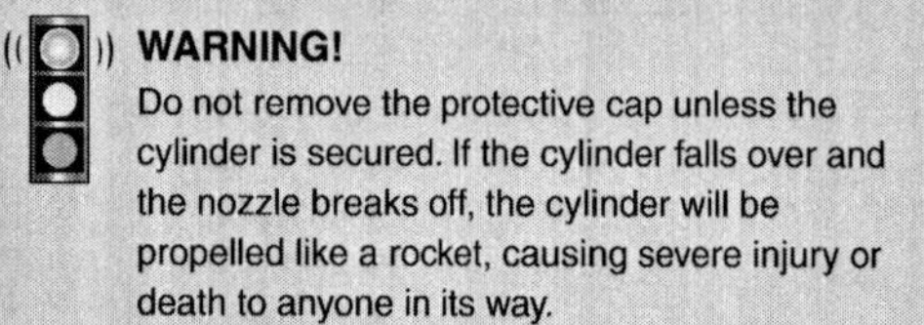

WARNING!

Do not remove the protective cap unless the cylinder is secured. If the cylinder falls over and the nozzle breaks off, the cylinder will be propelled like a rocket, causing severe injury or death to anyone in its way.

3.2.0 Acetylene

Acetylene gas (C2H2), a compound of carbon and hydrogen, is lighter than air. It is formed by dissolving calcium carbide in water. It has a strong, distinctive, garlic-like odor. In its gaseous form, acetylene is extremely unstable and explodes easily.

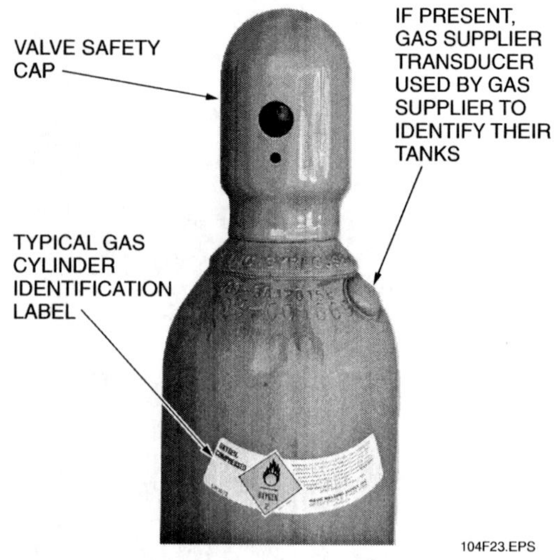

Figure 23 ◆ Oxygen cylinder with standard safety cap.

Because of this instability, it cannot be compressed at pressures of more than 15 psi when in its gaseous form. At higher pressures, acetylene gas breaks down chemically, producing heat and pressure that could result in a violent explosion. When combined with oxygen, acetylene creates a flame that burns hotter than 5,500°F, one of the hottest gas flames. Acetylene can be used for flame cutting, welding, heating, flame hardening, and stress relieving.

3.2.1 Acetylene Cylinders

Because of the explosive nature of acetylene gas, it cannot be stored above 15 psi in a hollow cylinder. To solve this problem, acetylene cylinders are specially constructed to store acetylene at higher pressures. The acetylene cylinder is filled with a porous material that creates a solid, instead of a hollow, cylinder. The porous material is soaked with acetone, which absorbs the acetylene, stabilizing it for storage at higher pressures. Because of the liquid acetone inside the cylinder, acetylene cylinders must always be used in an upright position. If the cylinder is tipped over, stand the cylinder upright and wait at least 30 minutes before using it. If liquid acetone is withdrawn from a cylinder, it will gum up the safety check valves and regulators. Always take care to withdraw acetylene gas from a cylinder at pressures less than 15 psig and at hourly rates that do not exceed one-tenth of the cylinder capacity. Higher rates may cause liquid acetone to be withdrawn along with the acetylene.

Instructor's Notes:

4.18 PIPEFITTING ◆ LEVEL ONE

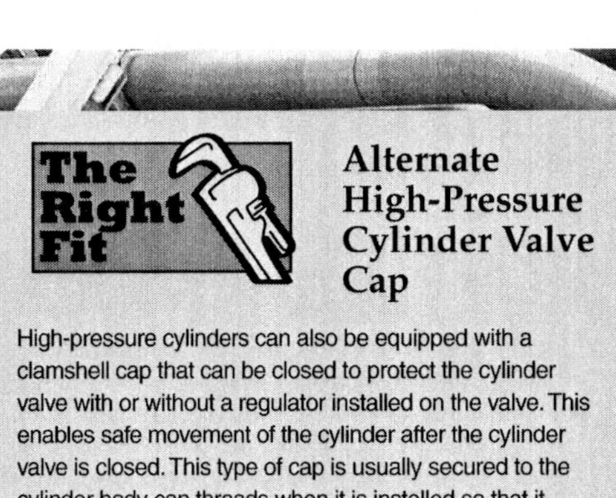

Alternate High-Pressure Cylinder Valve Cap

High-pressure cylinders can also be equipped with a clamshell cap that can be closed to protect the cylinder valve with or without a regulator installed on the valve. This enables safe movement of the cylinder after the cylinder valve is closed. This type of cap is usually secured to the cylinder body cap threads when it is installed so that it cannot be removed. When the clamshell is closed, it can also be padlocked to prevent unauthorized operation of the cylinder valve.

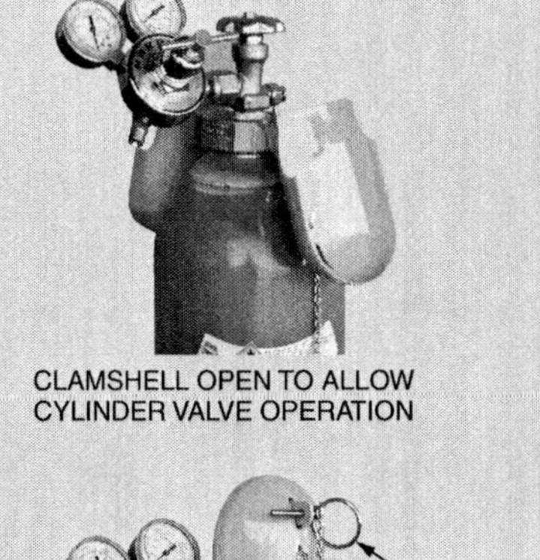

CLAMSHELL OPEN TO ALLOW CYLINDER VALVE OPERATION

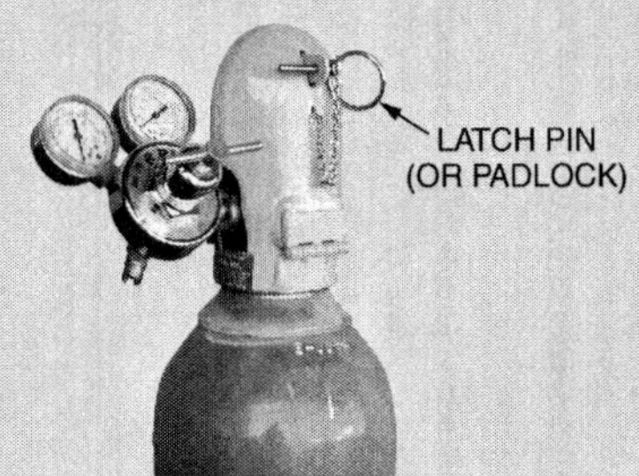

CLAMSHELL CLOSED FOR MOVEMENT OR PADLOCKED TO PREVENT OPERATION OF CYLINDER VALVE

CLAMSHELL CLOSED FOR TRANSPORT

104SA02.EPS

Acetylene cylinders have safety fuse plugs in the top and bottom of the cylinder that melt at 220°F (*Figure 24*). In the event of a fire, the fuse plugs will release the acetylene gas, preventing the cylinder from exploding.

Acetylene cylinders are available in a variety of sizes based on the cubic feet of acetylene that they can hold. The smallest standard cylinder holds about 10 cubic feet of gas. The largest standard cylinder holds about 420 cubic feet of gas. A cylinder that holds about 850 cubic feet is also available. *Figure 25* shows standard acetylene cylinder markings and sizes. Like oxygen cylinders, acetylene cylinders must be tested every 10 years.

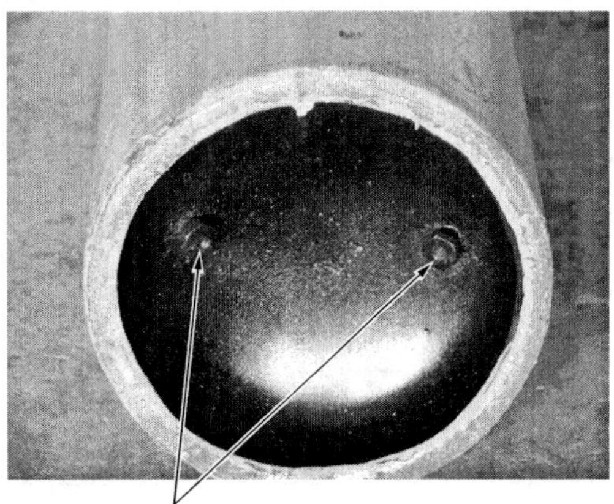

104F24.EPS

Figure 24 ◆ Acetylene cylinder valve and safety plugs.

Show Transparency 7 (Figure 25). Explain how acetylene cylinders are sized and marked.

MODULE 08104-06 ◆ OXYFUEL CUTTING 4.19

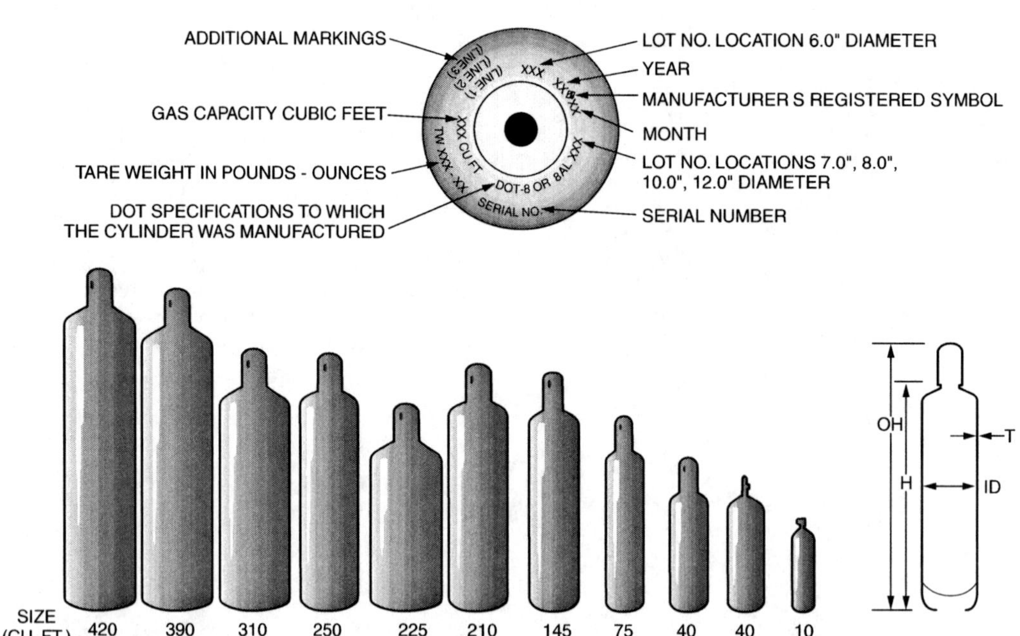

DOT SPECIFICATIONS	CAPACITY			NOMINAL DIMENSIONS (IN.)				ACETONE (LB. - OZ.)	APPROXIATE TARE WEIGHT WITH VALVE WITHOUT CAP (LB.)
	ACETYLENE	MIN. WATER							
	(FT.³)	(IN.³)	(LB.)	AVG. INSIDE DIAMETER ID "	HEIGHT W/OUT VALVE OR CAP "H"	HEIGHT W/VALVE AND CAP "OH"	MINIMUM WALL "T"		
8 AL[1]	10	125	4.5	3.83	13.1375	14.75	0.0650	1-6	8
8[1]	40	466	16.8	6.00	19.8000	23.31	0.0870	5-7	25
8[2]	40	466	16.8	6.00	19.8000	28.30	0.0870	5-7	28
8[3]	75	855	30.8	7.00	25.5000	31.25	0.0890	9-8	45
8	100	1055	38.0	7.00	30.7500	36.50	0.0890	12-2	55
8	145	1527	55.0	8.00	34.2500	40.00	0.1020	18-10	76
8	210	2194	79.0	10.00	32.2500	38.00	0.0940	25-13	105
8AL	225	2630	94.7	12.00	27.5000	32.75	0.1280	29-6	110
8	250	2606	93.8	10.00	38.0000	43.75	0.0940	30-12	115
8AL	310	3240	116.7	12.00	32.7500	38.50	0.1120	39-5	140
8AL	390	4151	150.0	12.00	41.0000	46.75	0.1120	49-14	170
8AL	420	4375	157.5	12.00	43.2500	49.00	0.1120	51-14	187
8	60	666	24.0	7.00	25.79 OH		0.0890	7-11	40
8	130	1480	53.3	8.00	36.00 OH		0.1020	17-2	75
8AL	390	4215	151.8	12.00	46.00 OH		0.1120	49-14	180

1. Tapped for 3/8" valve but are not equipped with valve protection caps.
2. Includes Valve protection cap.
3. Can be tared to hold 60 ft³ (1.7 m³) of acetylene gas.

Standard tapping (except cylinders tapped for 3/8") 3/4"-14 NGT.

Weight includes saturation gas, filler, paint, solvent, valve, fuse plugs. Does not include cap of 2 lb. (91 kg.)

Cylinder capacities are based upon commercially pure acetylene gas at 250 psi (17.5 kg/cm²), and 70°F (15°C).

104F25.EPS

Figure 25 ◆ Acetylene cylinder markings and sizes.

Instructor's Notes:

4.20 PIPEFITTING ◆ LEVEL ONE

Acetylene cylinders are usually equipped with a standard CGA 510 brass cylinder valve (see *Figure 25*). The handwheel of the valve controls the flow of acetylene from the cylinder to a regulator. Some acetylene cylinders are equipped with an alternate standard CGA 300 valve. Some obsolete valves still in use require a special long-handled wrench with a square socket end to operate the valve.

The smallest standard acetylene cylinder, which holds 10 cubic feet, is equipped with a CGA 200 small series valve, and 40 cubic foot cylinders use a CGA 520 small series valve. As with oxygen cylinders, place a protective valve cap on the acetylene cylinders during transport (*Figure 26*).

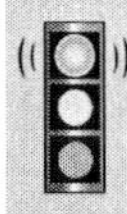

WARNING!
Do not remove the protective cap unless the cylinder is secured. If the cylinder falls over and the nozzle breaks off, the cylinder will release highly explosive gas.

104F26.EPS

Figure 26 ◆ Acetylene cylinder with standard valve safety cap.

3.3.0 Liquefied Fuel Gases

Many fuel gases other than acetylene are used for cutting. They include natural gas and liquefied fuel gases such as methylacetylene propadiene (MAPP®), propylene, and propane. Their flames are not as hot as acetylene, but they have higher British thermal unit (Btu) ratings and are cheaper and safer to use. The supervisor at your job site will determine which fuel gas to use.

Table 2 compares the flame temperatures of oxygen mixed with various fuel gases.

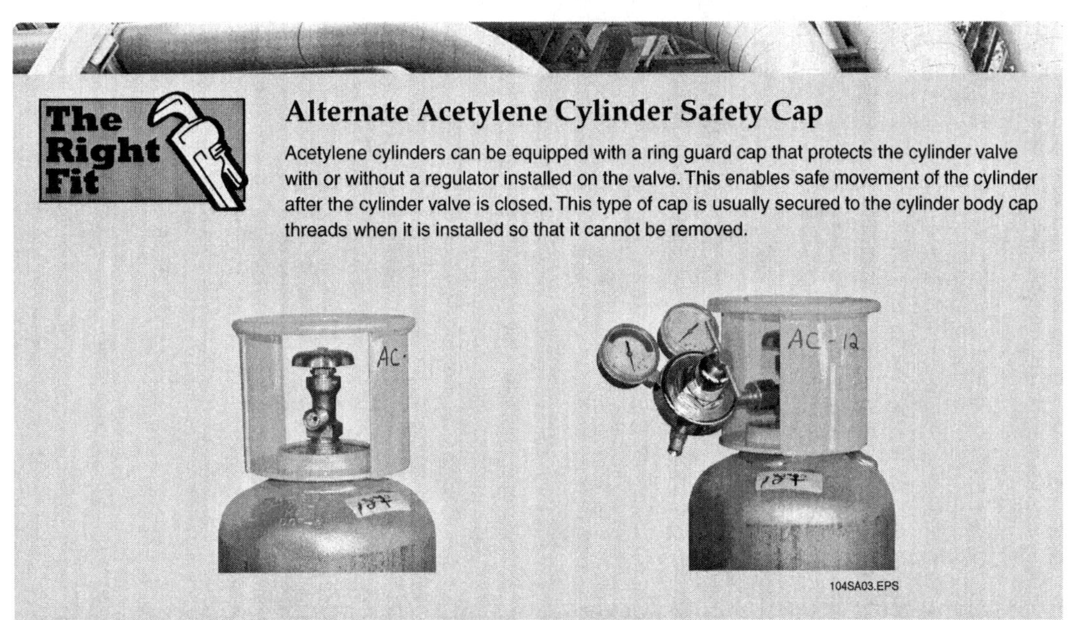

Alternate Acetylene Cylinder Safety Cap

Acetylene cylinders can be equipped with a ring guard cap that protects the cylinder valve with or without a regulator installed on the valve. This enables safe movement of the cylinder after the cylinder valve is closed. This type of cap is usually secured to the cylinder body cap threads when it is installed so that it cannot be removed.

104SA03.EPS

Explain how acetylene cylinders are regulated by a brass valve.

Emphasize the danger of removing protective caps from cylinders that are not secured.

Identify the various types of gases that are used for cutting.

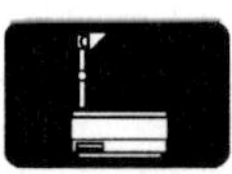

Show Transparency 8 (Table 2). Discuss the flame temperatures of oxygen with various fuel gases.

MODULE 08104-06 ◆ OXYFUEL CUTTING 4.21

Describe the various gases used as fuel gas.

Show Transparency 9 (Figure 27). Discuss the uses and precautions associated with liquefied fuel gas cylinders.

Tank Identification Transducers

Many gas suppliers mount transducers on their tanks so that they can readily identify the tanks. The transducer is electronically scanned, and a coded number is matched against the supplier's records for identification. This aids in quickly determining the purchaser or user of the tank as well as the required retesting date.

Table 2 Flame Temperatures of Oxygen with Various Fuel Gases

Type of Gas	Flame Temperature
Acetylene	More than 5,500°F
MAPP®	5,300°F
Propylene	5,190°F
Natural gas	4,600°F
Propane	4,580°F

MAPP® is a Dow Chemical Company product that is a chemical combination of acetylene and propane gases. MAPP® gas burns at temperatures almost as high as acetylene and has the stability of propane. Because of this stability, it can be used at pressures over 15 psi and is not as likely as acetylene to **backfire** or **flashback**. MAPP® also has an offensive odor that can be detected easily. MAPP® gas can be used for flame cutting, heating, stress relieving, brazing, soldering, and scarfing (cleaning cutting dross or other material from the workpiece).

Propylene mixtures are hydrocarbon-based gases that are stable and shock-resistant, making them relatively safe to use. They are purchased under trade names such as High Purity Gas (HPG™), Apachi™, and Prestolene™. These gases and others have distinctive odors to make leak detection easier. They burn at temperatures around 5,193°F, hotter than natural gas and propane. Propylene gases are used for flame cutting, scarfing, heating, stress relieving, brazing, and soldering.

Propane is also known as liquefied petroleum (LP) gas. It is stable and shock-resistant, and it has a distinctive odor for easy leak detection. It burns at 4,580°F, which is the lowest temperature of any fuel gas. It has a slight tendency toward backfire and flashback and is used quite extensively for cutting procedures.

Natural gas is delivered by pipeline rather than by cylinders. It burns at about 4,600°F. Natural gas is relatively stable and shock-resistant and has a slight tendency toward backfire and flashback. Because of its recognizable odor, leaks are easily detectable. Natural gas is used primarily for cutting on job sites with permanent cutting stations.

3.3.1 Liquefied Fuel Gas Cylinders

Liquefied fuel gases are shipped in hollow steel cylinders (*Figure 27*). When empty, they are much lighter than acetylene cylinders.

The liquefied fuel gas is stored in hollow steel cylinders of various sizes. They can hold from 30 to 225 pounds of fuel gas. As the cylinder valve is opened, the vaporized gas is withdrawn from the cylinder. The remaining liquefied gas absorbs heat and releases additional vaporized gas. The pressure of the vaporized gas varies with the outside temperature. The colder the outside temperature,

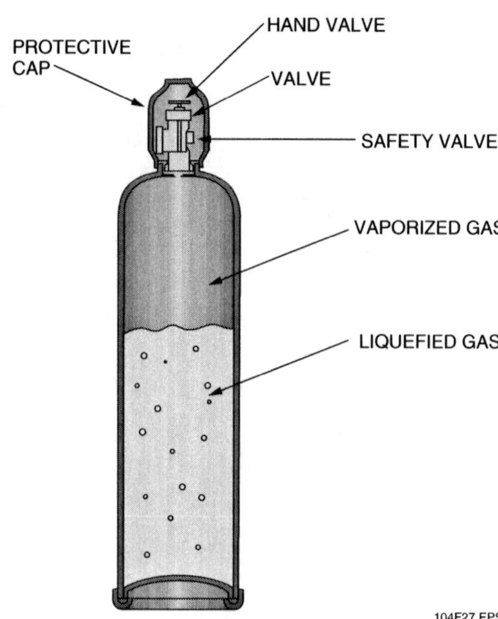

Figure 27 ◆ Liquefied fuel gas cylinder.

Instructor's Notes:

4.22 PIPEFITTING ◆ LEVEL ONE

the lower the vaporized gas pressure will be. If high volumes of gas are removed from a liquefied fuel gas cylinder, the pressure drops, and the temperature of the cylinder will also drop. A ring of frost can form around the base of the cylinder. If high withdrawal rates continue, the regulator may also start to ice up. If high withdrawal rates are required, special regulators with electric heaters should be used.

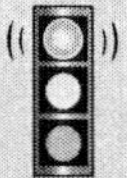

WARNING!

Never apply heat directly to a cylinder or regulator. This can cause excessive pressures, resulting in an explosion.

The pressure inside a liquefied fuel gas cylinder is not an indicator of how full or empty the cylinder is. The weight of a cylinder determines how much liquefied gas is left. Liquefied fuel gas cylinders are equipped with CGA 510, 350, or 695 valves, depending on the fuel and storage pressures.

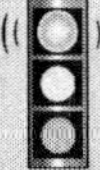

WARNING!

Do not remove the protective cap on liquefied fuel gas cylinders unless the cylinder is secured. If the cylinder falls over and the nozzle breaks off, the cylinder will release highly explosive gas. Cylinders containing a liquid such as propane must be kept in an upright position. If the valve is broken or is opened with the cylinder horizontal, the fuel can emerge as a liquid that will shoot a long distance before it vaporizes. If it is ignited, it will act like a flamethrower.

3.4.0 Regulators

Regulators (*Figure 28*) are attached to the cylinder valve. They reduce the high cylinder pressures to the required lower working pressures and maintain a steady flow of gas from the cylinder.

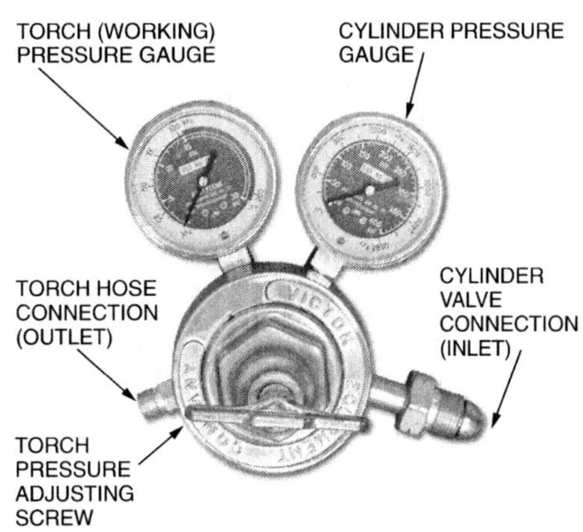

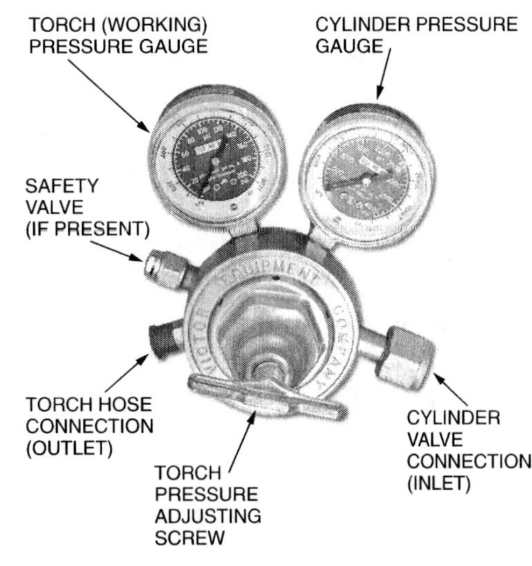

Figure 28 ◆ Oxygen and acetylene regulators.

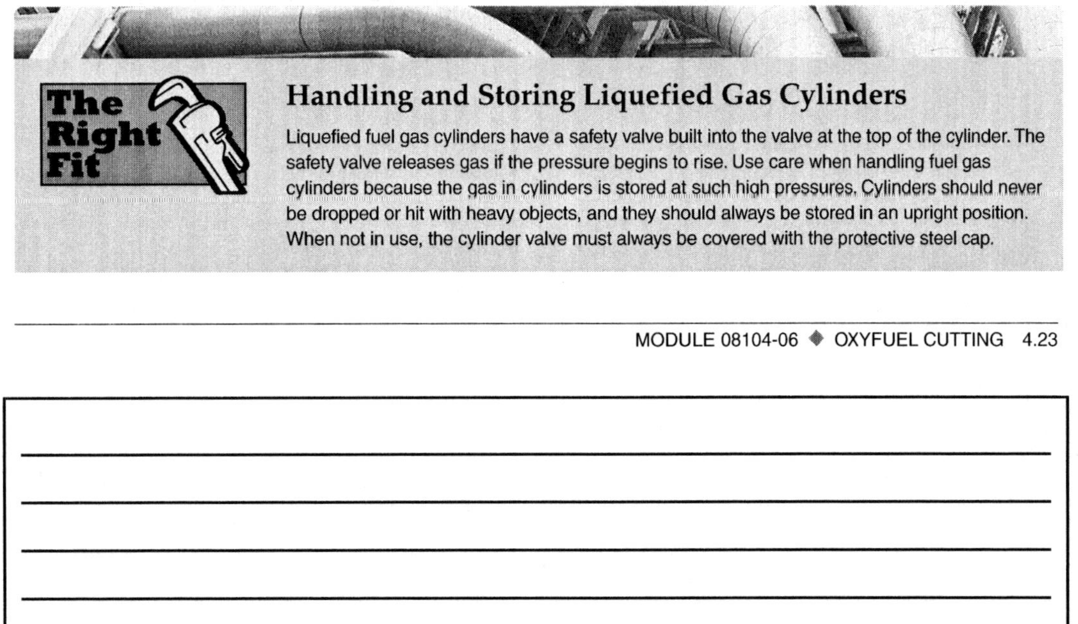

Handling and Storing Liquefied Gas Cylinders

Liquefied fuel gas cylinders have a safety valve built into the valve at the top of the cylinder. The safety valve releases gas if the pressure begins to rise. Use care when handling fuel gas cylinders because the gas in cylinders is stored at such high pressures. Cylinders should never be dropped or hit with heavy objects, and they should always be stored in an upright position. When not in use, the cylinder valve must always be covered with the protective steel cap.

Emphasize the danger of applying heat directly to a cylinder or regulator.

Explain that the weight of the cylinder, not pressure, is the measure of how much liquefied gas is in the cylinder.

Emphasize the danger of removing protective caps from cylinders that are not secured.

Describe a regulator.

Discuss the proper handling and storage of liquefied gas cylinders.

MODULE 08104-06 ◆ OXYFUEL CUTTING 4.23

Explain the purpose and use of regulators.

Identify the guidelines for preventing injury to personnel and damage to regulators.

Compare and contrast two types of regulators.

Discuss the purpose and use of check valves and flashback arrestors.

The pressure adjusting screw controls the gas pressure. Turned clockwise, it increases the flow of gas. Turned counterclockwise, it reduces the flow of gas. When turned counterclockwise until loose (released), it stops the flow of gas.

Most regulators contain two gauges. The high-pressure or cylinder-pressure gauge indicates the actual cylinder pressure; the low-pressure or working-pressure gauge indicates the pressure of the gas leaving the regulator.

Oxygen regulators differ from fuel gas regulators. Oxygen regulators are often painted green and always have right-hand threads on all connections. The oxygen regulator's high-pressure gauge generally reads up to 3,000 psi and includes a second scale that shows the amount of oxygen in the cylinder in terms of cubic feet. The low-pressure or working-pressure gauge may read 100 psi or higher.

Fuel gas regulators are often painted red and always have left-hand threads on all the connections. As a reminder that the regulator has left-hand threads, a V-notch may be cut around the nut. The fuel gas regulator's high-pressure gauge usually reads up to 400 psi. The low-pressure or working-pressure gauge may read up to 40 psi. Acetylene gauges, however, are always red-lined at 15 psi as a reminder that acetylene pressure should not be increased over 15 psi.

There are two types of regulators: single-stage and two-stage.

WARNING!

To prevent injury and damage to regulators, always follow these guidelines:

- Never subject regulators to jarring or shaking, as this can damage the equipment beyond repair.
- Always check that the adjusting screw is fully released before the cylinder valve is turned on and when the welding has been completed.
- Always open cylinder valves slowly and stand on the side of the cylinder opposite the regulator.
- Never use oil to lubricate a regulator. This can result in an explosion when the regulator is in use.
- Never use fuel gas regulators on oxygen cylinders or oxygen regulators on fuel gas cylinders.
- Never work with a defective regulator. If it is not working properly, shut off the gas supply and have the regulator repaired by someone who is qualified to work on it.
- Never use large wrenches, pipe wrenches, pliers, or slipjoint pliers to install or remove regulators.

3.4.1 Single-Stage Regulators

Single-stage, spring-compensated regulators reduce pressure in one step. As gas is drawn from the cylinder, the internal pressure of the cylinder decreases. A single-stage, spring-compensated regulator is unable to automatically adjust for this decrease in internal cylinder pressure. Therefore, it becomes necessary to adjust the spring pressure to periodically raise the output gas pressure as the gas in the cylinder is consumed. These regulators are the most commonly used because of their low cost and high flow rates.

3.4.2 Two-Stage Regulators

The two-stage, pressure-compensated regulator reduces pressure in two steps. It first reduces the input pressure from the cylinder to a predetermined intermediate pressure. The intermediate pressure is then adjusted by the pressure-adjusting screw. With this type of regulator, the delivery pressure to the torch remains constant, and no readjustment is necessary as the gas in the cylinder is consumed. Standard two-stage regulators are more expensive than single-stage regulators and have lower flow rates. There are also heavy-duty types with higher flow rates that are usually preferred for thick material and/or continuous-duty cutting operations.

3.4.3 Check Valves and Flashback Arrestors

Check valves and flashback arrestors (*Figure 29*) are safety devices for regulators, hoses, and torches. Check valves allow gas to flow in one direction only. Flashback arrestors stop fire.

Check valves consist of a ball and spring that open inside a cylinder. The valve allows gas to move in one direction but closes if the gas attempts to flow in the opposite direction. When a torch is first pressurized or when it is being shut off, back-pressure check valves prevent the entry and mixing of acetylene with oxygen in the oxygen hose or the entry and mixing of oxygen with acetylene in the acetylene hose.

Flashback arrestors prevent flashbacks from reaching the hoses and/or regulator. They have a flame-retarding filter that will allow heat, but not flames, to pass through. Most flashback arrestors also contain a check valve.

Add-on check valves and flashback arrestors are designed to be attached either to the torch handle connections or to the regulator outlets. As a minimum, flashback arrestors with check valves should be attached to the torch handle connections. Both devices have arrows on them to

Instructor's Notes:

4.24 PIPEFITTING ◆ LEVEL ONE

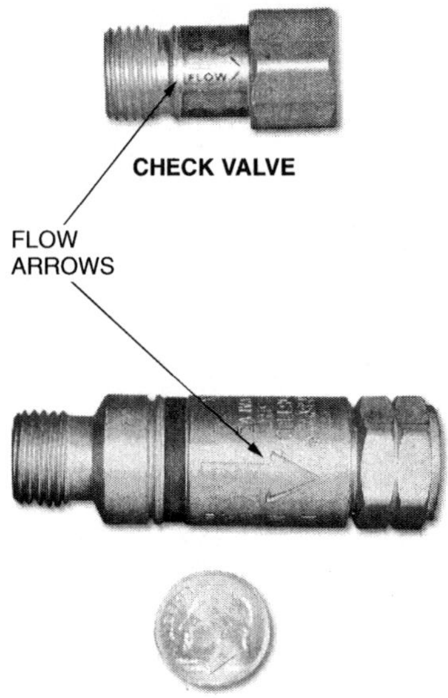

FLASHBACK ARRESTOR WITH INTERNAL CHECK VALVE

104F29.EPS

Figure 29 ◆ Torch handle add-on check valve or flashback arrestor.

indicate flow direction. When installing add-on check valves and flashback arrestors, be sure the arrow matches the desired gas flow direction.

3.5.0 Hoses

Hoses transport gases from the regulators to the torch. Oxygen hoses are usually green or black with right-hand threaded connections. Hoses for fuel gas are usually red and have left-hand threaded connections. The fuel gas connections may also be grooved as a reminder that they have left-hand threads.

Proper care and maintenance of the hose is important for maintaining a safe, efficient work area. Remember the following guidelines for hoses:

- Protect the hose from molten dross or sparks, which will burn the exterior. Although some hoses are flame retardant, they will burn.
- Remove the hoses from under the metal being cut. If the hot metal falls on the hose, the hose will be damaged.
- Frequently inspect and replace hoses that show signs of cuts, burns, worn areas, cracks, or damaged fittings.
- Never use pipe-fitting compounds or lubricants around hose connections. These compounds often contain oil or grease, which ignite and burn or explode in the presence of oxygen.

3.6.0 Cutting Torches

Cutting torches mix oxygen and fuel gas for the torch flame and control the stream of oxygen necessary for the cutting jet. Depending on the job site, you may use either a one-piece or a combination cutting torch.

3.6.1 One-Piece Hand Cutting Torch

The one-piece hand cutting torch, sometimes called a demolition torch, contains the fuel gas and oxygen valves that allow the gases to enter the chambers and then flow into the tip where they are mixed. The main body of the torch is called the handle. The torch valves control the fuel gas and oxygen used for preheating the metal to be cut. The cutting oxygen lever, which is spring-loaded, controls the jet of cutting oxygen. Hose connections are located at the end of the torch body behind the valves. *Figure 30* shows a three-tube one-piece positive-pressure hand cutting torch in which the preheat fuel and oxygen are mixed in the tip. These torches are designed for heavy-duty cutting. They have long supply tubes from the torch handle to the torch head to reduce radiated heat to the operator's hands. The torches are generally available in capacities for cutting steel up to 12 inches thick. Larger-capacity torches, with the ability to cut steel up to 36 inches thick, can also be obtained.

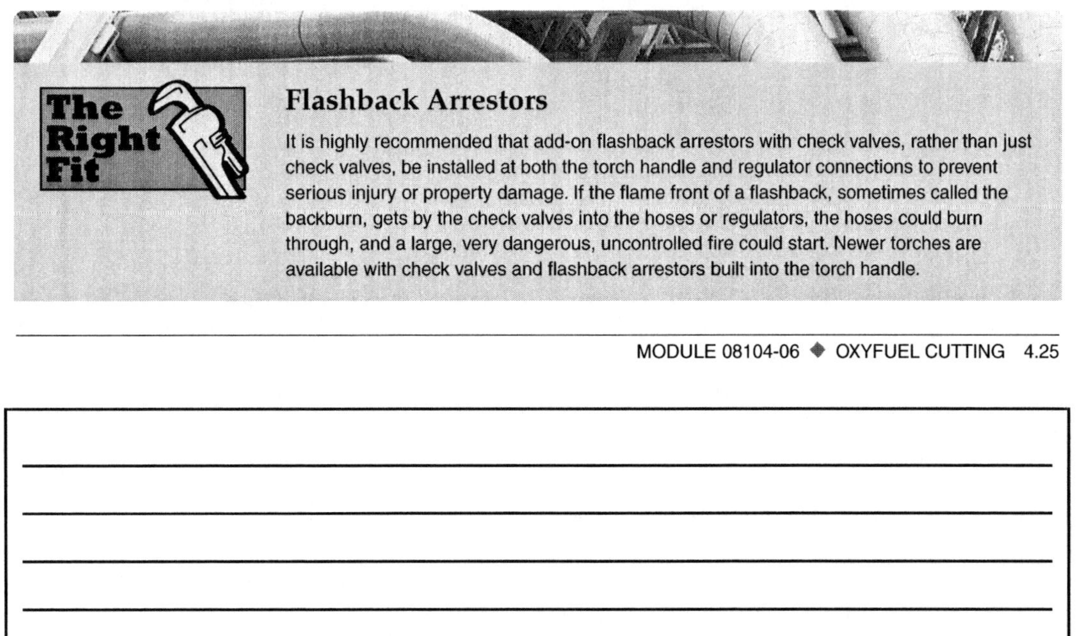

Flashback Arrestors

It is highly recommended that add-on flashback arrestors with check valves, rather than just check valves, be installed at both the torch handle and regulator connections to prevent serious injury or property damage. If the flame front of a flashback, sometimes called the backburn, gets by the check valves into the hoses or regulators, the hoses could burn through, and a large, very dangerous, uncontrolled fire could start. Newer torches are available with check valves and flashback arrestors built into the torch handle.

Describe the purpose and use of hoses.

Explain that there are two types of cutting torches.

Explain that the one-piece hand cutting torch is also called a demolition torch.

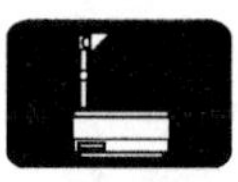

Show Transparency 10 (Figure 30). Describe a one-piece hand cutting torch.

Discuss the applications for a one-piece hand cutting torch.

Provide a one-piece hand cutting torch for the trainees to examine.

Describe a flashback arrestor and explain how they are used.

MODULE 08104-06 ◆ OXYFUEL CUTTING 4.25

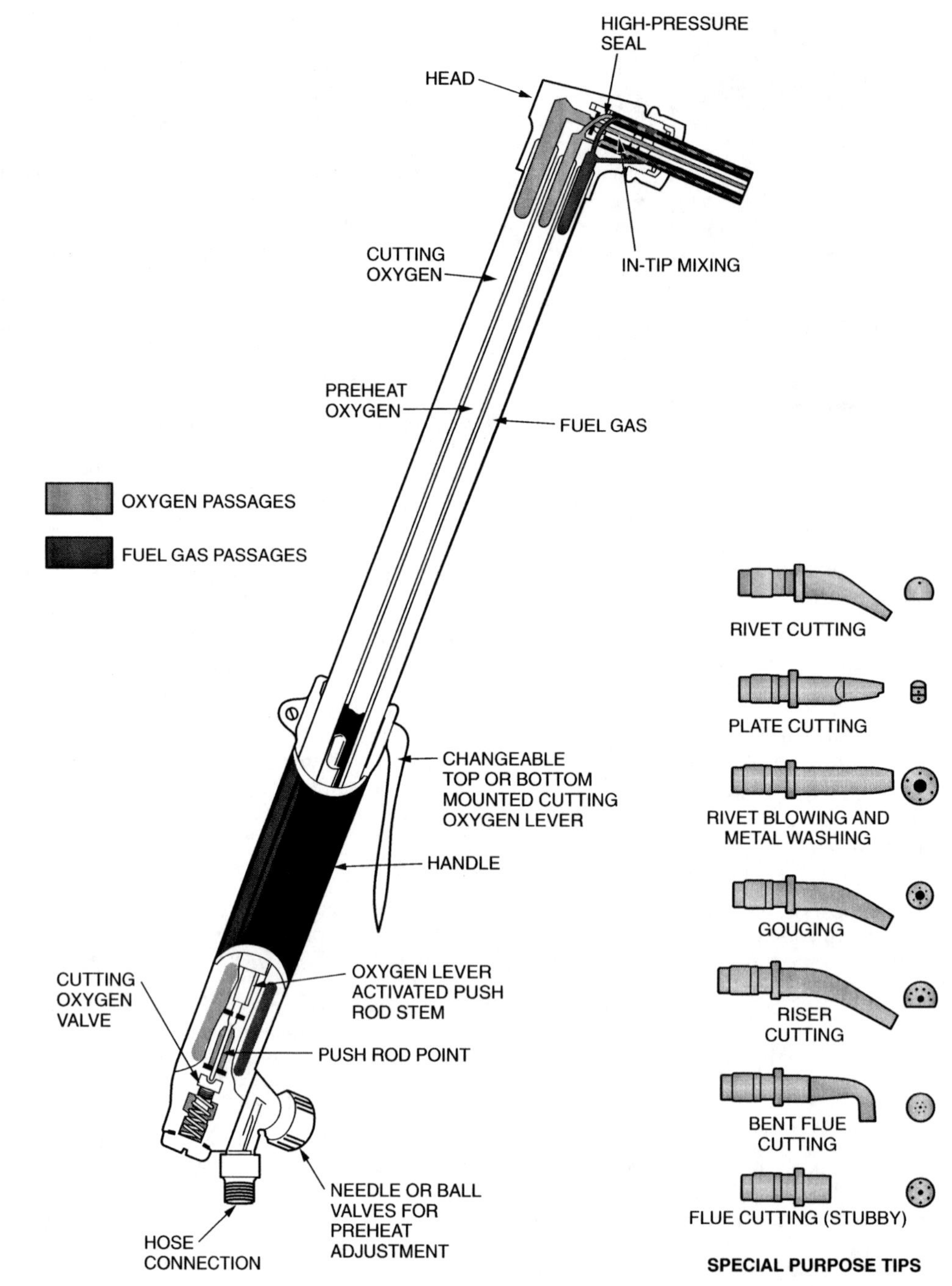

Figure 30 ◆ Heavy-duty three-tube one-piece positive-pressure hand cutting torch.

Instructor's Notes:

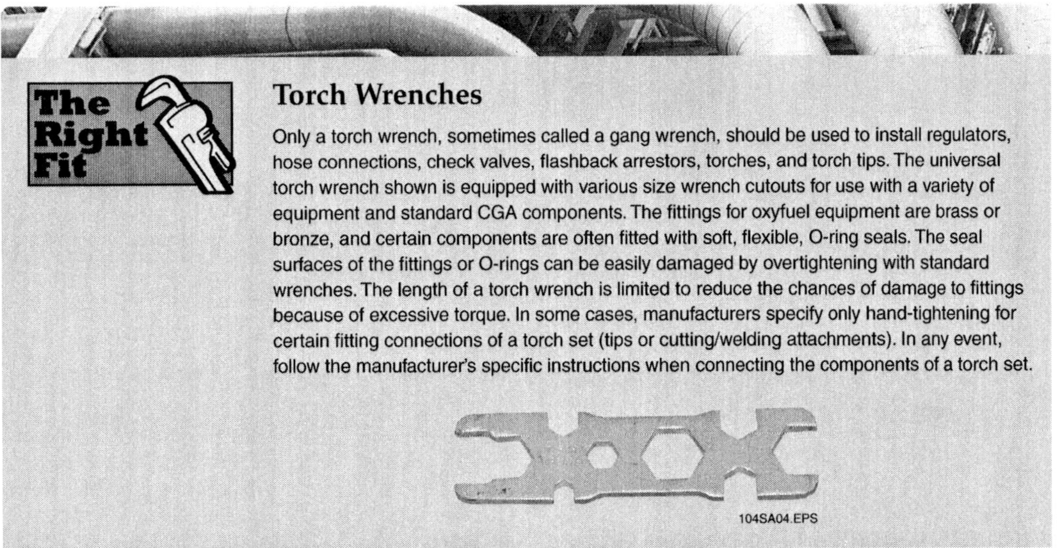

Torch Wrenches

Only a torch wrench, sometimes called a gang wrench, should be used to install regulators, hose connections, check valves, flashback arrestors, torches, and torch tips. The universal torch wrench shown is equipped with various size wrench cutouts for use with a variety of equipment and standard CGA components. The fittings for oxyfuel equipment are brass or bronze, and certain components are often fitted with soft, flexible, O-ring seals. The seal surfaces of the fittings or O-rings can be easily damaged by overtightening with standard wrenches. The length of a torch wrench is limited to reduce the chances of damage to fittings because of excessive torque. In some cases, manufacturers specify only hand-tightening for certain fitting connections of a torch set (tips or cutting/welding attachments). In any event, follow the manufacturer's specific instructions when connecting the components of a torch set.

Compare and contrast the two types of oxyfuel cutting torches in general use.

Compare combination torches to one-piece hand cutting torches.

Two different types of oxyfuel cutting torches are in general use. The positive-pressure torch is designed for use with fuel supplied through a regulator from pressurized fuel storage cylinders. The injector torch is designed to use a vacuum created by oxygen flow to draw the necessary amount of fuel from a very low-pressure fuel source, such as a natural gas line or acetylene generator. The injector torch, when used, is most often found in continuous-duty high-volume manufacturing applications. Both types may employ one of two different fuel-mixing methods:

- Torch-handle or supply-tube mixing
- Torch-head or tip mixing

The two methods can normally be distinguished by the number of supply tubes from the torch handle to the torch head. Torches that use three tubes (see *Figure 30*) from the handle to the head mix the preheat fuel and oxygen at the torch head or tip. This method tends to help eliminate any flashback damage to the torch head supply tubes and torch handle. Torches with two tubes usually mix the preheat fuel and oxygen in a mixing chamber in the torch body or in one of the supply tubes. Injector torches usually have the injector located in one of the supply tubes, and the mixing occurs in that tube from the injector to the torch head. Some older torches that have only two visible tubes are actually three-tube torches that mix the preheat fuel and oxygen in the torch head or tip. This is accomplished by using a separate preheat fuel tube inside a larger preheat oxygen tube.

3.6.2 Combination Torch

The combination torch consists of a cutting torch attachment that fits onto a welding torch handle. These torches are normally used in light-duty or medium-duty applications. Fuel gas and oxygen valves are on the torch handle. The cutting attachment has a cutting oxygen lever and another oxygen valve to control the preheat flame. When the cutting attachment is screwed onto the torch handle, the torch handle oxygen valve is opened all the way, and the preheat oxygen is controlled by an oxygen valve on the cutting attachment. When the cutting attachment is removed, welding and heating tips can be screwed onto the torch handle. *Figure 31* shows a two-tube combination torch in which the preheat mixing is accomplished in a supply tube. These torches are usually positive-pressure torches with mixing occurring in the attachment body, supply tube, head, or tip. These torches are also equipped with built-in flashback arrestors and check valves.

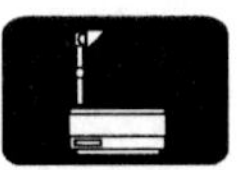

Show Transparency 11 (Figure 32). Describe one- and two-piece cutting torch tips.

3.7.0 Cutting Torch Tips

Cutting torch tips, or nozzles, fit into the cutting torch and are either screwed in or secured with a tip nut. There are one- and two-piece cutting tips (*Figure 32*).

One-piece cutting tips are made from a solid piece of copper. Two-piece cutting tips have a separate external sleeve and internal section.

Torch manufacturers supply literature explaining the appropriate torch tips and gas pressures to

Provide a combination torch and cutting torch tips for the trainees to examine.

MODULE 08104-06 ◆ OXYFUEL CUTTING 4.27

Show Transparency 12 (Table 3). Explain how to use a cutting tip chart to determine tip size and gas pressures.

Warn trainees not to mix cutting tips and cutting tip charts from different manufacturers.

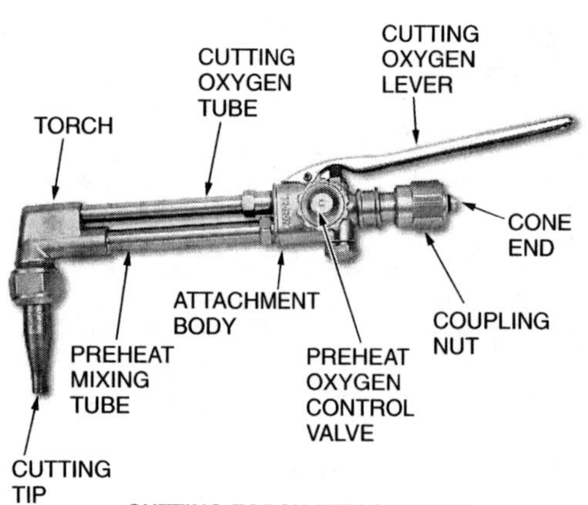

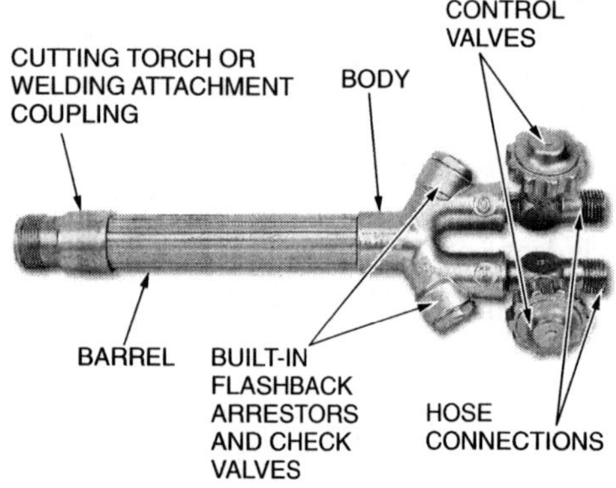

Figure 31 ◆ Typical combination torch.

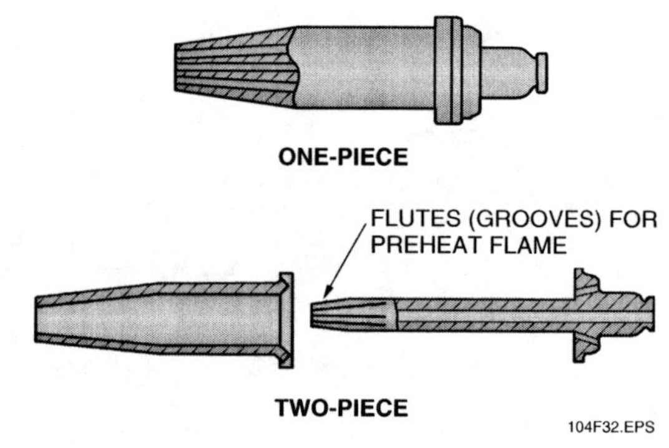

Figure 32 ◆ One- and two-piece cutting tips.

be used for various applications. *Table 3* shows a sample cutting tip chart that lists recommended tip sizes and gas pressures for use with acetylene fuel gas and a specific manufacturer's torch and tips.

The cutting torch tip to be used depends on the base metal thickness and fuel gas being used. Special-purpose tips are also available for use in such operations as gouging and grooving.

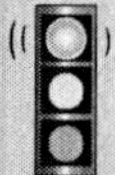

WARNING!

Do not use the cutting tip chart from one manufacturer for the cutting tips of another manufacturer. The gas flow rate of the tips may be different, resulting in excessive flow rates. Different gas pressures may also be required.

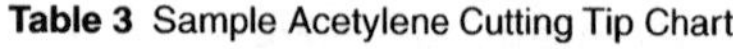

Table 3 Sample Acetylene Cutting Tip Chart

Cutting Tip Series 1-101, 3-101, and 5-101						
Metal Thickness (in)	**Tip Size**	**Cutting Oxygen Pressure (psig)**	**Preheat Oxygen (psig)**	**Acetylene Pressure (psig)**	**Speed (in/min)**	**Kerf Width**
⅛	000	20/25	3/5	3/5	20/30	.04
¼	00	20/25	3/5	3/5	20/28	.05
⅜	0	25/30	3/5	3/5	18/26	.06
½	0	30/35	3/6	3/5	16/22	.06
¾	1	30/35	4/7	3/5	15/20	.07
1	2	35/40	4/8	3/6	13/18	.09
2	3	40/45	5/10	4/8	10/12	.11
3	4	40/50	5/10	5/11	8/10	.12
4	5	45/55	6/12	6/13	6/9	.15
6	6	45/55	6/15	8/14	4/7	.15
10	7	45/55	6/20	10/15	3/5	.34
12	8	45/55	7/25	10/15	3/4	.41

Instructor's Notes:

4.28 PIPEFITTING ◆ LEVEL ONE

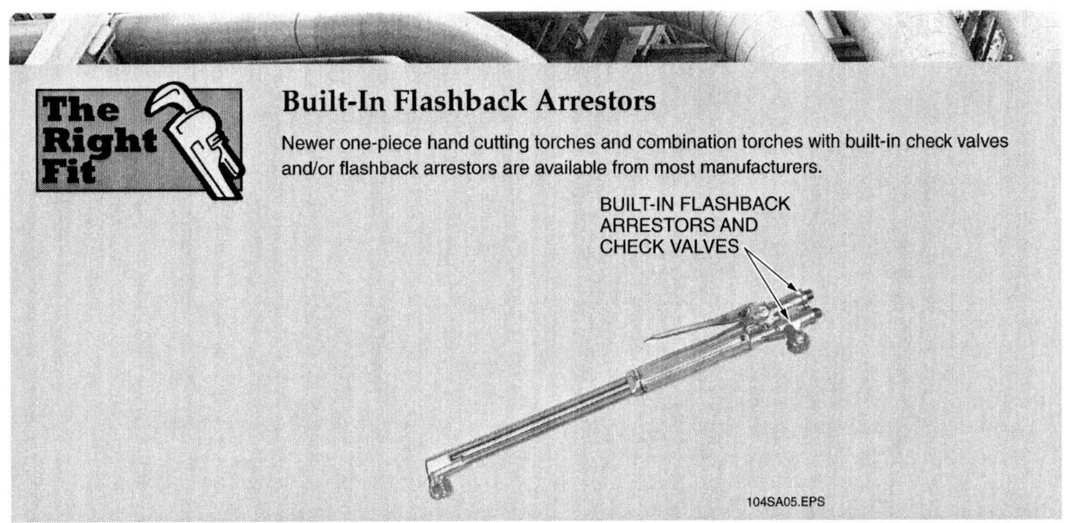

The Right Fit

Built-In Flashback Arrestors

Newer one-piece hand cutting torches and combination torches with built-in check valves and/or flashback arrestors are available from most manufacturers.

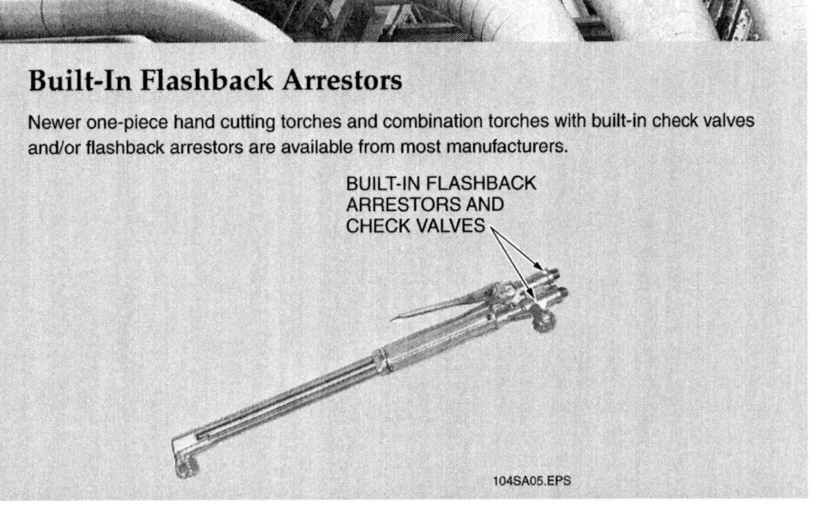

104SA05.EPS

Show Transparencies 13 and 14 (Figures 33 and 34). Describe cutting torch tips used for acetylene and liquefied fuel gas cutting torches.

Provide acetylene and liquefied fuel gas cutting torch tips for the trainees to examine.

3.7.1 Cutting Tips for Acetylene

One-piece torch tips are usually used with acetylene cutting because of the high temperatures involved. They can have four, six, or eight preheat holes in addition to the single cutting hole. *Figure 33* shows typical acetylene torch cutting tips.

3.7.2 Cutting Tips for Liquefied Fuel Gases

Tips used with liquefied fuel gases must have at least six preheat holes. Because fuel gases burn at lower temperatures than acetylene, more holes are necessary for preheating. Tips used with liquefied fuel gases can be one- or two-piece cutting tips. *Figure 34* shows typical cutting tips used with liquefied fuel gases.

The Right Fit

Torch Tip Styles

Nearly all manufacturers use different tip-to-torch mounting designs, sealing surfaces, and diameters. In addition, tip sizes and flow rates are usually not the same between manufacturers even though the number designations may be the same. This makes it impossible to safely interchange cutting tips between torches from different manufacturers. Even though some tips from different manufacturers may appear to be the same, do not interchange them. The sealing surfaces are very precise, and serious leaks may occur that could result in a dangerous fire or flashback.

104SA06.EPS

MODULE 08104-06 ◆ OXYFUEL CUTTING 4.29

Show Transparency 15 (Figure 35). Describe special-purpose cutting torch tips and explain how they are used.

Provide special-purpose cutting torch tips for the trainees to examine.

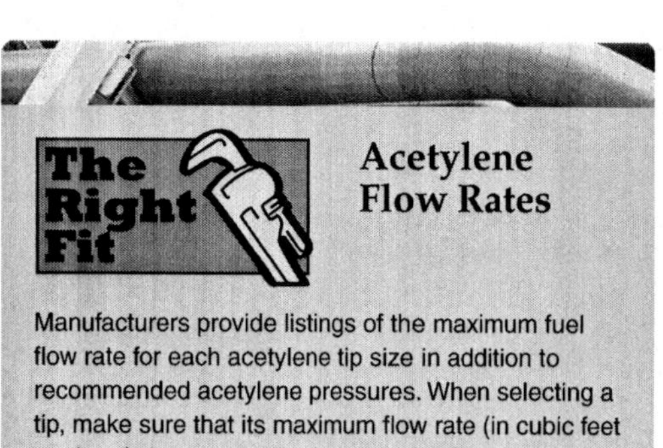

Acetylene Flow Rates

Manufacturers provide listings of the maximum fuel flow rate for each acetylene tip size in addition to recommended acetylene pressures. When selecting a tip, make sure that its maximum flow rate (in cubic feet per hour) does not exceed one-tenth of the total fuel capacity (in cubic feet) for the acetylene cylinder in use. Multiple cylinders must be manifolded together if the flow rate exceeds the cylinder(s) in use in order to prevent withdrawal of acetone along with acetylene.

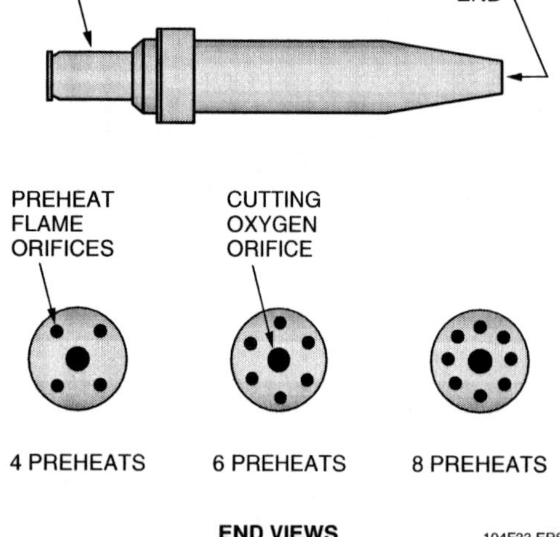

Figure 33 ◆ Typical acetylene torch cutting tips.

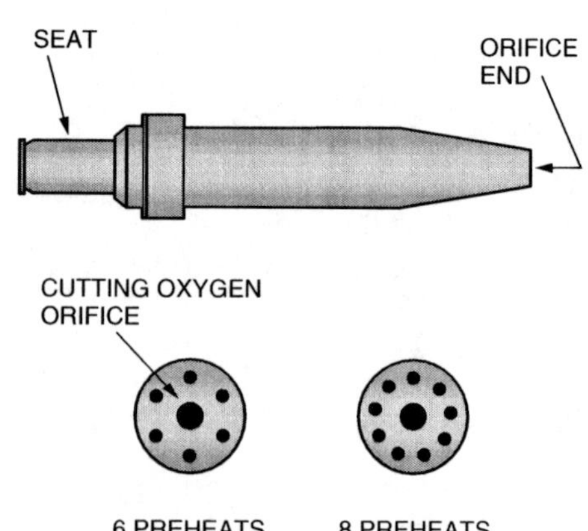

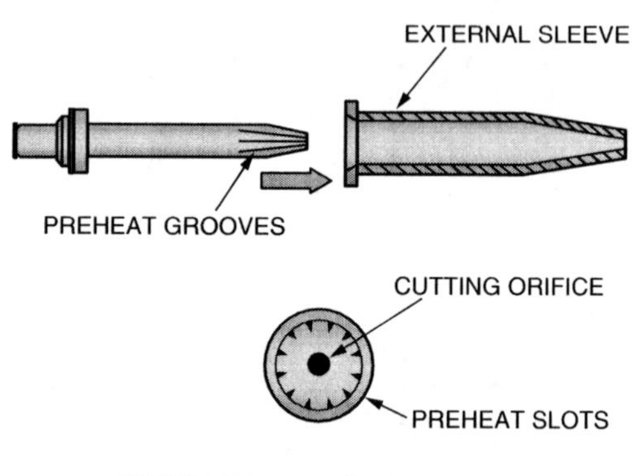

Figure 34 ◆ Typical cutting tips for liquefied fuel gases.

3.7.3 Special-Purpose Tips

Special-purpose tips are available for special cutting jobs. These jobs include cutting sheet metal, rivets, risers, and flues, as well as washing and gouging. *Figure 35* shows special-purpose torch cutting tips.

- The sheet metal cutting tip has only one preheat hole. This minimizes the heat and prevents distortion in the sheet metal. These tips are normally used with a motorized carriage but can also be used for hand cutting.
- Rivet cutting tips are used to cut off rivet heads, bolt heads, and nuts.
- Rosebuds are used to provide large amounts of heat to loosen frozen parts or to expand parts prior to pressing them together for a machine fit. Care must be taken when using a rosebud as they produce much more heat than a standard cutting or welding tip, and as such, the metal being worked will heat up much faster.
- Riser cutting tips are similar to rivet cutting tips and can also be used to cut off rivet heads, bolt heads, and nuts. They have extra preheat holes to cut risers, flanges, or angle legs faster. They can be used for any operation that requires a cut close to and parallel to another surface, such as in removing a metal backing.

Instructor's Notes:

4.30 PIPEFITTING ◆ LEVEL ONE

- Rivet blowing and metal washing tips are heavy-duty tips designed to withstand high heat. They are used for coarse cutting and for removing such items as clips, angles, and brackets.
- Flue cutting tips are designed to cut flues inside boilers. They also can be used for any cutting operation in tight quarters where it is difficult to get a conventional tip into position.

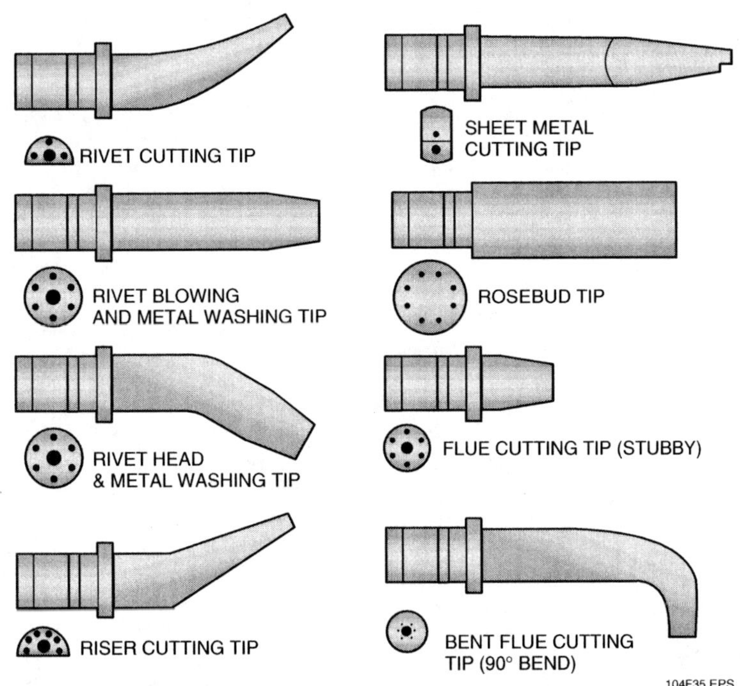

Figure 35 ◆ Special-purpose torch cutting tips.

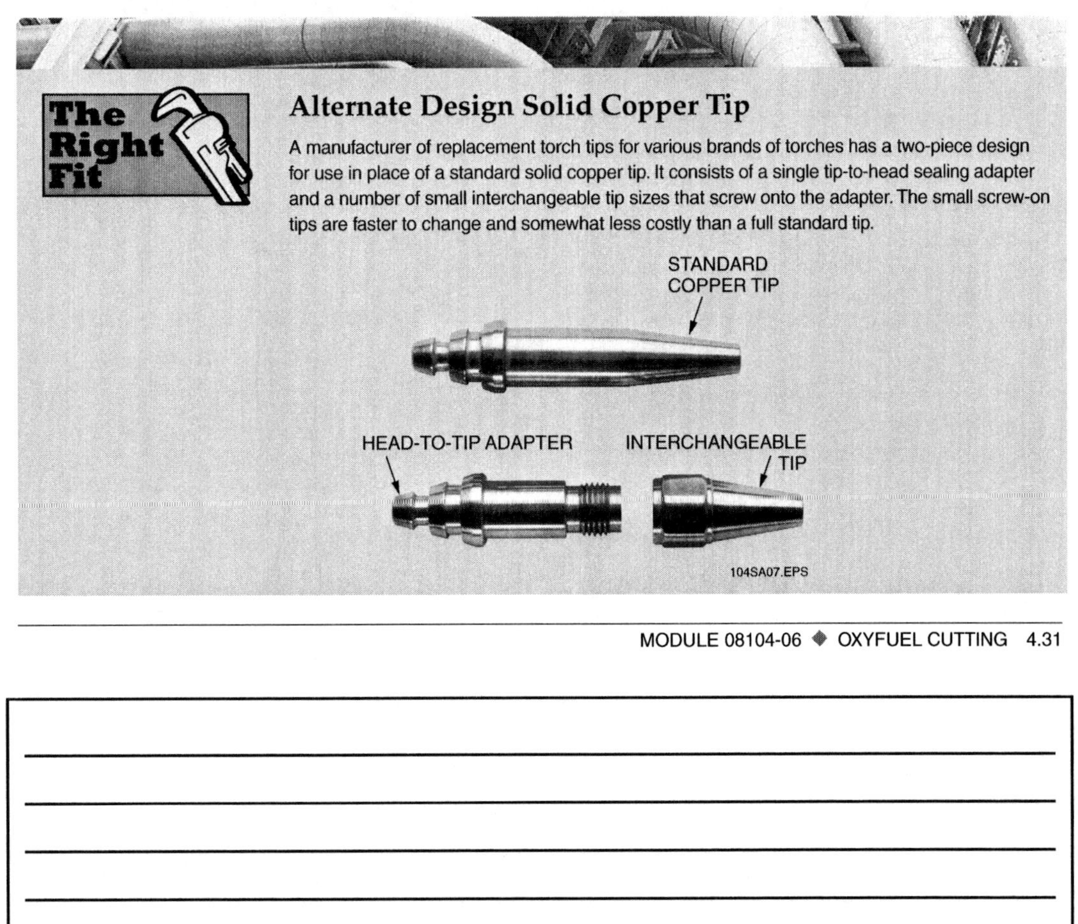

Alternate Design Solid Copper Tip

A manufacturer of replacement torch tips for various brands of torches has a two-piece design for use in place of a standard solid copper tip. It consists of a single tip-to-head sealing adapter and a number of small interchangeable tip sizes that screw onto the adapter. The small screw-on tips are faster to change and somewhat less costly than a full standard tip.

MODULE 08104-06 ◆ OXYFUEL CUTTING 4.31

Explain the purpose of tip cleaners and tip drills. Point out that improper use may damage tips.

Describe a friction lighter and explain that they must be used to light a torch.

Emphasize that matches and gas-filled lighters must never be used to light torches. Point out that friction lighters should always be used.

Show trainees how to use a cup-type striker.

Describe a cylinder cart and explain its use.

Provide tip cleaners, tip drills, a friction lighter, and a cylinder cart for the trainees to examine.

3.8.0 Tip Cleaners and Tip Drills

With use, cutting tips become dirty. Carbon and other impurities build up inside the holes, and molten metal often sprays and sticks onto the surface of the tip. A dirty tip will result in a poor-quality cut with an uneven **kerf** and excessive dross buildup. To ensure good cuts with straight kerfs and minimal dross buildup, clean cutting tips with tip cleaners or tip drills (*Figure 36*).

Tip cleaners are small round files. They usually come in a set with files to match the diameters of the various tip holes. In addition, each set usually includes a file that can be used to lightly recondition the face of the cutting tip. Tip cleaners are inserted into the tip hole and moved back and forth a few times to remove deposits from the hole.

Tip drills are used for major cleaning and for holes that are plugged. Tip drills are tiny drill bits that are sized to match the diameters of tip holes. The drill fits into a drill handle for use. The handle is held, and the drill bit is turned carefully inside the hole to remove debris. They are more brittle than tip cleaners, making them more difficult to use.

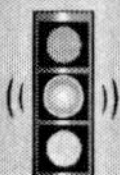

CAUTION

Tip cleaners and tip drills are brittle. If you are not careful, they may break off inside a hole. Broken tip cleaners are difficult to remove. Improper use of tip cleaners or tip drills can enlarge the tip, causing improper burning of gases. If this occurs, tips must be discarded. If the end of the tip has been partially melted or deeply gouged, do not attempt to cut it off or file it flat. The tip should be discarded and replaced with a new tip. This is because some tips have tapered preheat holes, and if a significant amount of metal is removed from the end of the tip, the preheat holes will become too large.

3.9.0 Friction Lighters

Always use a friction lighter (*Figure 37*), also known as a striker or spark-lighter, to ignite the cutting torch. The friction lighter works by rubbing a piece of flint on a steel surface to create sparks.

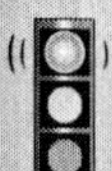

WARNING!

Do not use a match or a gas-filled lighter to light a torch. This could result in severe burns and/or could cause the lighter to explode.

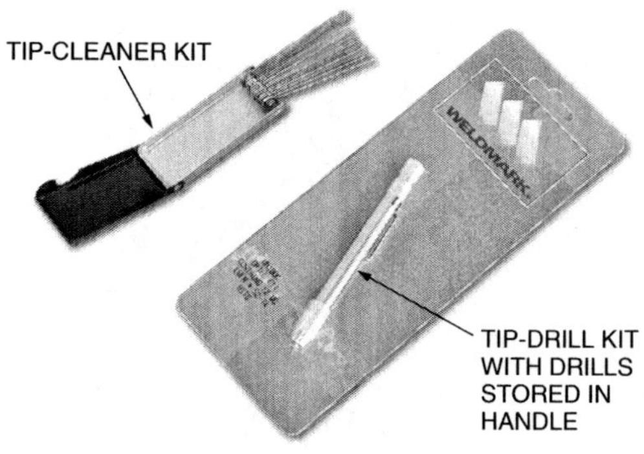

Figure 36 ◆ Tip cleaner and drill kits.

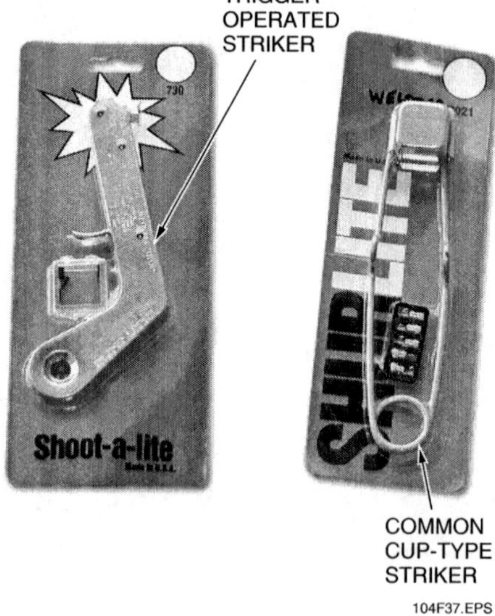

Figure 37 ◆ Typical friction lighters.

3.10.0 Cylinder Cart

The cylinder cart, or bottle cart, is a modified hand truck that has been equipped with seats and chains to hold cylinders firmly in place. Bottle carts help ensure the safe transportation of gas cylinders. *Figure 38* shows a typical cylinder cart for a welding/cutting rig. Some carts are equipped with tool/accessory trays or boxes as well as rod holders.

Instructor's Notes:

Show the trainees a soapstone and explain how to use it.

Provide a soapstone marker for the trainees to examine.

Cup-Type Striker

When using a cup-type striker to ignite a welding torch, hold the cup of the striker slightly below and to the side of the tip, parallel with the fuel gas stream from the tip. This prevents the ignited gas from deflecting back toward you from the cup and reduces the amount of carbon soot in the cup. Note that the flint in a striker can be replaced.

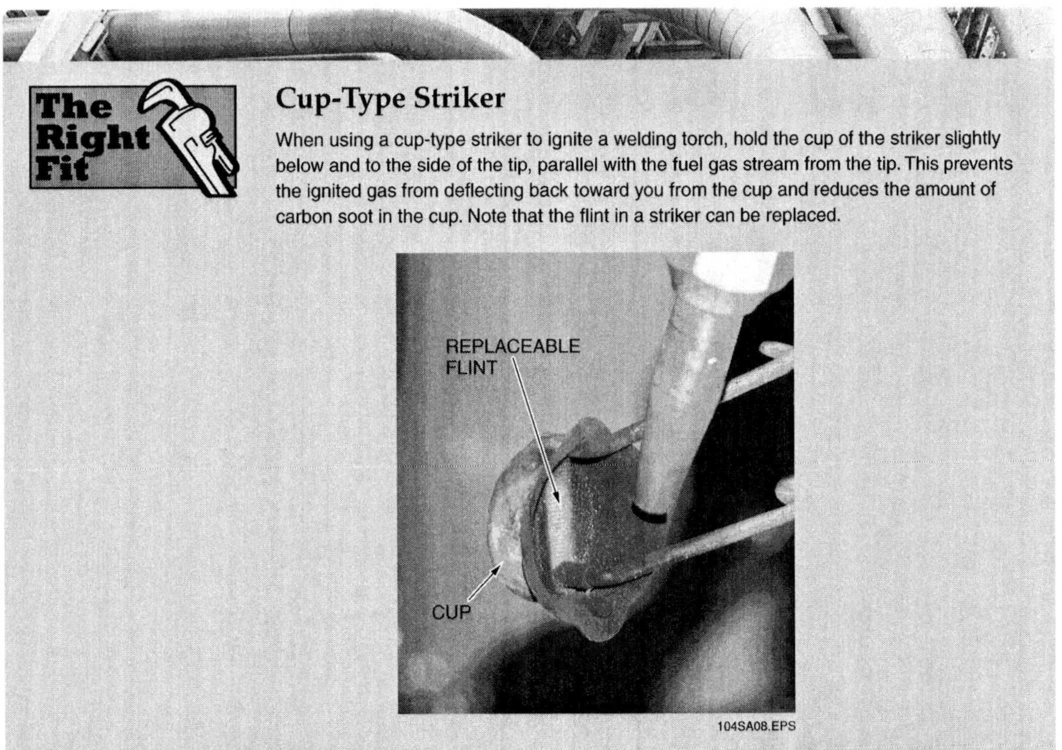

Figure 38 ◆ Typical cutting/welding rig cylinder cart.

3.11.0 Soapstone Markers

Because of the heat involved in welding and cutting operations, along with the tinted lenses that are required, ordinary pen or pencil marking for cutting lines or welding locations is not effective. The oldest and most common material used for marking is **soapstone** in the form of sticks or cylinders (*Figure 39*). Soapstone is soft and feels greasy and slippery. It is actually steatite, a dense, impure form of talc that is heat resistant. It also shows up well through a tinted lens under the illumination of an electric arc or gas welding/cutting flame. Some welders prefer to use silver-graphite pencils for marking dark materials and red-graphite pencils for aluminum or other bright metals. Graphite is also highly heat resistant (*Figure 39*). A few manufacturers also market heat-resistant paint/dye markers for welding.

3.12.0 Specialized Cutting Equipment

In addition to the common hand cutting torches, other types of equipment are used in oxyfuel

MODULE 08104-06 ◆ OXYFUEL CUTTING 4.33

Show trainees how to sharpen a soapstone with a penknife.

Describe the use of a mechanical guide.

Provide a mechanical guide for the trainees to examine.

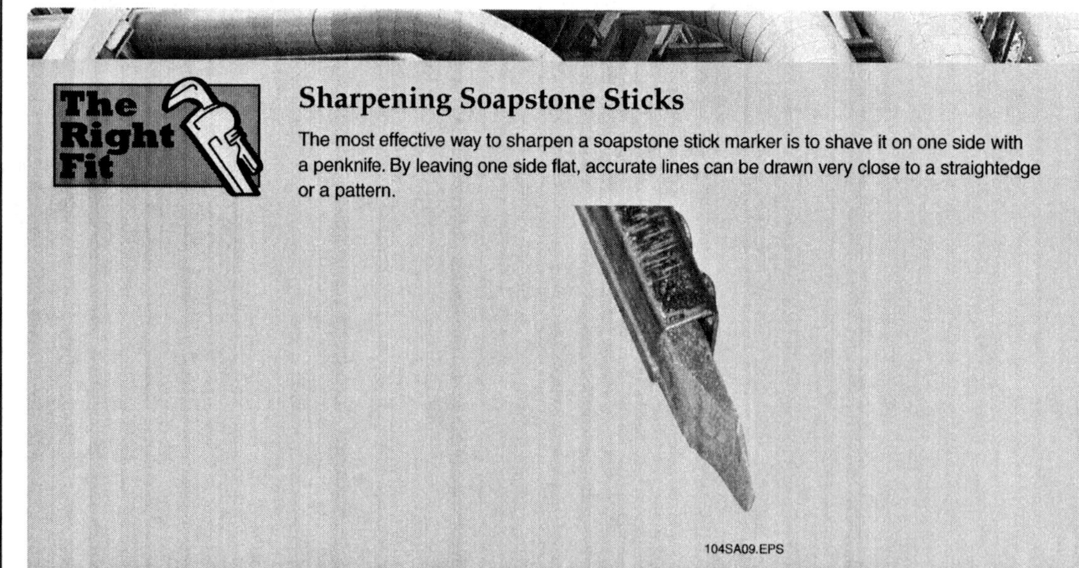

Sharpening Soapstone Sticks

The most effective way to sharpen a soapstone stick marker is to shave it on one side with a penknife. By leaving one side flat, accurate lines can be drawn very close to a straightedge or a pattern.

104SA09.EPS

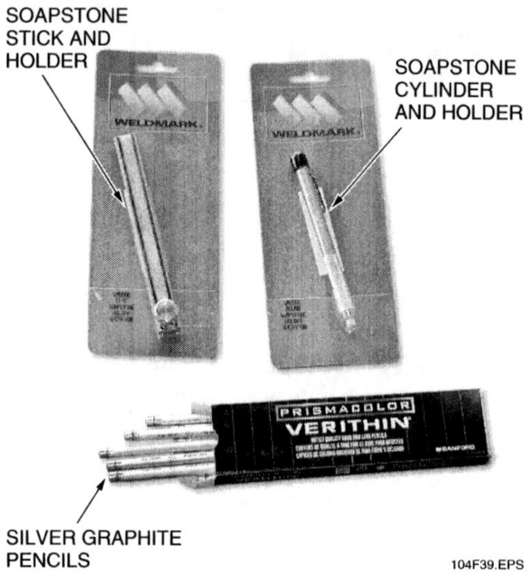

Figure 39 ◆ Typical soapstone and graphite markers.

cutting applications. This equipment includes mechanical guides used with a hand cutting torch, various types of motorized cutting machines, and oxygen lances. All of the motorized units use special straight body machine cutting or welding torches with a gear rack attached to the torch body to set the tip distance from the work.

3.12.1 Mechanical Guides

On long, circular, or irregular cuts, it is very difficult to control and maintain an even kerf with a hand cutting torch. Mechanical guides can help maintain an accurate and smooth kerf along the cutting line. For straight line or curved cuts, use a one- or two-wheeled accessory that clamps on the torch tip in a fixed position. The wheeled accessory maintains the proper tip distance while the tip is guided by hand along the cutting line. The fixed, two-wheeled accessory is similar to the rotating-mount wheeled unit used for a circle cutter but without the radius bar (*Figure 40*).

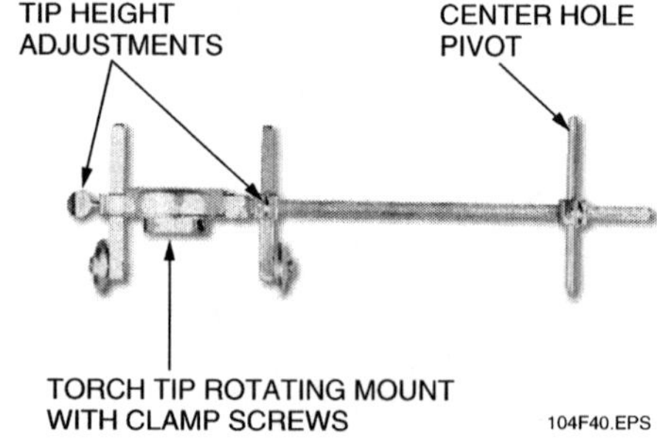

Figure 40 ◆ Circle cutting accessory.

Perform arc or circular cuts with the circle cutting accessory shown in *Figure 40*. The torch tip fits through and is secured to a rotating mount

Instructor's Notes:

4.34 PIPEFITTING ◆ LEVEL ONE

between the two small metal wheels. The wheel heights are adjustable so that the tip distance from the work can be set. The radius of the circle is set by moving the pivot point on a radius bar. After a starting hole is cut (if needed), the torch tip is placed through and secured to the circle cutter rotating mount. Then the pivot point is placed in a drilled hole or a magnetic holder at the center of the circle. When the cut is restarted, the torch can be moved in a circle around the cut, guided by the circle cutter.

When large work with an irregular pattern must be cut, a template is often used. The torch is drawn around the edges of the template to trace the pattern as the cut is made. If multiple copies must be cut, a metal pattern, held in place and spaced for tip distance from the work by stacked magnets, is usually used. For a one- or two-time copy, a heavily weighted Masonite or aluminum template that is spaced off the workpiece could be carefully used and discarded.

3.12.2 Motor-Driven Equipment

A variety of fixed and portable motorized cutting equipment is available for straight and curved cutting/welding. The computer-controlled gantry cutting machine (*Figure 41*) and the optical pattern-tracing machine (*Figure 42*) are fixed-location machines used in industrial manufacturing applications. The computer-controlled machine can be programmed to **pierce** and then cut any pattern from flat metal stock. The optical pattern-tracing machine follows lines on a drawing using a light beam and an optical detector. They both can be rigged to cut multiple items using multiple torches operated in parallel. Both units have a motor-driven gantry that travels the length of a table and a transverse motor-driven beam or torch head that moves back and forth across the table. Both units are also equipped to use both oxyfuel cutting and plasma cutting torches. The size of the patterns can also be adjusted.

104F41.EPS

Figure 41 ◆ Computer-controlled gantry cutting machine.

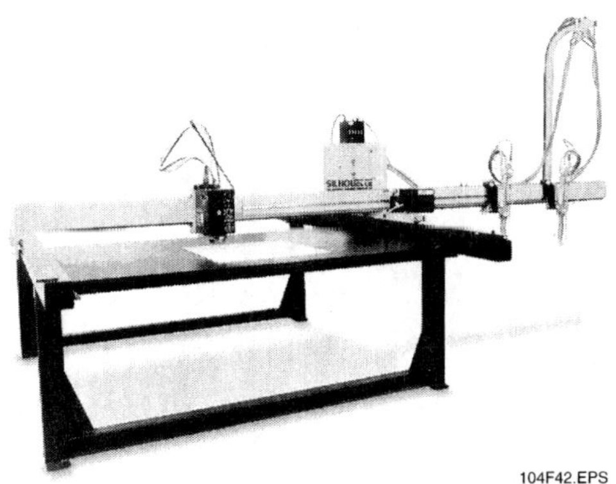

104F42.EPS

Figure 42 ◆ Optical pattern-tracing machine.

Explain how to make long, circular, or irregular cuts using a mechanical guide.

Discuss shop-made cutting guides.

Discuss the various motor-driven cutting machines available.

Compare and contrast computer-controlled gantry cutting machines and optical pattern-tracing machines.

See the Teaching Tip for Section 3.12.0 at the end of this module.

The Right Fit

Shop-Made Straight Line Cutting Guide

A simple solution for straight line cutting is to clamp a piece of angle iron to the work and use a band clamp around the cutting torch tip to maintain the cutting tip distance from the work. When the cut is started, the band clamp rests on the top of the vertical leg of the angle iron, and the torch is drawn along the length of the angle iron at the correct cutting speed.

104SA10.EPS

MODULE 08104-06 ◆ OXYFUEL CUTTING 4.35

Compare and contrast portable track cutting machines and portable motor-driven and hand-cranked ring gear cutters and bevelers.

Other types of pattern-tracing machines use metal templates that are clamped in the machine. A follower wheel traces the pattern from the template. The pattern size can be increased or decreased by electrical or mechanical linkage to a moveable arm holding one or more cutting torches that cut the pattern from flat metal stock.

Portable track cutting machines (track burners) can be used in the field for straight or curved cutting and beveling. *Figures 43* and *44* show units driven by a variable-speed motor. The unit shown in *Figure 43* is available with track extensions for any length of straight cutting or beveling, along with a circle cutting attachment. The unit shown in *Figure 44* uses a somewhat flexible magnetic track for both flat straight-line or large-diameter object cutting or beveling. It is shown equipped with an optional plasma machine torch. Both units can be adapted to metal inert gas (MIG) or tungsten inert gas (TIG) welding.

A portable, motor-driven band track or hand-cranked ring gear cutter/beveler can be set up in the field for cutting and beveling pipe with oxyfuel or plasma machine torches (*Figures 45* and *46*). The stainless steel band track cutter uses a chain and motor sprocket drive to rotate the machine cutting torch around the pipe a full 360 degrees.

104F43.EPS

Figure 43 ◆ Track burner with an oxyfuel machine torch.

104F44.EPS

Figure 44 ◆ Track burner with a plasma machine torch.

104F45.EPS

Figure 45 ◆ Band track pipe cutter/beveler.

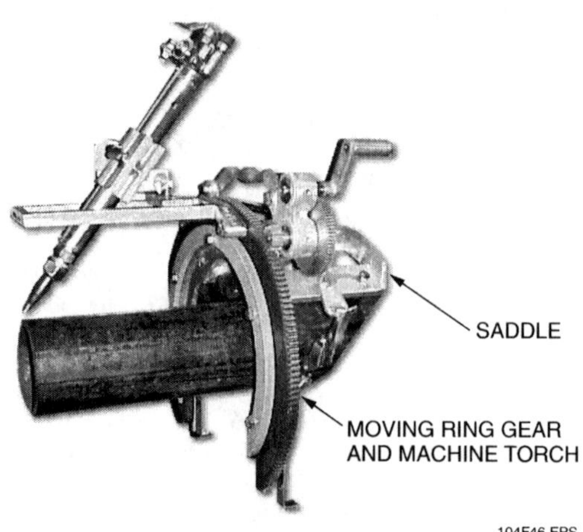

104F46.EPS

Figure 46 ◆ Ring gear pipe cutter/beveler.

Instructor's Notes:

4.36 PIPEFITTING ◆ LEVEL ONE

Describe the operation of exothermic oxygen lances.

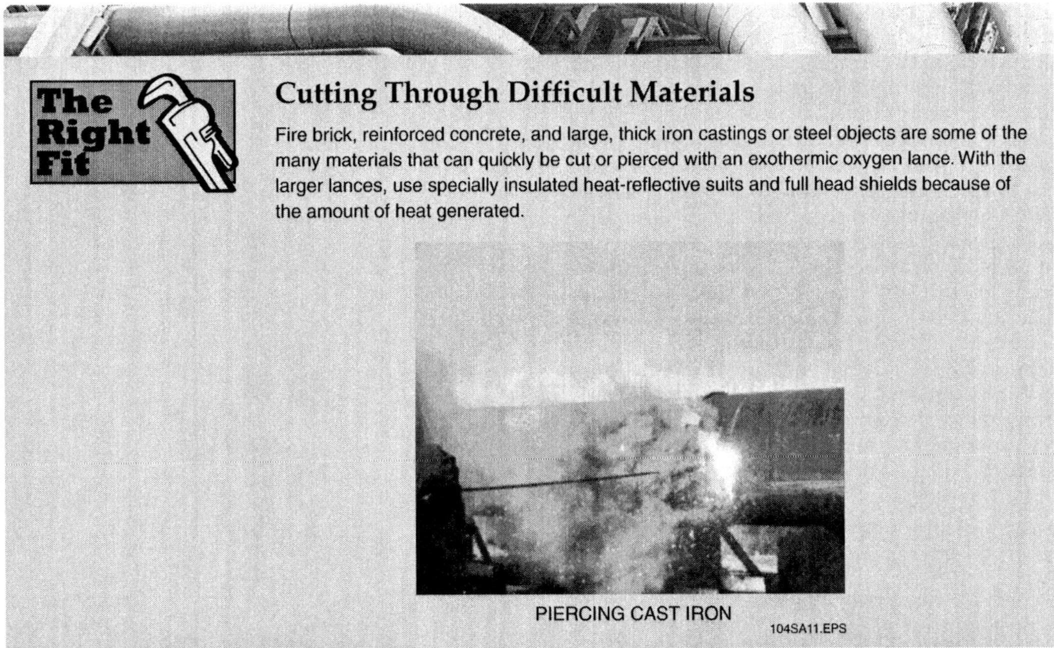

Cutting Through Difficult Materials

Fire brick, reinforced concrete, and large, thick iron castings or steel objects are some of the many materials that can quickly be cut or pierced with an exothermic oxygen lance. With the larger lances, use specially insulated heat-reflective suits and full head shields because of the amount of heat generated.

PIERCING CAST IRON

104SA11.EPS

The all-aluminum ring gear type of cutter/ beveler is positioned on the pipe, and then the saddle is clamped in place. In operation, the ring gear and the cutting torch rotate at different rates around the saddle for a full 360-degree cut.

A hand-guided oxyfuel cutting torch with an integral-precision, variable-speed motor drive (*Figure 47*), which can be used for straight line and curved cutting to achieve machine-quality cuts, is available. A circle cutting accessory is also available for the unit.

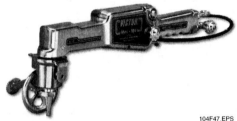

104F47.EPS

Figure 47 ◆ Hand-guided motorized oxyfuel cutting torch.

3.12.3 Exothermic Oxygen Lances

Exothermic (combustible) oxygen lances are a special oxyfuel cutting tool usually used in heavy industrial applications and demolition work. The lance is a steel pipe that contains magnesium- and aluminum-cored powder or rods (fuel). In operation, the lance is clamped into a holder (*Figure 48*) that seals the lance to a fitting that supplies oxygen to the lance through a hose at pressures of 75 to 80 psi. With the oxygen turned on, the end of the lance is ignited with an acetylene torch or flare. As long as the oxygen is applied, the lance will burn and consume itself. The oxygen-fed flame of the burning magnesium, aluminum, and steel pipe creates temperatures approaching 10,000°F. At this temperature, the lance will rapidly cut or pierce any material, including steel, metal alloys, and cast iron, even under water.

OXYGEN INLET AND VALVE

CLAMPING ASSEMBLY

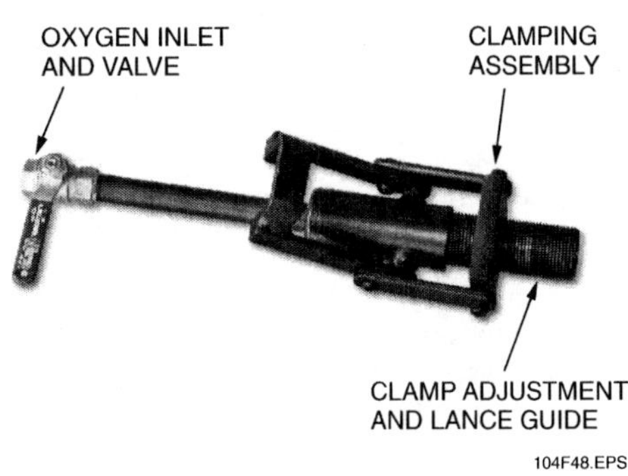

CLAMP ADJUSTMENT AND LANCE GUIDE

104F48.EPS

Figure 48 ◆ Oxygen lance holder.

MODULE 08104-06 ◆ OXYFUEL CUTTING 4.37

Explain the uses of exothermic oxygen lances.

Have trainees review Sections 4.0.0–4.12.3.

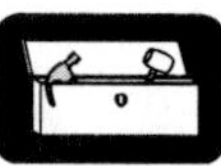

Ensure that you have everything required for teaching this session.

Explain that specific procedures must be followed to set up oxyfuel cutting equipment.

Review the procedure for transporting cylinders.

Show trainees how to transport cylinders safely.

Emphasize the need to handle cylinders carefully.

The lances for the holder shown in *Figure 48* are 10 feet long and range in size from ⅜ to 1 inch in diameter. The larger sizes can be coupled to obtain a longer lance. A small pistol-grip heat-shielded unit that can, if desired, be used with an electric welder is also available. The arc, combined with an oxygen lance, can create temperatures exceeding 10,000°F. This small unit uses lances from ¼ to ⅜ inch in diameter that are 22 to 36 inches long and that cut very rapidly at a maximum burning time of 60 to 70 seconds. The small unit is primarily used to burn out large frozen pins and frozen headless bolts or rivets. Like a large lance, it can be used to cut any material, including concrete-lined pipe. It also can be used to remove hard-surfacing material that has been applied to wear surfaces. Both units are relatively inexpensive and can be set up in the field with only an oxygen cylinder, hose, and ignition device.

4.0.0 ◆ SETTING UP OXYFUEL EQUIPMENT

When setting up oxyfuel equipment, you must follow certain procedures to ensure that the equipment operates properly and safely. The following sections explain the procedures for setting up oxyfuel equipment.

4.1.0 Transporting and Securing Cylinders

Follow these steps to transport and secure cylinders:

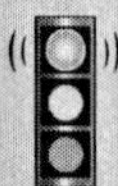

WARNING!

Always handle cylinders with care. They are under high pressure and should never be dropped, knocked over, or exposed to excessive heat. When moving cylinders, always be certain that the valve caps are in place. Use a cylinder cage to lift cylinders. Never use a sling or electromagnet.

Step 1 Transport cylinders to the workstation in the upright position on a hand truck or bottle cart (*Figure 49*).

Step 2 Secure the cylinders at the workstation.

Step 3 Remove the protective cap from each cylinder and inspect the outlet nozzles to ensure that the seat and threads are not damaged. Place the protective caps where they will not be lost and where they will be available when the cylinders are empty.

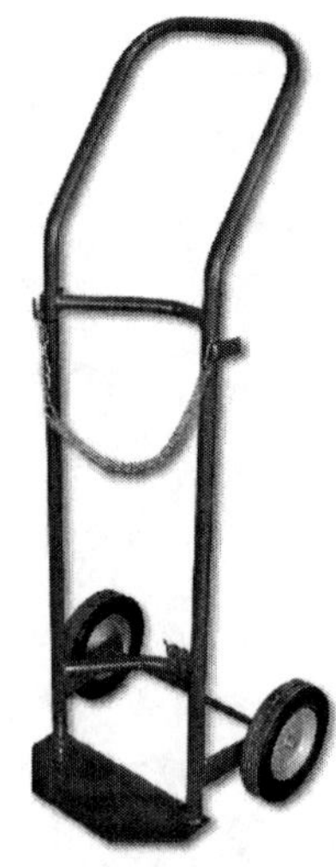

SINGLE-CYLINDER HAND CART

HEAVY-DUTY TWIN-CYLINDER HAND CART

104F49.EPS

Figure 49 ◆ Carts for transporting cylinders.

CAUTION

Do not transport or immediately use an acetylene cylinder found resting on its side. Stand it upright and wait at least 30 minutes to allow the acetone to settle before using it.

Instructor's Notes:

4.38 PIPEFITTING ◆ LEVEL ONE

4.2.0 Cracking Cylinder Valves

Follow these steps to crack cylinder valves:

Step 1 Crack open the cylinder valve momentarily to remove any dirt from the valves (*Figure 50*).

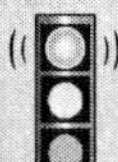

WARNING!
Always stand to one side of the valves when opening them to avoid injury from dirt that may be lodged in the valve.

Figure 50 ◆ Cracking a cylinder valve.

Step 2 Wipe out the connection seat of the valves with a clean cloth. Dirt frequently collects in the outlet nozzle of a cylinder valve and must be cleaned out to keep it from entering the regulator when pressure is turned on.

WARNING!
Be sure the cloth used does not have any oil or grease on it. Oil or grease mixed compressed oxygen will cause an explosion.

4.3.0 Attaching Regulators

Follow these steps to attach the regulators:

Step 1 Check that the regulator is closed (adjustment screw loose).

Step 2 Check the regulator fittings to ensure that they are free of oil and grease (*Figure 51*).

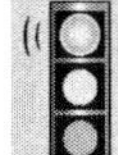

WARNING!
Do not work with a regulator that shows signs of damage, such as cracked gauges, bent thumbscrews, or worn threads. Set it aside for repairs. Never attempt to repair regulators yourself. When tightening the connections, always use a torch wrench.

Figure 51 ◆ Checking connection fittings.

Step 3 Connect and tighten the oxygen regulator to the oxygen cylinder using a torch wrench (*Figure 52*).

Step 4 Connect and tighten the fuel gas regulator to the fuel gas cylinder. Remember that all fuel gas fittings have left-hand threads.

Step 5 Crack the cylinder valve slightly and open the regulator to expel any debris from the outlet. Shut the cylinder valve and close the regulator (*Figure 53*).

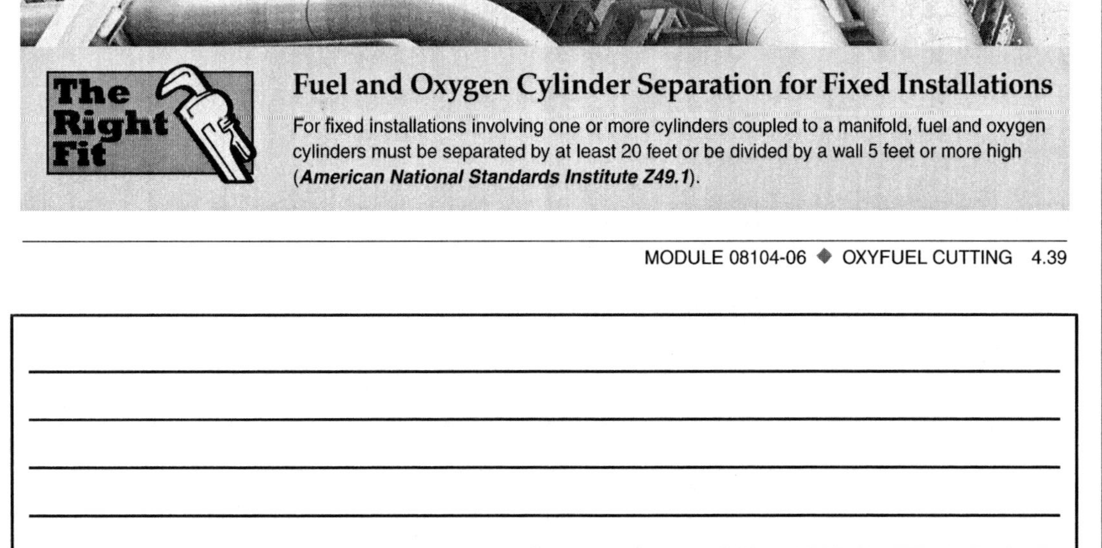

Fuel and Oxygen Cylinder Separation for Fixed Installations

For fixed installations involving one or more cylinders coupled to a manifold, fuel and oxygen cylinders must be separated by at least 20 feet or be divided by a wall 5 feet or more high (***American National Standards Institute Z49.1***).

Review the procedure for cracking the cylinder valve.

Warn trainees to stand aside when opening valves.

Point out that cloths used to wipe valve seats must be free of oil and grease.

Review the procedure for attaching a regulator.

Show trainees how to crack the cylinder valve and attach a regulator.

Warn trainees not to work with damaged regulators.

MODULE 08104-06 ◆ OXYFUEL CUTTING 4.39

Review the procedure for installing a flash-back arrestor or check valve.

Show trainees how to install a flashback ar-restor or check valve.

Discuss the dangers of flashback.

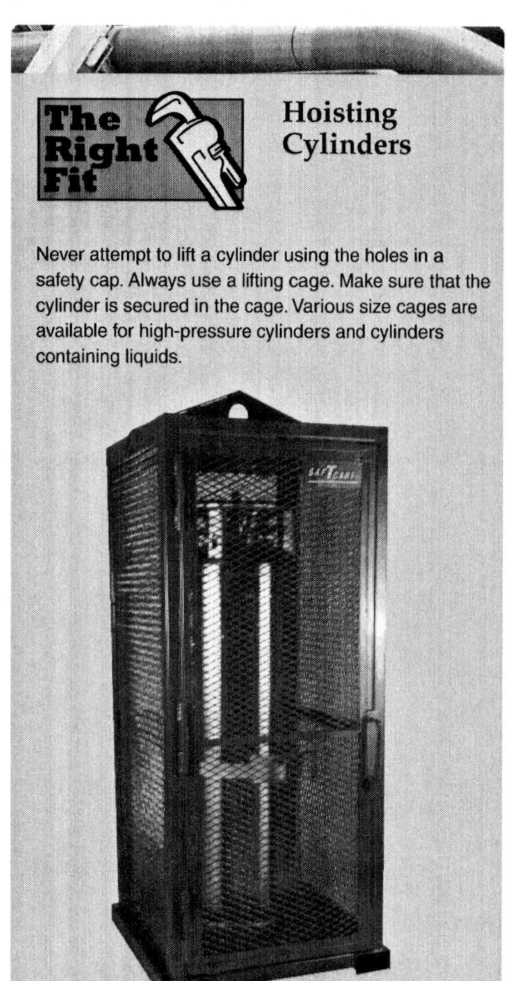

Hoisting Cylinders

Never attempt to lift a cylinder using the holes in a safety cap. Always use a lifting cage. Make sure that the cylinder is secured in the cage. Various size cages are available for high-pressure cylinders and cylinders containing liquids.

104SA12.EPS

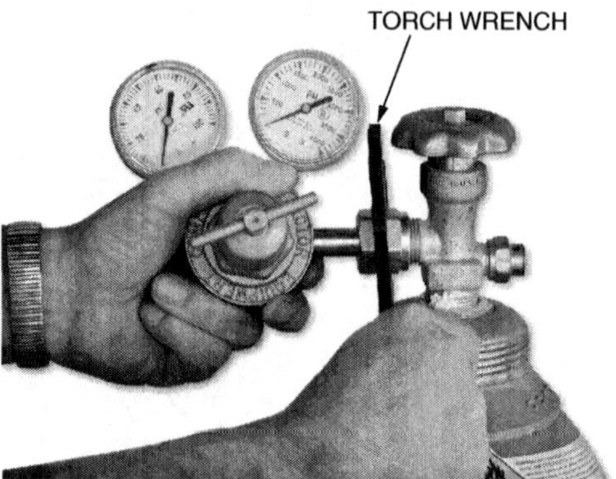

104F52.EPS

Figure 52 ◆ Tightening regulator connection.

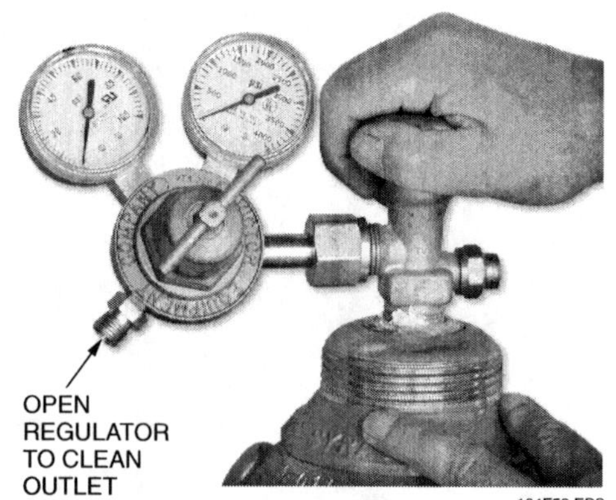

104F53.EPS

Figure 53 ◆ Cleaning the regulator.

4.4.0 Installing Flashback Arrestors or Check Valves

Follow these steps to install flashback arrestors or check valves:

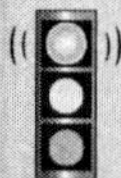

WARNING!

At least one flashback arrestor must be used with each hose. The absence of a flashback arrestor could result in flashback. Flashback arrestors can be attached either to the regulator, the torch, or both; however, flashback arrestors installed at the torch handle are preferred if only one is being used.

Step 1 Attach a flashback arrestor or check valve to the hose connection on the oxygen regulator (*Figure 54*) and tighten with a torch wrench.

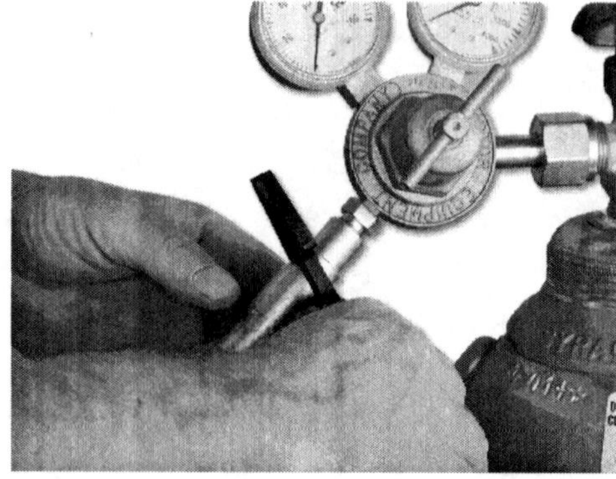

104F54.EPS

Figure 54 ◆ Attaching a flashback arrestor.

Instructor's Notes:

Step 2 Attach and tighten a flashback arrestor or check valve to the hose connection on the fuel gas regulator. Keep in mind that all fuel gas fittings have left-hand threads.

4.5.0 Connecting Hoses to Regulators

New hoses contain talc and loose bits of rubber. These materials must be blown out of the hoses using an inert gas such as nitrogen or argon before the torch is connected. If they are not blown out, they will clog the torch needle valves.

> **WARNING!**
> Never blow out hoses with compressed air, fuel gas, or oxygen. Compressed air often contains some oil that could explode or cause a fire when compressed in the hose with oxygen. Using fuel gas or oxygen creates a fire and explosion hazard.
>
> Check that used hoses are not cracked, cut, damaged, or contaminated with oil or grease. Replace the hoses if these conditions exist.

Follow these steps to connect the hoses to the regulators:

Step 1 Inspect both the oxygen and fuel gas hoses for any damage, burns, cuts, or fraying.

Step 2 Replace any damaged hoses.

Step 3 Connect the green oxygen hose to the oxygen regulator flashback arrestor or check valve (*Figure 55*).

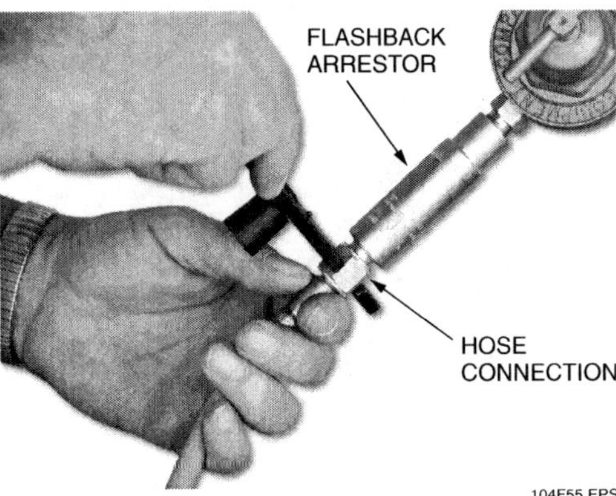

Figure 55 ◆ Connecting hose to regulator flashback arrestor.

Step 4 Connect the red or black fuel gas hose to the fuel gas regulator flashback arrestor or check valve. Keep in mind that all fuel gas fittings have left-hand threads.

4.6.0 Attaching Hoses to the Torch

Follow these steps to attach the hoses to the torch:

Step 1 Attach flashback arrestors to the oxygen and fuel gas hose connections on the torch body unless the torch has built-in flashback arrestors and check valves. Keep in mind that all fuel gas fittings have left-hand threads.

Step 2 Attach and tighten the green oxygen hose to the oxygen fitting on the flashback arrestor or torch (*Figure 56*).

Step 3 Attach and tighten the red or black hose to the fuel gas fitting on the flashback arrestor or torch. Remember that all fuel gas fittings have left-hand threads.

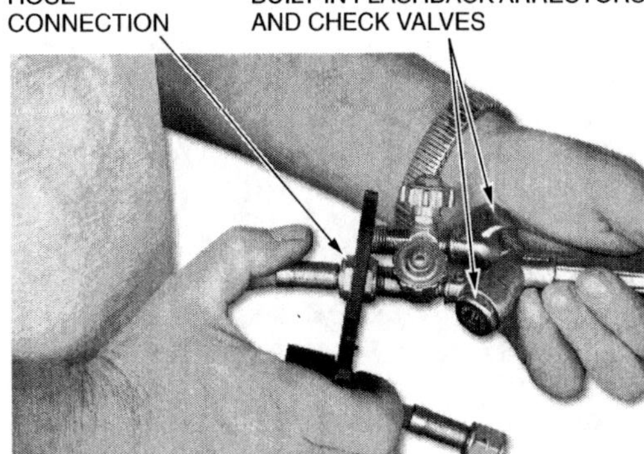

Figure 56 ◆ Connecting hoses to torch body.

4.7.0 Connecting Cutting Attachments (Combination Torch Only)

If a combination torch is being used, connect cutting attachments as follows:

Step 1 Check the torch manufacturer's instructions for the correct method of installing the attachment.

Step 2 Connect the attachment and tighten with a torch wrench or by hand as required (*Figure 57*).

Explain that new hoses contain material that can clog torch needle valves.

Review the procedure for connecting hoses to the regulator and the torch.

Emphasize the danger of blowing out hoses with compressed air, fuel gas, or oxygen.

Show trainees how to connect hoses to the regulators and the torch.

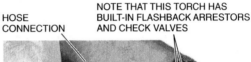

Review the procedure for installing cutting attachments and tips on torches.

Show trainees how to install cutting attachments and tips on torches.

Discuss the dangers associated with excessive acetylene flow rates.

List the steps that must be followed to close torch valves.

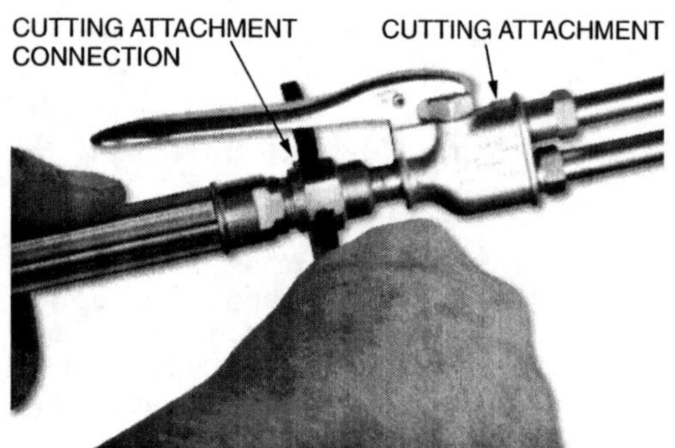

Figure 57 ◆ Connecting a cutting attachment.

4.8.0 Installing Cutting Tips

Follow these steps to install a cutting tip in the cutting torch:

Step 1 Identify the thickness of the material to be cut.

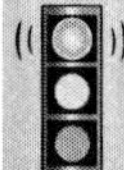

WARNING!

If acetylene fuel is being used, make sure that the maximum fuel flow rate per hour of the tip does not exceed one-tenth of the fuel cylinder capacity. If a purplish flame is observed when the torch is operating, the fuel rate is too high and acetone is being withdrawn from the acetylene cylinder along with the acetylene gas.

Step 2 Identify the proper size cutting tip from the manufacturer's recommended tip size chart for the fuel being used.

Step 3 Inspect the cutting tip sealing surfaces and orifices for damage or plugged holes. If the sealing surfaces are damaged, discard the tip. If the orifices are plugged, clean them with a tip cleaner or drill.

Step 4 Check the torch manufacturer's instructions for the correct method of installing cutting tips.

Step 5 Install the cutting tip, securing it with a torch wrench or by hand as required (*Figure 58*).

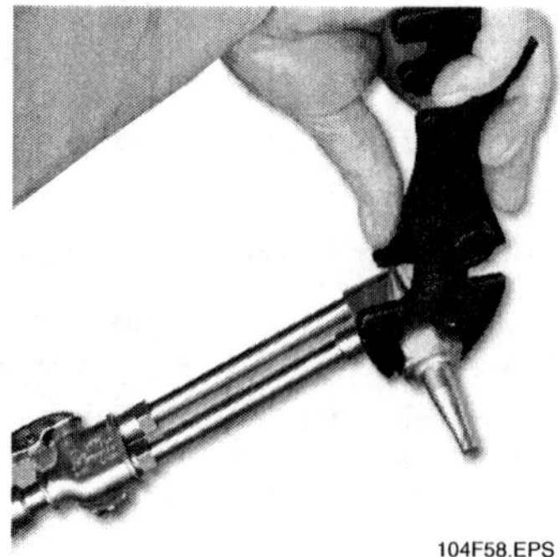

Figure 58 ◆ Installing a cutting tip.

4.9.0 Closing Torch Valves and Loosening Regulator Adjusting Screws

Follow these steps to close the torch valves and loosen the regulator adjusting screws (*Figure 59*):

Step 1 Check the fuel and oxygen valves on the torch to be sure they are closed. Closing the torch gas valves prevents gases from backing up inside the torch.

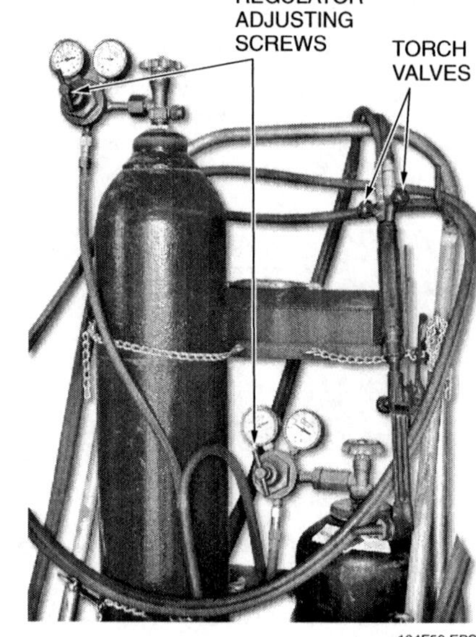

Figure 59 ◆ Torch valves and regulator adjusting screws.

Instructor's Notes:

4.42 PIPEFITTING ◆ LEVEL ONE

CAUTION

Loosening regulator adjusting screws closes the regulators and prevents damage to the regulator diaphragms when the cylinder valves are opened.

Step 2 Check the oxygen regulator adjusting screw to be sure it is loose (backed out).

Step 3 Check the fuel gas regulator adjusting screw to be sure it is loose.

4.10.0 Opening Cylinder Valves

Follow these steps to open cylinder valves (*Figure 60*):

WARNING!

Never stand directly in front of or behind a regulator. The regulator adjusting screw can blow out, causing serious injury. Always open the cylinder valve gradually. Quick openings can damage a regulator or gauge or even cause a gauge to explode.

Oxygen cylinder valves must be opened all the way until the valve seats at the top. Seating the valve at the fully open position prevents high-pressure leaks at the valve stem.

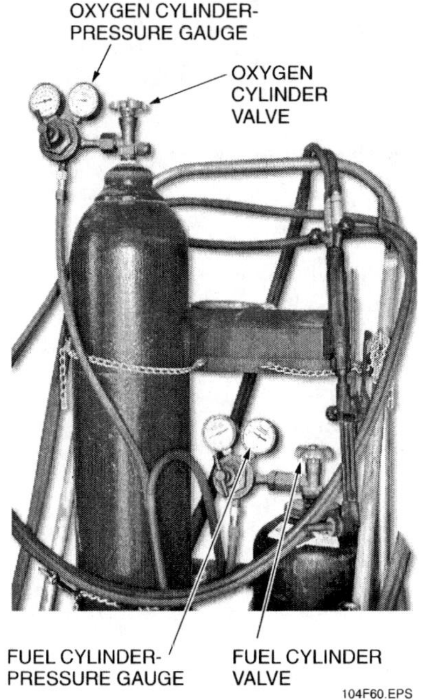

Figure 60 ◆ Cylinder valves and gauges.

Step 1 Standing on the cylinder valve side of the oxygen regulator, slowly open the oxygen cylinder valve all the way, allowing the pressure in the cylinder pressure gauge to rise gradually until the gauge indicates the oxygen cylinder pressure.

Step 2 Standing on the cylinder valve side of the fuel gas regulator, slowly open the fuel gas cylinder valve a quarter turn or until the cylinder pressure gauge indicates the cylinder pressure. Opening the cylinder valve a quarter turn allows it to be quickly closed in case of a fire.

4.11.0 Purging the Torch and Setting the Working Pressures

Follow these steps to purge the torch and set the working pressures (*Figure 61*):

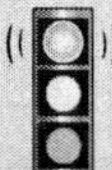

WARNING!

The working pressure gauge readings on single-stage regulators will rise after the torch valves are turned off. This is normal. However, if acetylene is being used as the fuel gas, ensure that the static pressure does not rise to 15 psig. Make sure that equipment is purged and leak tested in a well-ventilated area to avoid creating an explosive concentration of gases.

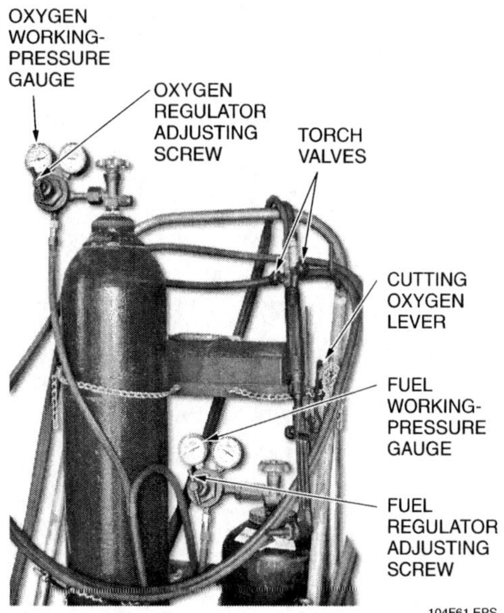

Figure 61 ◆ Typical points for purging the equipment and setting the working pressures.

Discuss the importance of loosening regulator adjusting screws.

List the steps that must be followed to open cylinder valves.

Warn trainees to stand aside when opening cylinder valves. Emphasize the need to open these valves gradually.

List the steps that must be followed to purge the torch and set the working pressure.

Warn trainees that explosive concentrations of gases may accumulate if equipment is not purged and leak tested in a well-ventilated area.

MODULE 08104-06 ◆ OXYFUEL CUTTING 4.43

Show trainees how to close the torch valves, open cylinder valves, purge the torch, and set the working pressure.

Emphasize the importance of leak testing. Review the areas to check for leaks.

Emphasize that detergent used for leak testing must not contain oil.

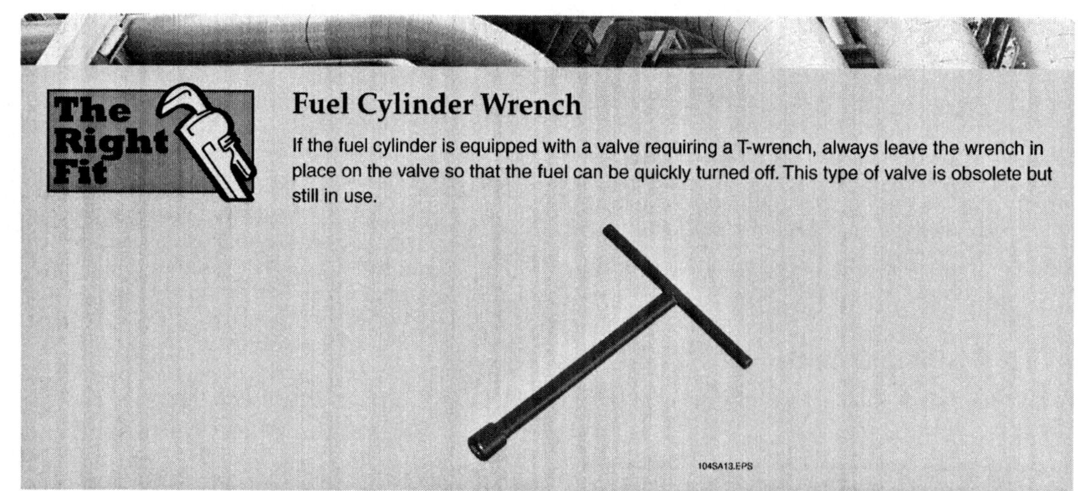

Fuel Cylinder Wrench

If the fuel cylinder is equipped with a valve requiring a T-wrench, always leave the wrench in place on the valve so that the fuel can be quickly turned off. This type of valve is obsolete but still in use.

104SA13.EPS

Step 1 Fully open the oxygen valve on the torch. Then depress and hold or lock open the cutting oxygen lever.

Step 2 Tighten the oxygen regulator adjusting screw until the working pressure gauge shows the correct oxygen gas working pressure with the gas flowing. Allow the gas to flow for five to ten seconds to purge the torch and hoses of air or fuel gas.

Step 3 At the torch, release the cutting lever and close the oxygen valve.

Step 4 Open the fuel valve on the torch about ⅛ of a turn.

Step 5 Tighten the fuel regulator adjusting screw until the working pressure gauge shows the correct fuel gas working pressure with the gas flowing. Allow the gas to flow for five to ten seconds to purge the hoses and torch of air.

Step 6 At the torch, close the fuel valve. If acetylene is used, check that the acetylene static pressure does not rise to 15 psig. If it does, immediately open the torch fuel valve and reduce the regulator output pressure as required.

4.12.0 Testing for Leaks

Equipment must be tested for leaks immediately after it is set up and periodically thereafter. The torch should be checked for leaks each time before use. Leaks could cause a fire or explosion if undetected. To test for leaks, brush a commercially prepared leak-testing formula or a solution of detergent and water on the following points. If bubbles form, a leak is present.

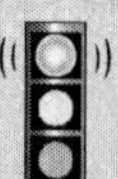

WARNING!

Make sure the detergent contains no oil. In the presence of oxygen, oil can cause fires or explosions.

Leak points include the following:

- Oxygen cylinder valve
- Fuel gas cylinder valve
- Oxygen regulator and regulator inlet and outlet connections
- Fuel gas regulator and regulator inlet and outlet connections
- Hose connections at the regulators, check valves/flashback arrestors, and torch

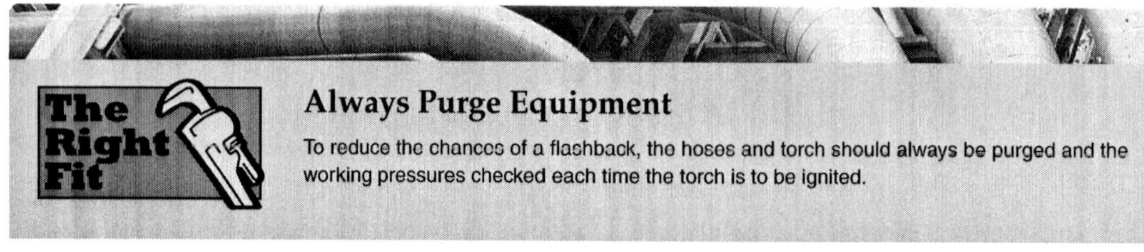

Always Purge Equipment

To reduce the chances of a flashback, the hoses and torch should always be purged and the working pressures checked each time the torch is to be ignited.

Instructor's Notes:

- Torch valves and cutting oxygen lever valve
- Cutting attachment connection (if used)
- Cutting tip

If there is a leak at the fuel gas cylinder valve stem, attempt to stop it by tightening the packing gland. If this does not stop the leak, mark and remove the cylinder and notify the supplier. For other leaks, tighten the connections slightly with a wrench. If this does not stop the leak, turn off the gas pressure, open all connections, and inspect the screw threads.

CAUTION

Use care not to overtighten connections. The brass connections used will strip if overtightened.

Repair or replace equipment that does not seal properly. If the torch valves leak, try lightly tightening the packing nut at the valves. If a cutting attachment or torch tip leaks, disassemble and check the sealing surfaces and/or O-rings for damage. If the equipment or torch tip is new and the sealing surfaces do not appear damaged, overtighten the connection slightly to seat the sealing surfaces; then loosen and retighten normally. If the leaks persist or the equipment is visibly damaged, replace the torch, cutting attachment (if used), or cutting tip as necessary.

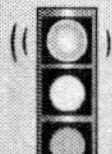

WARNING!

Make sure that equipment is purged and leak tested in a well-ventilated area to avoid creating an explosive concentration of gases.

4.12.1 Initial and Periodic Leak Testing

Perform the following steps at initial equipment setup and periodically thereafter:

Step 1 Set the equipment to the correct working pressures with the torch valves turned off.

Step 2 Using a leak-test solution, check for leaks at the cylinder valves, regulator relief ports, and regulator gauge connections (*Figure 62*). Also, check for leaks at hose connections, regulator connections, and check valve/flame arrestor connections up to the torch.

4.12.2 Leak-Down Testing of Regulators, Hoses, and Torch

Perform the following steps to quickly leak test the regulators, hoses, and torch before the torch is ignited.

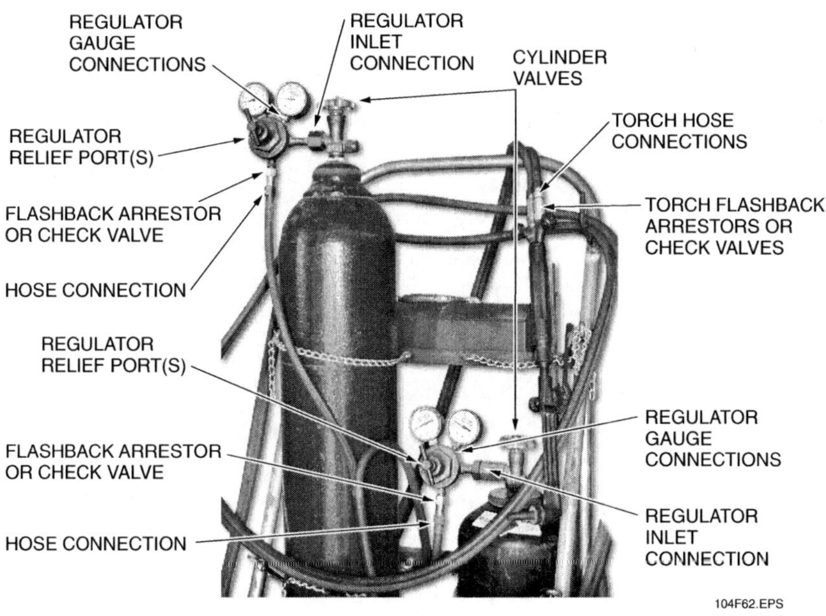

Figure 62 ◆ Typical initial and periodic leak-test points.

Explain how to stop a leak.

Discuss the harmful effects of overtightening connections.

Warn trainees that explosive concentrations of gases may accumulate if equipment is not purged and leak tested in a well-ventilated area.

Review the procedure for initial and ongoing leak tests for cutting equipment.

See the Teaching Tip for Section 4.12.0

Emphasize the need to leak test cutting and welding torches.

Review the procedures for leak testing for regulators, torches, and hoses.

Review the procedures for full leak testing for a torch.

Warn trainees that explosive concentrations of gases may accumulate if equipment is not purged and leak-tested in a well-ventilated area.

Demonstrate how to set up oxyfuel equipment.

Have trainees practice setting up oxyfuel equipment. Note the proficiency of each trainee. This laboratory corresponds to Performance Task 1.

Have trainees review Sections 5.0.0–8.0.0.

Always Leak Test a Cutting/Welding Torch

Always take the time to perform a leak test on regulators, hoses, and the torch before you use them the first time that day. Always perform a leak test on a torch after you change tips or if you convert the torch from welding to cutting or vice versa. Leaks, especially fuel leaks, can cause a fire or explosion after a cutting or welding operation begins. A dangerous fire occurring at or near the torch may not be immediately noticed by the welder due to limited visibility caused by the tinted lenses in the welding goggles or face shield.

Step 1 Set the equipment to the correct working pressures with the torch valves turned off. Then loosen both regulator adjusting screws. Check the working pressure gauges after a minute or two to see if the pressure drops. If the pressure drops, check the hose connection and regulators for leaks; otherwise, proceed to Step 2.

Step 2 Place a thumb over the cutting tip orifices and press to block the orifices.

Step 3 Turn on the torch oxygen valve and then depress and hold the cutting oxygen lever down.

Step 4 After the gauge pressure drops slightly, observe the oxygen working pressure gauge for a minute to see if the pressure continues to drop. If the pressure keeps dropping, perform the leak test described in the following section to determine the source of the leak. If the pressure does not change, close the torch oxygen valve and release the pressure at the cutting tip.

Step 5 With the tip blocked, turn on the torch fuel valve. After the gauge pressure drops slightly, carefully observe the fuel working pressure gauge for a minute. If the pressure continues to drop, perform the leak test described in the following section to determine the source of the leak. If the pressure does not change, close the torch fuel valve and release the pressure at the cutting tip.

Step 6 If no leaks are apparent, set the equipment to the correct working pressures.

4.12.3 Full Leak Testing of a Torch

Perform the following steps to test for and isolate torch leaks (*Figure 63*).

Step 1 Set the equipment to the correct working pressures with the torch valves turned off.

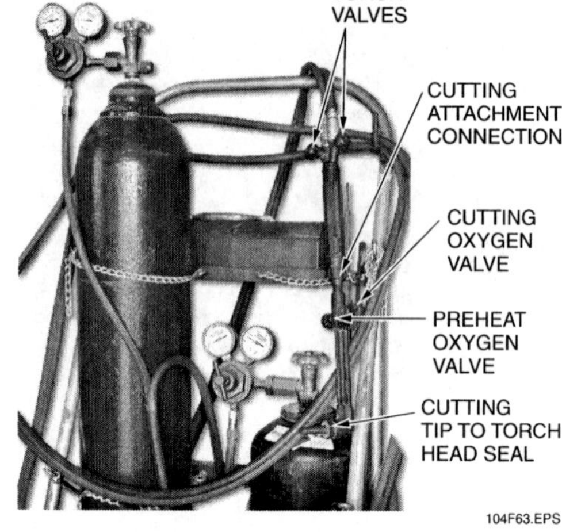

Figure 63 ◆ Torch leak-test points.

Step 2 Place a thumb over the cutting tip orifices and press to block the orifices.

Step 3 Turn on the torch oxygen valve and then depress and lock the cutting oxygen lever down.

Step 4 With the cutting tip blocked, check for leaks using a leak-test solution at the torch oxygen valve, cutting oxygen lever valve, cutting attachment connection (if used), preheat oxygen valve (if used), and cutting tip seal at the torch head.

Step 5 Release the cutting oxygen lever and close the torch oxygen valve. Release the pressure at the cutting tip.

Step 6 With the cutting tip blocked, open the torch fuel valve.

Step 7 Using a leak-test solution, check for leaks at the torch fuel valve, cutting attachment (if used), and cutting tip seal at the torch head.

Step 8 Close the torch fuel valve and remove thumb from the cutting tip.

Instructor's Notes:

4.46 PIPEFITTING ◆ LEVEL ONE

5.0.0 ◆ CONTROLLING THE OXYFUEL TORCH FLAME

To be able to safely use a cutting torch, the operator must understand the flame and be able to adjust it and react to unsafe conditions. The following sections will explain the oxyfuel flame and how to control it safely.

5.1.0 Oxyfuel Flames

There are three types of oxyfuel flames: **neutral flame**, **carburizing flame**, and **oxidizing flame**.

- *Neutral flame* – A neutral flame burns proper proportions of oxygen and fuel gas. The inner cones will be light blue in color, surrounded by a darker blue outer flame envelope that results when the oxygen in the air combines with the super-heated gases from the inner cone. A neutral flame is used for all but special cutting applications.
- *Carburizing flame* – A carburizing flame has a white feather created by excess fuel. The length of the feather depends on the amount of excess fuel present in the flame. The outer flame envelope is longer than that of the neutral flame, and it is much brighter in color. The excess fuel in the carburizing flame (especially acetylene) produces large amounts of carbon. The carbon will combine with red-hot or molten metal, making the metal hard and brittle. The carburizing flame is cooler than a neutral flame and is never used for cutting. It is used for some special heating applications.
- *Oxidizing flame* – An oxidizing flame has an excess of oxygen. The inner cones are shorter, much bluer in color, and more pointed than a neutral flame. The outer flame envelope is very short and often fans out at the ends. An oxidizing flame is the hottest flame. A slightly oxidizing flame is recommended with some special fuel gases, but in most cases it is not used. The excess oxygen in the flame can combine with many metals, forming a hard, brittle, low-strength oxide. However, the preheat flames of a properly adjusted cutting torch will be slightly oxidizing when the cutting oxygen is shut off.

Figure 64 shows the various flames that occur at a cutting tip for both acetylene and LP gas.

5.2.0 Backfires and Flashbacks

When the torch flame goes out with a loud pop or snap, a backfire has occurred. Backfires are usually caused when the tip or nozzle touches the work surface or when a bit of hot dross briefly interrupts the flame. When a backfire occurs, you can relight the torch immediately. Sometimes the torch even relights itself. If a backfire recurs without the tip making contact with the base metal, shut off the torch and find the cause. Possible causes are:

- Improper operating pressures
- A loose torch tip
- Dirt in the torch tip seat or a bad seat

When the flame goes out and burns back inside the torch with a hissing or whistling sound, a flashback is occurring. Immediately shut off the oxygen valve on the torch; the flame is burning inside the torch. If the flame is not extinguished quickly, the end of the torch will melt off. The flashback will stop as soon as the oxygen valve is closed. Therefore, quick action is crucial. Flashbacks can cause fires and explosions within the cutting rig and, therefore, are very dangerous. Flashbacks can be caused by:

- Equipment failure
- Overheated torch tip
- Dross or spatter hitting and sticking to the torch tip
- Oversized tip (tip is too large for the gas flow rate being used)

After a flashback has occurred, wait until the torch has cooled. Then, blow oxygen (not fuel gas) through the torch for several seconds to remove soot that may have built up in the torch during the flashback before relighting it. If you hear the hissing or whistling after the torch is reignited or if the flame does not appear normal, shut off the torch immediately and have the torch serviced by a qualified technician. If the torch or tip has been damaged, replace the flashback arrestors.

5.3.0 Igniting the Torch and Adjusting the Flame

After the cutting equipment has been properly set up and purged, the torch can be ignited and the

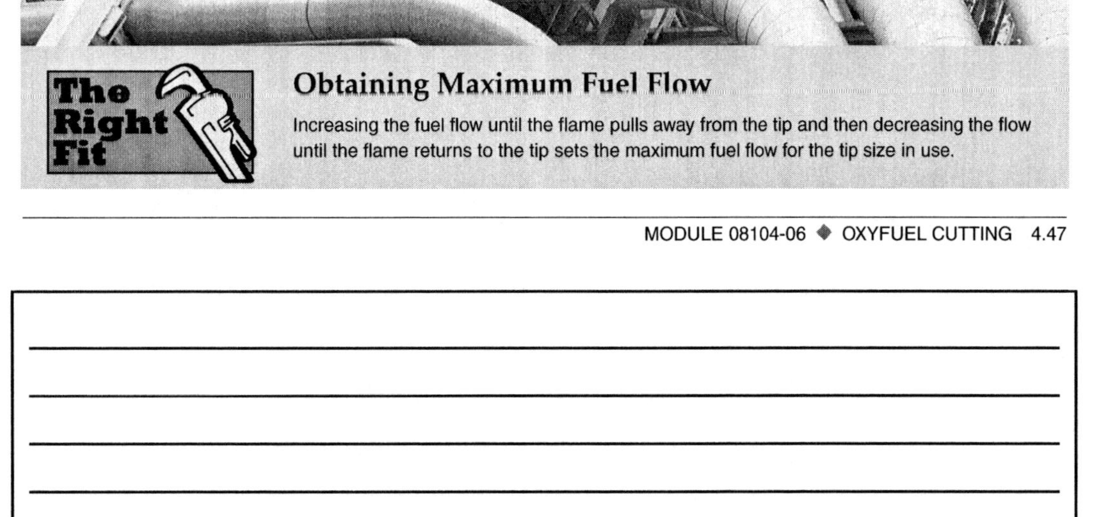

The Right Fit

Obtaining Maximum Fuel Flow

Increasing the fuel flow until the flame pulls away from the tip and then decreasing the flow until the flame returns to the tip sets the maximum fuel flow for the tip size in use.

MODULE 08104-06 ◆ OXYFUEL CUTTING 4.47

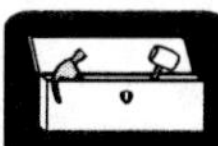

Ensure that you have everything required for teaching this session.

Compare and contrast the three types of oxyfuel flames.

Define backfire and discuss possible causes.

Discuss possible causes of flashback.

Review the steps that must be followed for lighting torches.

Remind trainees to check the manufacturer's charts when selecting tips and igniting torches.

Point out that the tip manual may contain recommendations for proper igniting.

Acetylene Burning in Atmosphere
Open fuel gas valve until smoke clears from flame.

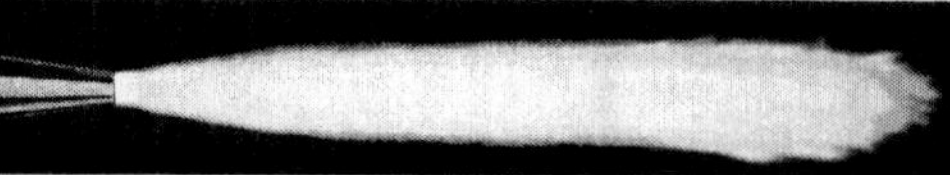

Carburizing Flame
(Excess acetylene with oxygen) Preheat flames require more oxygen.

Neutral Flame
(Acetylene with oxygen) Temperature 5589°F (3087°C). Proper preheat adjustment when cutting.

Neutral Flame with Cutting Jet Open
Cutting jet must be straight and clean. If it flares, the pressure is too high for the tip size.

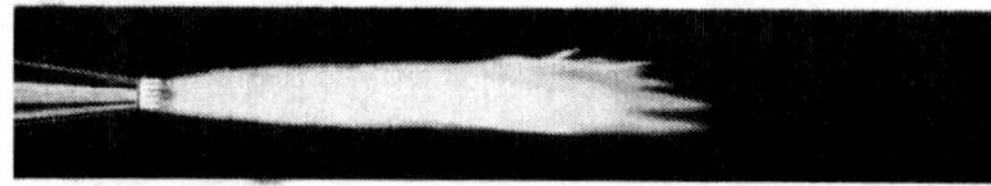

Oxidizing Flame
(Acetylene with excess oxygen) Not recommended for average cutting. However, if the preheat flame is adjusted for neutral with the cutting oxygen on, then this flame is normal after the cutting oxygen is off.

104F64A.EPS

LP Gas Burning in Atmosphere
Open fuel gas valve until flame begins to leave tip end.

Reducing Flame
(Excess LP-gas with oxygen) Not hot enough for cutting.

Neutral Flame
(LP-gas with oxygen) For preheating prior to cutting.

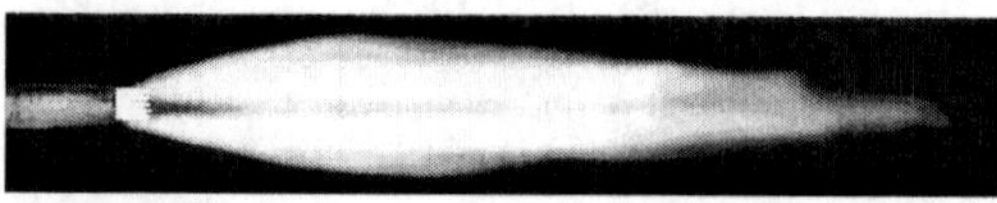

Oxidizing Flame with Cutting Jet Open
Cutting jet stream must be straight and clean.

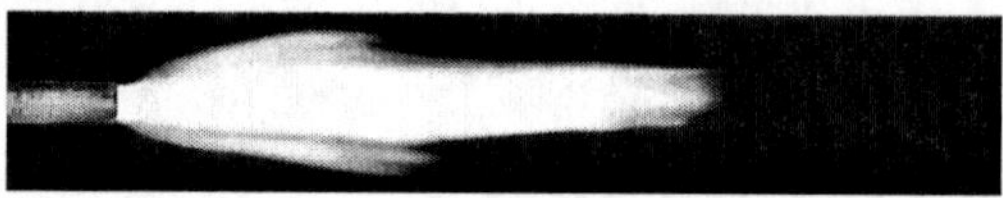

Oxidizing Flame without Cutting Jet Open
(LP-gas with excess oxygen) The highest temperature flame for fast starts and high cutting speeds.

104F64B.EPS

Figure 64 ◆ Acetylene and LP gas flames.

flame adjusted for cutting. Follow these steps to ignite the torch:

Step 1 Choose the appropriate cutting torch tip according to the base metal thickness you will be cutting and fuel gas you are using.

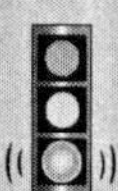

NOTE
Refer to the manufacturer's charts. You may have to readjust the oxygen and fuel gas pressure depending on the tip selected.

Step 2 Inspect the tip sealing surfaces and orifices. Attach the tip to the cutting torch or cutting attachment by placing it on the end of the torch and tightening the nut.

NOTE
Some manufacturers recommend tightening the nut with a torch wrench. Others recommend tightening the nut by hand. Check the tip manual to see if the manufacturer recommends that the nut be tightened manually or with a torch wrench.

Step 3 Put on proper protective clothing, gloves, and eye/face protection.

Step 4 Raise tinted eye protection and/or face shield.

Step 5 Release the oxygen cutting lever. If present, close the preheat oxygen valve and open the torch oxygen valve fully.

Instructor's Notes:

Step 6 Open the fuel gas valve on the torch handle about one-quarter turn.

Step 7 Holding the friction lighter near the side and to the front of the torch tip, ignite the torch.

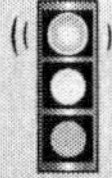

WARNING!

Hold the friction lighter near the side of the tip when igniting the torch to prevent deflecting the ignited gas back toward you. Always use a friction lighter. Never use matches or cigarette lighters to light the torch because this could result in severe burns and/or could cause the lighter to explode. Always point the torch away from yourself, other people, equipment, and flammable material.

Step 8 Once the torch is lit, adjust the torch fuel gas flame by adjusting the flow of fuel gas with the fuel gas valve. Increase the flow of fuel gas until the flame stops smoking or pulls slightly away from the tip. Decrease the flow until the flame returns to the tip.

Step 9 Open the preheat oxygen valve (if present) or the oxygen torch valve very slowly and adjust the torch flame to a neutral flame.

Step 10 Press the cutting oxygen lever all the way down and observe the flame. It should have a long, thin, high-pressure oxygen cutting jet up to 8 inches long extending from the cutting oxygen hole in the center of the tip. If it does not:

- Check that the working pressures are set as recommended on the manufacturer's chart.
- Clean the cutting tip. If this does not clear up the problem, change the cutting tip.

Step 11 With the cutting oxygen on, observe the preheat flame. If it has changed slightly to a carburizing flame, increase the preheat oxygen until the flame is neutral. After this adjustment, the preheat flame will change slightly to an acceptable oxidizing flame when the cutting oxygen is shut off.

5.4.0 Shutting Off the Torch

Follow these steps to shut off the torch after a cutting operation is completed:

WARNING!

Always turn off the oxygen flow first to prevent a possible flashback into the torch.

Step 1 Release the cutting oxygen lever. Then close the torch or preheat oxygen valves.

Step 2 Close the torch fuel gas valve quickly to extinguish the flame.

6.0.0 ◆ SHUTTING DOWN OXYFUEL CUTTING EQUIPMENT

When a cutting job is completed and the oxyfuel equipment is no longer needed, it must be shut down. Follow these steps to shut down oxyfuel cutting equipment (*Figure 65*):

Step 1 Close the fuel gas and oxygen cylinder valves.

Step 2 Open the fuel gas and oxygen torch valves to allow gas to escape. Do not proceed to Step 3 until all pressure is released and all regulator gauges read zero.

Step 3 Back out the fuel gas and oxygen regulator adjusting screws until they are loose.

Step 4 Close the fuel gas and oxygen torch valves.

Step 5 Coil up the hose and secure the torch to prevent damage.

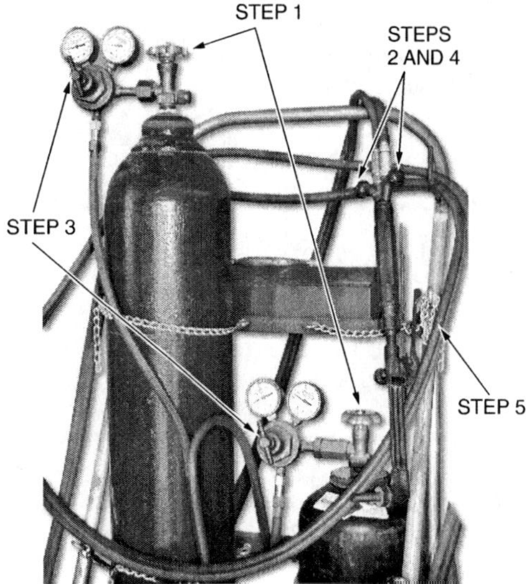

Figure 65 ◆ Shutting down oxyfuel cutting equipment.

Explain how to set the maximum fuel flow for the tip size in use.

Discuss the hazards and safety precautions needed when lighting torches.

Describe the steps used to shut down oxyfuel equipment.

Warn trainees to turn off oxygen flow first to prevent flashback.

Demonstrate how to light, adjust, and shut down an oxyfuel cutting torch.

Have trainees practice lighting, adjusting, and shutting down an oxyfuel cutting torch. Note the proficiency of each trainee. This laboratory corresponds to Performance Tasks 2 and 3.

Describe the steps used to disassemble oxyfuel equipment. Explain how to change an empty cylinder.

Discuss the proper handling and transportation of cylinders. Point out that oxygen and fuel cylinders must never be stored together.

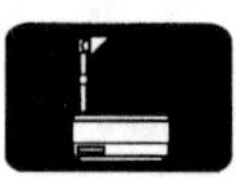

Show Transparency 16 (Figure 66). Explain how to mark and store an empty cylinder.

Demonstrate how to disassemble oxyfuel equipment and change an empty cylinder.

Have trainees practice disassembling oxyfuel equipment and changing an empty cylinder. Note the proficiency of each trainee. This laboratory corresponds to Performance Tasks 4 and 5.

Have trainees review Sections 9.0.0–10.4.0.

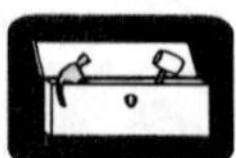

Ensure that you have everything required for teaching this session.

7.0.0 ◆ DISASSEMBLING OXYFUEL EQUIPMENT

Follow these steps if the oxyfuel equipment must be disassembled after use:

Step 1 Check to be sure the equipment has been properly shut down. This includes checking that:

- The cylinder valves are closed.
- All pressure gauges read zero.

Step 2 Remove both hoses from the torch.

Step 3 Remove both hoses from the regulators.

Step 4 Remove both regulators from the cylinder valves.

Step 5 Replace the protective caps on the cylinders.

Step 6 Return the oxygen cylinder to its proper storage place.

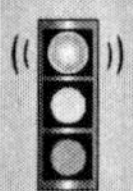

WARNING!
Always transport and store gas cylinders in the upright position. Be sure they are properly secured (chained) and capped.

Step 7 Return the fuel gas cylinder to its proper storage place.

WARNING!
Regardless if the cylinders are empty or full, never store fuel gas cylinders and oxygen cylinders together. Storing cylinders together is a violation of OSHA and local fire regulations and could result in a fire and explosion.

8.0.0 ◆ CHANGING EMPTY CYLINDERS

Follow these procedures to change a cylinder when it is empty:

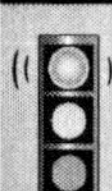

WARNING!
When moving cylinders, always be certain that they are in the upright position and the valve caps are secured in place. Never use a sling or electromagnet to lift cylinders. To lift cylinders, use a cylinder cage.

Step 1 Check to be sure equipment has been properly shut down. This includes checking that:

- The cylinder valves are closed.
- All pressure gauges read zero.

Step 2 Remove the regulator from the empty cylinder.

Step 3 Replace the protective cap on the empty cylinder.

Step 4 Transport the empty cylinder from the workstation to the storage area.

Step 5 Mark MT (empty) and the date (or the accepted site notation for indicating an empty cylinder) near the top of the cylinder using soapstone (*Figure 66*).

Step 6 Place the empty cylinder in the empty cylinder section of the cylinder storage area for the type of gas in the cylinder.

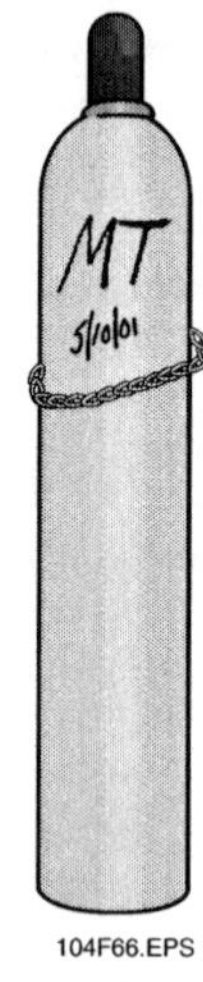

Figure 66 ◆ Typical empty cylinder marking.

9.0.0 ◆ PERFORMING CUTTING PROCEDURES

The following sections explain how to recognize good and bad cuts, how to prepare for cutting operations, and how to perform straight-line cutting, piercing, bevel cutting, washing, and gouging.

9.1.0 Inspecting the Cut

Before attempting to make a cut, you must be able to recognize good and bad cuts and know what causes bad cuts. This is explained in the following list and illustrated in *Figure 67*:

Instructor's Notes:

4.50 PIPEFITTING ◆ LEVEL ONE

Marking and Tagging Cylinders

Do not use permanent markers on cylinders; use soapstone or another temporary marker. If a cylinder is defective, place a warning tag on it.

DIRECTION OF TRAVEL

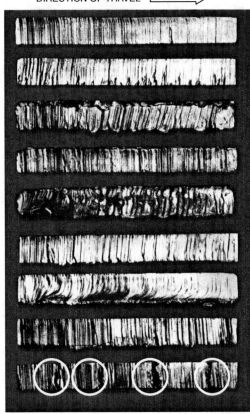

GOOD CUT

PREHEAT INSUFFICIENT

TOO MUCH PREHEAT

CUTTING PRESSURE TOO LOW

OXYGEN PRESSURE TOO HIGH AND UNDERSIZE TIP

TRAVEL SPEED TOO SLOW

TRAVEL SPEED TOO FAST

TORCH HELD OR MOVED UNSTEADILY

CUT NOT RESTARTED CAREFULLY, CAUSING GOUGES AT RESTARTING POINTS (CIRCLED)

104F67.EPS

Figure 67 ◆ Examples of good and bad cuts.

- A good cut features a square top edge that is sharp and straight, not ragged. The bottom edge can have some dross adhering to it but not an excessive amount. What dross there is should be easily removable with a chipping hammer. The **drag lines** should be near vertical and not very pronounced.
- When preheat is insufficient, bad gouging results at the bottom of the cut because of slow travel speed.
- Too much preheat will result in the top surface melting over the cut, an irregular cut edge, and an excessive amount of dross.
- When the cutting oxygen pressure is too low, the top edge will melt over because of the resulting slow cutting speed.
- Using cutting oxygen pressure that is too high will cause the operator to lose control of the cut, resulting in an uneven kerf.
- A travel speed that is too slow results in bad gouging at the bottom of the cut and irregular drag lines.
- When the travel speed is too fast, there will be gouging at the bottom of the cut, a pronounced break in the drag line, and an irregular kerf.

List the various cutting procedures typically performed with oxyfuel equipment.

Describe the features of good and bad cuts.

Discuss the causes of bad cuts.

Emphasize the importance of properly marking and tagging cylinders.

Explain how to prepare the metal to make a cut.

- A torch that is held or moved unsteadily across the metal being cut can result in a wavy and irregular kerf.
- When a cut is lost and then not restarted carefully, bad gouges will result at the point where the cut is restarted.

Show Transparency 17 (Figure 68). Review the procedure for cutting thin steel.

NOTE

The following tasks are designed to develop your skills with a cutting torch. Practice each task until you are thoroughly familiar with the procedure. As you complete each task, take it to your instructor for evaluation. Do not proceed to the next task until your instructor tells you to continue.

Emphasize the dangers of holding the tip upright when cutting thin steel.

9.2.0 Preparing for Oxyfuel Cutting with a Hand Cutting Torch

Before metal can be cut, the equipment must be set up and the metal prepared. One important step is to properly lay out the cut by marking it with soapstone or punch marks. The few minutes this takes will result in a quality job, reflecting craftsmanship and pride in your work. Follow these steps to prepare to make a cut:

Step 1 Prepare the metal to be cut by cleaning any rust, scale, or other foreign matter from the surface.

Step 2 If possible, position the work so you will be comfortable when cutting.

Step 3 Mark the lines to be cut with soapstone or a punch.

Step 4 Select the correct cutting torch tip according to the thickness of the metal to be cut, the type of cut to be made, the amount of preheat needed, and the type of fuel gas to be used.

Step 5 Ignite the torch.

Step 6 Use the procedures outlined in the following sections for performing particular types of cutting operations.

9.3.0 Cutting Thin Steel

Thin steel is 3/16 inch thick or less. A major concern when cutting thin steel is distortion caused by the heat of the torch and the cutting process. To minimize distortion, move as quickly as you can without losing the cut. Follow these steps to cut thin steel:

Step 1 Prepare the metal surface.

Step 2 Light the torch.

Step 3 Hold the torch so that the tip is pointing in the direction the torch is traveling at a 15- to 20-degree angle. Make sure that a preheat orifice and the cutting orifice are centered on the line of travel next to the metal (*Figure 68*).

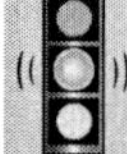

CAUTION

Holding the tip upright when cutting thin steel will overheat the metal, causing distortion.

Step 4 Preheat the metal to a dull red. Use care not to overheat thin steel because this will cause distortion.

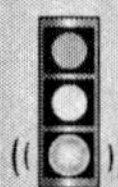

NOTE

The edge of the tip can be rested on the surface of the metal being cut and then slid along the surface when making the cut.

Step 5 Press the cutting oxygen lever to start the cut, and then move quickly along the line to be cut. To minimize distortion, move as quickly as you can without losing the cut.

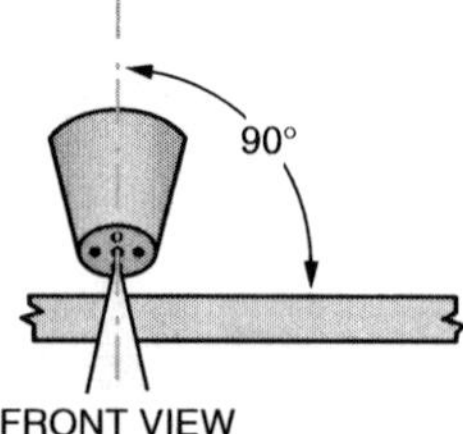

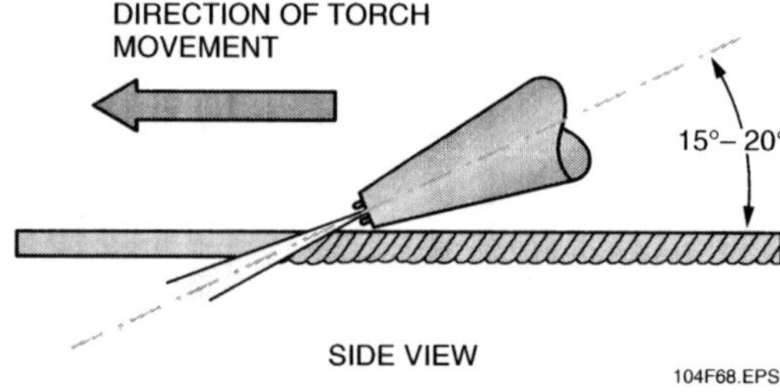

Figure 68 ◆ Cutting thin steel.

Instructor's Notes:

4.52 PIPEFITTING ◆ LEVEL ONE

9.4.0 Cutting Thick Steel

Most oxyfuel cutting will be on steel that is more than 3/16 inch thick. Whenever heat is applied to metal, distortion is a problem, but as the steel gets thicker, it becomes less of a problem. Follow these steps to cut thick steel with a hand cutting torch:

Step 1 Prepare the metal surface and torch. Light the torch.

Step 2 Ignite and adjust the torch flame.

> **NOTE**
> The torch can be moved from either right to left or left to right. Choose the direction that is the most comfortable for you.

Step 3 Follow the number sequence shown in *Figure* 69 to perform the cut.

> **NOTE**
> When cutting begins, the tips of the preheat flame should be held 1/16 to 1/8 inch above the workpiece. For steel up to 3/8 inch thick, the first and third procedures can usually be omitted.

9.5.0 Piercing a Plate

Before holes or slots can be cut in a plate, the plate must be pierced. Piercing puts a small hole through the metal where the cut can be started. Because more preheat is necessary on the surface of a plate than at the edge, choose the next-larger cutting tip than is recommended for the thickness to be pierced. When piercing steel that is more than 3 inches thick, it may help to preheat the bottom side of the plate directly under the spot to be pierced. Follow these steps to pierce a plate for cutting:

Step 1 Prepare the metal surface and torch.

Step 2 Ignite the torch and adjust the flame.

Step 3 Hold the torch tip 1/4 to 5/16 inch above the spot to be pierced until the surface is a bright cherry red (*Figure 70*).

Step 4 Slowly press the cutting oxygen lever. As the cut starts, raise the tip about 1/2 inch above the metal surface and tilt the torch slightly so that molten metal does not blow back onto the tip. The tip should be raised and tipped before the cutting oxygen lever is fully depressed.

Step 5 Maintain the tipped position until a hole burns through the plate. Then rotate the tip vertically.

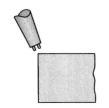

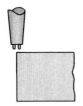

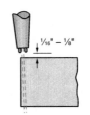

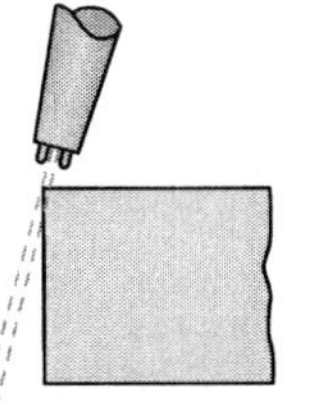
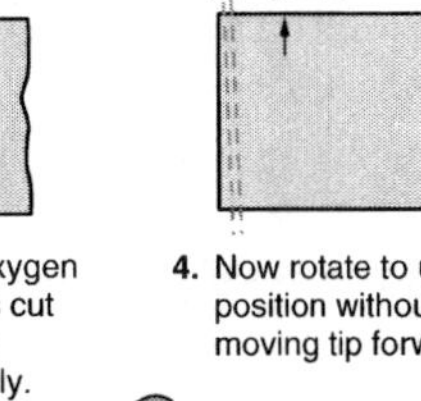
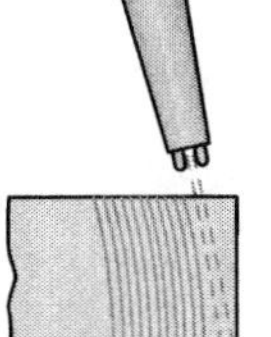
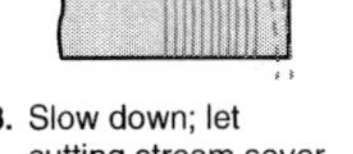
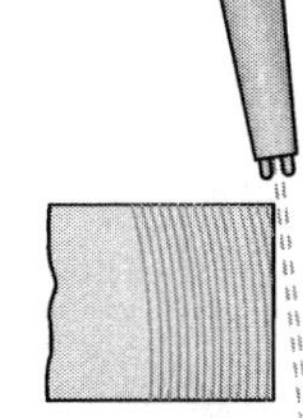

1/16" – 1/8"
1. Start to preheat; point tip at angle on edge of plate.
2. Rotate tip to upright position.
3. Press cutting oxygen valve slowly; as cut starts, rotate tip backward slightly.
4. Now rotate to upright position without moving tip forward.
5. Rotate tip more to point slightly in direction of cut.
6. Advance as fast as good cutting action will permit.
7. Do not jerk; maintain slight leading angle toward direction of cut.
8. Slow down; let cutting stream sever corner edge at bottom.
9. Continue steady forward motion until tip has cleared end.
104F69.EPS

Figure 69 ◆ Procedures for flame cutting with a hand torch.

Show Transparency 18 (Figure 69). Review the procedure for cutting thick steel.

Show trainees how to cut thin and thick steel.

Have trainees practice straight line and square shape cutting. Note the proficiency of each trainee. This laboratory corresponds to Performance Task 6.

Have the trainees adjust a torch incorrectly, use the wrong size tip, or use the wrong travel speed or cutting angle in order to observe the effects of these mistakes in cutting thin and thick steel.

Show Transparency 19 (Figure 70). Review the procedures for piercing a plate.

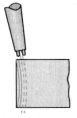

Show trainees how to pierce a steel plate and cut a slot.

MODULE 08104-06 ◆ OXYFUEL CUTTING 4.53

Have trainees practice piercing and slot cutting. Note the proficiency of each trainee. This laboratory corresponds to Performance Task 7.

Show Transparency 20 (Figure 71). Explain how to cut a bevel.

Show trainees how to make beveled cuts.

Have trainees practice making beveled cuts. Note the proficiency of each trainee. This laboratory corresponds to Performance Task 8.

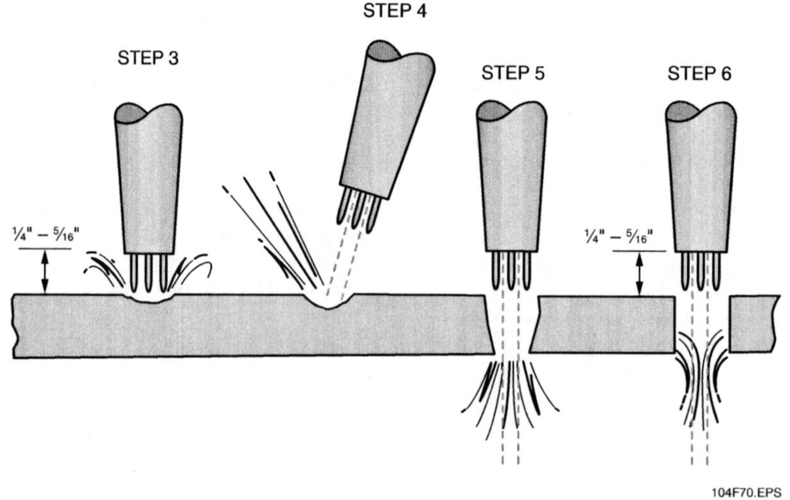

Figure 70 ◆ Steps for piercing steel.

Step 6 Lower the torch tip to about 3⁄16 inch above the metal surface and continue to cut outward from the original hole to the edge of the line to be cut.

9.6.0 Cutting Bevels

Bevel cutting is often performed to prepare the edge of steel plate for welding. Follow these steps to perform bevel cutting (*Figure 71*):

Step 1 Prepare the metal surface and the torch.

Step 2 Ignite the torch and adjust the flame.

Step 3 Hold the torch so that the tip faces the metal at the desired bevel angle.

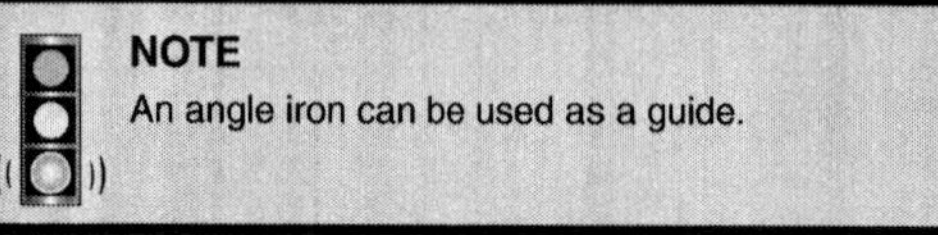

NOTE
An angle iron can be used as a guide.

Step 4 Preheat the edge to a bright cherry red.

Step 5 Press the cutting oxygen lever to start the cut.

Step 6 As cutting begins, move the torch tip at a steady rate along the line to be cut. Pay particular attention to the torch angle to ensure it is uniform along the entire length of the cut.

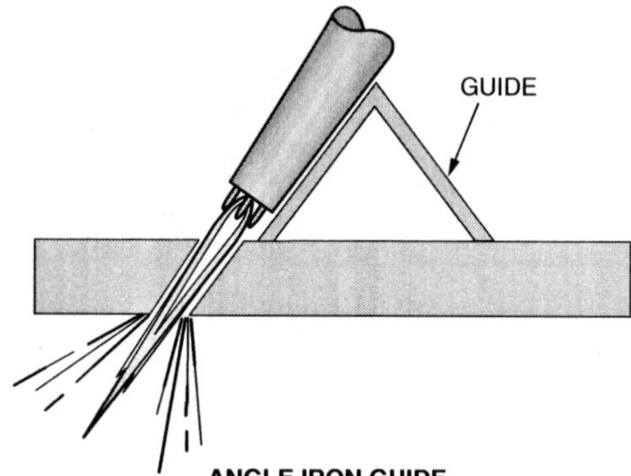

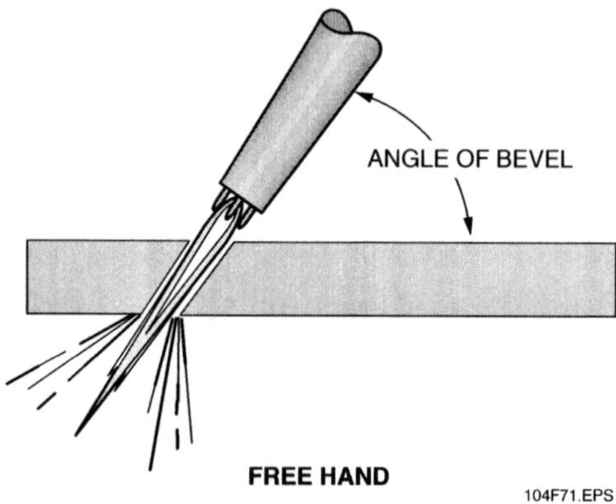

Figure 71 ◆ Cutting a bevel.

Instructor's Notes:

9.7.0 Washing

Washing is a term used to describe the process of cutting out bolts or rivets. Washing operations use a special tip with a large cutting hole that produces a low-velocity stream of oxygen. The low-velocity oxygen stream helps prevent cutting into the surrounding base metal. Washing tips can also be used to remove items such as blocks, angles, or channels that are welded onto a surface. Follow these steps to perform washing (*Figure 72*):

Step 1 Prepare the metal surface and torch.

Step 2 Ignite the torch and adjust the flame.

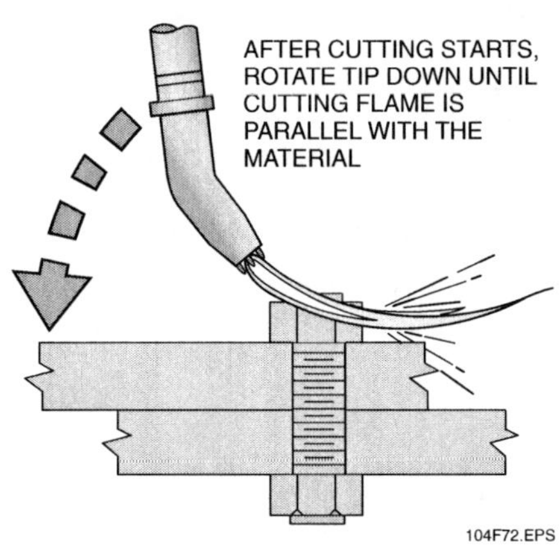

Figure 72 ◆ Washing.

Step 3 Preheat the metal to be cut until it is a bright cherry red.

Step 4 Move the cutting torch at a 55-degree angle to the metal surface.

Step 5 At the top of the material, press the cutting oxygen lever to cut the material to be removed. Continue moving back and forth across the material while rotating the tip to a position parallel with the material. Move the tip back and forth and down to the surrounding metal. Use care not to cut into the surrounding metal.

CAUTION

As the surrounding metal heats up, there is a greater danger of cutting into it. Try to complete the washing operation as quickly as possible. If the surrounding metal gets too hot, stop and let it cool down.

10.0.0 ◆ PORTABLE OXYFUEL CUTTING MACHINE OPERATION

As explained previously, machine oxyfuel gas cutters or track burners are basic guidance systems driven by a variable speed electric motor to enable the operator to cut or bevel straight lines at any desired speed. The device (*Figure 73*) is usually mounted on a track or used with a circle-cutting

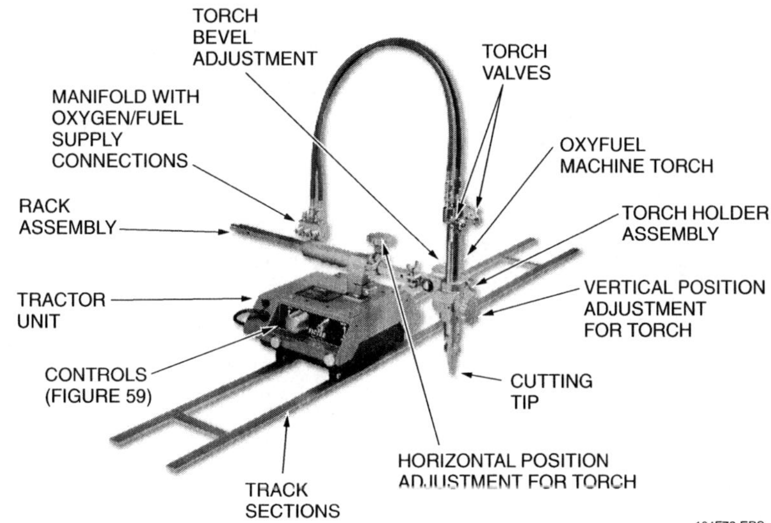

Figure 73 ◆ Victor track burner.

Show Transparency 21 (Figure 72). Describe how washing is performed.

Show trainees how to perform washing.

Have trainees practice washing. Note the proficiency of each trainee. This laboratory corresponds to Performance Task 9.

Describe portable oxyfuel cutting equipment.

Discuss the operation of portable oxyfuel cutting equipment.

Show Transparency 22 (Figure 74). Describe the controls for portable oxyfuel cutting machines.

Explain how to adjust the torch on an oxyfuel cutting machine.

Review the procedure for straight line cutting with an oxyfuel cutting machine.

attachment to enable the operator to cut various diameters from 4 to 96 inches. It consists of a heavy-duty tractor unit fitted with an adjustable torch mount and gas hose attachments. It is also equipped with an ON/OFF switch, a reversing switch, a clutch, and a speed-adjusting dial calibrated in feet/meters per minute.

The device shown in *Figure 73* offers the following operational features:

- Makes straight-line cuts of any length
- Makes circle cuts up to 96 inches in diameter
- Makes bevel or chamfer cuts
- Has an infinitely variable cutting speed from 1 to 60 inches per minute
- Has dual speed and clutch controls to enable operation of the machine from either end

10.1.0 Machine Controls

Figure 74 shows the location of the following controls:

- *Directional control* – Set the machine direction by toggling the FWD-OFF-REV toggle switch located next to the power cord.
- *Speed control* – Turn the large knob on either end of the machine to position the speed indicator at the desired cutting speed.
- *Clutch operation* – Engage the clutch by rotating one of the two clutch levers, located on either end of the machine, to the DRIVE position. Place the clutch lever in the FREE position to permit easy manual positioning of the machine prior to or after the actual cutting operation.

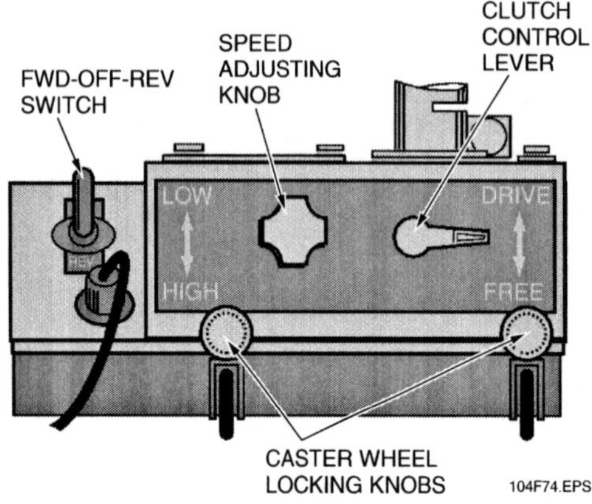

Figure 74 ◆ Victor track burner controls.

10.2.0 Torch Adjustment

The rack assembly permits the torch holder assembly to move toward or away from the tractor unit. The torch holder allows vertical positioning of the torch. The torch bevel adjustment allows torch positioning at any angle from 190 to 290 degrees in a plane perpendicular to the track. After adjusting the torch to the desired position, tighten all clamping screws to prevent the torch from making any unexpected movements.

10.3.0 Straight Line Cutting

Cut straight lines using the following procedure:

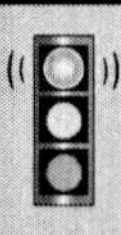

WARNING!

Most cutting machines are not designed to detect the end of their track or workpiece. Take care that an unattended machine does not fall from an elevated workpiece.

Step 1 Place the machine track on the workpiece and line it up before placing the machine on the track.

Step 2 Be sure the track is long enough for the cut to be made. If not, install additional track. Connect track sections carefully. Extend the track on both sides of the cut and support the track. When properly connected, the machine should travel smoothly from one track section to the next. If the cut is long, the track may have to be clamped at both ends beyond the cut to keep the track from moving during the cut.

Step 3 Place the machine on the track. Place the clutch lever in the FREE position. Be sure that the supply gas hoses and the power lines are long enough and free to move with the machine so that it can complete the cut properly.

Step 4 Move the machine to the approximate point where the cut will start. Set the drive speed control to the desired cutting speed. Set the FWD-OFF-REV switch to the OFF position. Plug the power cord into a 115 alternating current (AC), 60 Hertz (Hz) power outlet.

Instructor's Notes:

Step 5 Ensure that all clamping screws are properly tightened. Ignite and properly adjust the torch, then preheat the start of the cut. Set the FWD-OFF-REV switch to the desired direction of travel. Simultaneously turn on the cutting oxygen and set the clutch lever to the DRIVE position.

Step 6 When the cut is completed, stop the machine and shut off the torch.

10.4.0 Bevel Cutting

Perform bevel cutting operations using the following procedure:

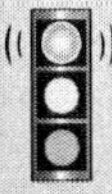

WARNING!
Most cutting machines are not designed to detect the end of their track or workpiece. Take care that an unattended machine does not fall from an elevated workpiece.

Step 1 Place the machine track on the workpiece and line it up before placing the machine on the track.

Step 2 Be sure the track is long enough for the cut to be made. If not, install additional track. Connect track sections carefully. Extend the track on both sides of the cut and support the track. When properly connected, the machine should travel smoothly from one track section to the next. If the cut is long, the track may have to be clamped at both ends beyond the cut to keep the track from moving during the cut.

Step 3 Place the machine on the track. Place the clutch lever in the FREE position. Be sure that the supply gas hoses and the power lines are long enough and free to move with the machine so that it can complete the cut properly.

Step 4 Loosen the bevel adjusting knob, set the torch angle to the desired bevel angle, and then tighten the bevel adjusting knob.

Step 5 Move the machine to the approximate point where the cut will start. Set the drive speed control to the desired cutting speed. Set the FWD-OFF-REV switch to the OFF position. Plug the power cord into a 115AC, 60Hz power outlet.

Step 6 Ensure that all clamping screws are properly tightened. Ignite and properly adjust the torch, then preheat the start of the cut. Set the FWD-OFF-REV switch to the desired direction of travel. Simultaneously turn on the cutting oxygen and set the clutch lever to the DRIVE position.

Step 7 When the cut is completed, stop the machine and shut off the torch.

Review the procedure for bevel cutting with an oxyfuel cutting machine.

Point out that unattended machines can fall from elevated workpieces.

Show trainees how to perform straight line and bevel cutting with an oxyfuel cutting machine.

MODULE 08104-06 ◆ OXYFUEL CUTTING 4.57

Summarize the major concepts presented in the module.

Summary

Oxyfuel cutting has many uses on job sites. It can be used to cut plate and shapes to size, prepare joints for welding, clean metals or welds, and disassemble structures. Because of the high pressures and flammable gases involved, there is a danger of fire and explosion when using oxyfuel equipment. However, these risks can be minimized when the oxyfuel cutting operator is well trained and knowledgeable. Be sure you know and understand the safety precautions and equipment presented in this module before using oxyfuel equipment.

Notes

Instructor's Notes:

4.58 PIPEFITTING ◆ LEVEL ONE

Have the trainees complete the Review Questions, and go over the answers prior to administering the Module Examination.

Review Questions

1. To prevent fume hazards, there must be at least _____ cubic feet of air for each welder.
   a. 2,000
   b. 4,000
   c. 5,000
   d. 10,000

2. A good safety practice is _____.
   a. welding near an open window
   b. welding over wooden floors
   c. to use cardboard to deflect welding/grinding sparks away from others
   d. writing *HOT* on hot metal before leaving it unattended

3. Cutting operations should never be performed without a _____ in the area.
   a. bucket of sand
   b. bucket of water
   c. fire watch
   d. fire hose

4. When pure oxygen is combined with fuel gases, the pure oxygen produces _____.
   a. argon
   b. hydrogen
   c. a colorless, odorless, and tasteless gas
   d. a high-temperature flame needed for flame cutting

5. The smallest standard oxygen cylinder holds about _____ cubic feet of oxygen.
   a. 40
   b. 65
   c. 80
   d. 85

6. Acetylene gas must be withdrawn from a cylinder at an hourly rate that does *not* exceed _____ of the cylinder capacity.
   a. 1⁄10
   b. 1⁄5
   c. 1⁄3
   d. 1⁄2

7. Safety fuse plugs in the top and bottom of an acetylene cylinder are designed to _____.
   a. release acetylene gas in the event of a fire
   b. prevent the release of acetylene gas
   c. prevent the withdrawal of acetone
   d. release acetone from the cylinder

8. Methylacetylene propadiene (MAPP®) gas, a liquefied fuel used in oxyfuel cutting, burns at temperatures almost as high as acetylene and has the stability of _____.
   a. natural gas
   b. propylene
   c. propane
   d. oxygen

9. The amount of liquefied gas remaining in a cylinder is determined by the _____ of the cylinder.
   a. color
   b. heat
   c. weight
   d. pressure

10. The regulators used on fuel gas cylinders are often painted red and always have _____ on all connections.
    a. right-hand threads
    b. left-hand threads
    c. metric threads
    d. safety latches

11. The attachment on the top of a fuel gas cylinder that allows the gas to flow only in one direction is called a _____.
    a. single-stage regulator
    b. two-stage regulator
    c. flashback arrestor
    d. check valve

12. The cutting tips used with liquefied fuel gases must have at least _____ preheat holes.
    a. four
    b. five
    c. six
    d. seven

13. When lifting oxyfuel cutting cylinders, always use a(n) _____.
    a. sling cable
    b. sling strap
    c. cylinder cage
    d. electromagnet

MODULE 08104-06 ◆ OXYFUEL CUTTING 4.59

Review Questions

14. To avoid injury from dirt that may be lodged in the valve and regulator seat of a gas cylinder, always stand _____ the valve when opening the valve to clear the regulator seat.
    a. to the side of
    b. in front of
    c. behind
    d. above

15. When clearing debris from new oxyfuel cutting equipment hoses, blow the hoses out with _____.
    a. nitrogen
    b. propane
    c. propylene
    d. compressed air

16. The first step in installing a cutting tip is to _____.
    a. inspect the cutting tip sealing surfaces and orifices for damage
    b. determine the size of cutting tip to use
    c. determine the kind of gas being used
    d. identify the thickness of the material to be cut

17. Before opening cylinder valves, verify that the adjusting screws on the oxygen and fuel gas regulators are _____.
    a. tight
    b. fully clockwise
    c. loose
    d. fully counterclockwise

18. When a cutting flame has an excess of fuel, the flame is called a(n) _____ flame.
    a. cold
    b. neutral
    c. oxidizing
    d. carburizing

19. When disassembling oxyfuel equipment, verify that all pressure gauges read _____ before starting to take the equipment apart.
    a. –1.0
    b. 0
    c. within 0.3 of 0
    d. within 0.2 of 0

20. When inspecting a completed cut made with oxyfuel cutting equipment, the drag lines of the cut should be near _____ and *not* very pronounced.
    a. 3 degrees
    b. 45 degrees
    c. horizontal
    d. vertical

Instructor's Notes:

Trade Terms Quiz

Fill in the blank with the correct trade term that you learned from your study of this module.

1. During the cutting process, _________ is blown from the cut by the jet of cutting oxygen.
2. When the torch flame goes out and then burns back inside the torch with a hissing sound, it is called a(n) _________ and can be caused when the torch tip is too hot.
3. Steel and cast iron are common _________.
4. A flame with a white feather has too much fuel and is called a(n) _________.
5. Before a plate can be cut, the oxyfuel torch must _________ it.
6. If the torch tip touches the work, the torch flame could go out with a loud snap or pop called a(n) _________.
7. The soft white stone used to mark welding locations is called _________.
8. A torch flame burning with the correct proportions of oxygen and fuel gas is called a(n) _________.
9. Flux and nonmetallic impurities can form _________ during some welding processes.
10. A flame with an excess of oxygen is called a(n) _________.
11. A dirty cutting tip will result in an uneven _________.
12. In a good cut, the _________ on the kerf should be near vertical.

Trade Terms

Backfire
Carburizing flame
Drag line
Dross
Ferrous metals
Flashback
Kerf
Neutral flame
Oxidizing flame
Pierce
Slag
Soapstone

Have the trainees complete the Trade Terms Quiz, and go over the answers prior to administering the Module Examination.

Administer the Module Examination. Record the results on Craft Training Report Form 200, and submit the results to the Training Program Sponsor.

Administer the Performance Test, and fill out Performance Profile Sheets for each trainee. If desired, trainee proficiency noted during laboratory sessions may be used to complete the Performance Test. Record the results on Craft Training Report Form 200, and submit the results to the Training Program Sponsor.

MODULE 08104-06 ◆ OXYFUEL CUTTING 4.61

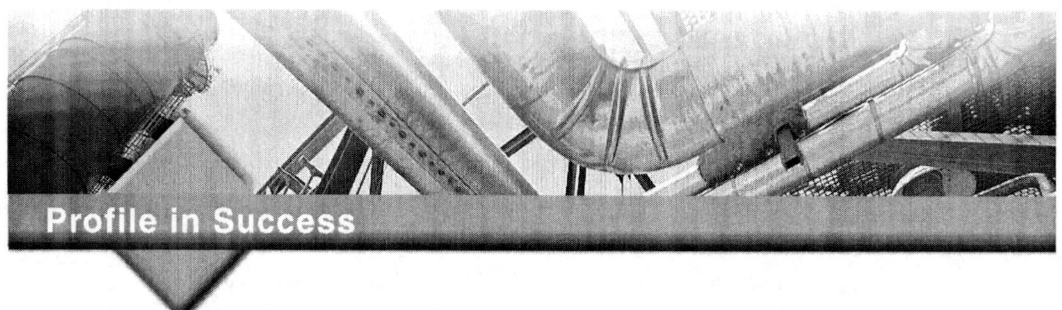

James M. Blair

Stracon International
Pipefitter Foreman

James "Matt" Blair, started out as a helper in his hometown in Pennsylvania. Matt teamed up with a pipe welder to travel as contract welder/fitters. His parents loaned him the money to start this new venture and they traveled around the United Sates working various jobs. Through one of his employers, he took NCCER courses in pipefitting. After completing his coursework, he participated in the Associated Builders and Contractors (ABC) Craft Championships and won first place in pipefitting. With this success, he was given an opportunity to become a pipefitting foreman and instructor.

How did you choose a career in the pipefitting field?
A friend of mine was a pipe welder. When he decided to travel, I jumped at the chance to be his fitter. It was a little scary borrowing money from my Mom and Dad to take off half way across America. However, several good things happened to me during this time. I met my future wife and made several good friends.

What types of training have you been through?
At first, my training was limited to on-the-job training and it went rather slowly. Then I met my future father-in-law, Jonny Millaway, who taught me the basics. Two years later, I was introduced to NCCER while I was on a job site. When I went to work for Starcon International, I became NCCER certified as a pipefitter and pipefitter instructor.

What kinds of work have you done in your career?
I started my career as a helper in Pennsylvania on a paper machine. From there, I worked all around the country for several contractors. I've worked on power houses and corn processing, pharmaceutical, water treatment, and petrochemical plants. My experience working in different environments helped get me to where I am today.

Tell us about your present job.
I am currently working for Starcon International in Illinois. Starcon has been very good to me. They gave me the opportunity to compete in the 2005 National Craft Championship after I completed my NCCER training. I placed first in the pipefitting division at the competition. My achievement was recognized by the company on the local and national levels.

After the competition, I earned a chance to prove myself as a Piping Foreman. From there, I have been able to run several small projects, which has let me further my education. I am currently also employed as an NCCER Instructor by Starcon where I get to work with several good apprentices.

What advice would you give to those new to pipefitter field?
Be on time and do a great job. If you want to be successful, you have to have good work habits and do quality work. Success breeds more success. If you give your company your best, they will give you opportunities to advance.

Learn everything you can. Find a good instructor and learn the tricks of the trade. Ask questions and soak up all the knowledge .

Instructor's Notes:

Trade Terms Introduced in This Module

Backfire: A loud snap or pop as a torch flame is extinguished.

Carburizing flame: A flame burning with an excess amount of fuel; also called a reducing flame.

Drag lines: The lines on the kerf that result from the travel of the cutting oxygen stream into, through, and out of the metal.

Dross: The material (oxidized and molten metal) that is expelled from the kerf when cutting using a thermal process.

Ferrous metals: Metals containing iron.

Flashback: The flame burning back into the tip, torch, hose, or regulator, causing a high-pitched whistling or hissing sound.

Kerf: The edge of the cut.

Neutral flame: A flame burning with correct proportions of fuel gas and oxygen.

Oxidizing flame: A flame burning with an excess amount of oxygen.

Pierce: To penetrate through metal plate with an oxyfuel cutting torch.

Slag: A nonmetallic product resulting from the mutual dissolution of flux and nonmetallic impurities in some welding and brazing processes.

Soapstone: Soft, white stone used to mark metal.

MODULE 08104-06 ◆ OXYFUEL CUTTING 4.63

Additional Resources

This module is intended to be a thorough resource for task training. The following reference works are suggested for further study. These are optional materials for continued education rather than for task training.

Safety in Welding, Cutting, and Allied Processes, ANSI Z49.1-99, 1999. Miami, FL: American Welding Society.

Welder's Handbook, Richard Finch, 1997. New York, NY: The Berkley Publishing Group, Inc.

Figure Credits

Controls Corporation of America, 104F01

Topaz Publications, Inc., 104F03–104F05, 104SA01–104SA04, 104F17, 104F20, 104F22–104F24, 104F26, 104F28, 104F29, 104F36–104F39, 104SA08–104SA10, 104F48–104F63, 104SA13, 104F65

Nederman, Inc., 104F06

3M, 104F07, 104F08, 104F11

Scott Health and Safety, 104F09, 104F10, 104F12

Brad Krauel, 104F14

Sellstrom Manufacturing Co., Inc., 104F15

J.R. Yochum, 104F19

Thermadyne, Inc., 104SA05, 104F31, 104F43, 104F47, 104F69, 104F73, 104F74

American Torch Tip Company, 104SA06, 104SA07

Lenco d/b/a NLC, Inc., 104F40

Koike Aronson, Inc., 104F41

ESAB Welding and Cutting Products, 104F42

Bug-O Systems, Inc., 104F44

H&M Pipe Beveling Machine Company, Inc., 104F45

Magnatech Limited Partnership, 104F46

American Welding Society, 104SA11, 104F67

Saf-T-Cart, 104SA12

Smith Equipment, 104F64

Zachary Construction Corporation, Title Page

Instructor's Notes:

4.64 PIPEFITTING ◆ LEVEL ONE

MODULE 02304-06 — TEACHING TIPS

The following are suggested activities or instructional methods to help you teach the material in this AIG.

General

When you call on someone to answer a question, the rest of the class relaxes or even tunes out because they expect that the question and answer will take place only between you and the trainee you called on. Instead, use this technique to involve more trainees in answering questions and to keep them on their toes.

1. Ask trainees to define a term or explain a concept.
2. After one trainee has answered, ask a trainee seated nearby if the answer is right. Then ask whether a trainee in the back of the room agrees.
3. Ask trainees to explain why they think an answer is right or wrong.
4. Use the session to clear up incorrect ideas and encourage trainees to learn from their mistakes.

Section 2.0.0 *Oxyfuel Cutting Equipment Safety*

Pipefitters must be able to operate oxyfuel cutting equipment safely. This exercise will familiarize trainees with various aspects of welding safety and safety equipment. Trainees will need pencils and paper. Obtain a safety video or arrange for a safety professional to give a presentation on oxyfuel cutting equipment safety. Allow 30 to 45 minutes for this exercise.

Obtain one of the following or another safety training video:

Welding Safety: Safe Work With Hotwork. 18 minutes. Coastal Training Technologies Corp. www.coastal.com. Rental $95.

Welding Safety. 13 minutes. National Safety Compliance. www.osha-safety-training.net. $195.

Cutting Torch Safety. 14 minutes. The Training Network. www.safetytrainingnetwork.com. $64.98.

1. Introduce the video or speaker who will give a presentation on oxyfuel cutting equipment safety.
2. Have the trainees take notes and write down questions during the video or presentation.
3. After the video or presentation answer any questions the trainees may have.

Section 3.12.0 *Specialized Cutting Equipment*

Pipefitters must be familiar with different methods used to cut steel. This exercise will familiarize the trainees with specialized equipment used to cut pipe. Trainees will need appropriate personal protective equipment, pencils, and paper. Arrange for an opportunity to tour an industrial metal fabricator or have a manufacturer's representative give a presentation to the class. Allow 1 to 1½ hours for this exercise.

1. Before the tour or presentation, discuss specialized cutting equipment.
2. Have trainees observe the operations of various types of cutting equipment, including gantry cutting machines, pipe cutters/bevelers, and exothermic oxygen lances. Have trainees take notes and write down questions during the tour or presentation.
3. After the tour or presentation answer any questions the trainees may have.

Section 4.12.0 *Leaking Cylinders*

Pipefitters work with various types of cylinders that hold gases for cutting operations. This exercise will familiarize trainees with cylinder and regulator repair. Trainees will need appropriate personal protective equipment, pencils, and paper. Arrange for an opportunity to tour a welding supply repair facility or have a manufacturer's representative give a presentation to the class. Allow 30 to 45 minutes for this exercise.

1. Before the tour or presentation, discuss leak testing for cylinders and regulators.
2. Have trainees observe how various types of cylinders, regulators, and torches are tested and repaired. Have trainees take notes and write down questions during the tour or presentation.
3. After the tour or presentation answer any questions the trainees may have.

MODULE 02304-06 — ANSWERS TO REVIEW QUESTIONS

Answer	Section
1. d	2.2.0
2. d	2.5.0
3. c	2.6.0
4. d	3.1.0
5. d	3.1.1
6. a	3.2.1
7. a	3.2.1
8. c	3.3.0
9. c	3.3.1
10. b	3.4.0
11. d	3.4.3
12. c	3.7.2
13. c	4.1.0
14. a	4.2.0
15. a	4.5.0
16. d	4.8.0
17. c	4.9.0
18. d	5.1.0
19. b	7.0.0
20. d	9.1.0

Answers to Trade Terms Quiz

1. Dross
2. Flashback
3. Ferrous metals
4. Carburizing flame
5. Pierce
6. Backfire
7. Soapstone
8. Neutral flame
9. Slag
10. Oxidizing flame
11. Kerf
12. Drag lines

CONTREN® LEARNING SERIES — USER UPDATE

NCCER makes every effort to keep these textbooks up-to-date and free of technical errors. We appreciate your help in this process. If you have an idea for improving this textbook, or if you find an error, a typographical mistake, or an inaccuracy in NCCER's Contren® textbooks, please write us, using this form or a photocopy. Be sure to include the exact module number, page number, a detailed description, and the correction, if applicable. Your input will be brought to the attention of the Technical Review Committee. Thank you for your assistance.

Instructors – If you found that additional materials were necessary in order to teach this module effectively, please let us know so that we may include them in the Equipment/Materials list in the Annotated Instructor's Guide.

Write: Product Development and Revision
National Center for Construction Education and Research
P.O. Box 141104, Gainesville, FL 32614-1104

Fax: 352-334-0932

E-mail: curriculum@nccer.org

Craft ______________________ Module Name ______________________

Copyright Date __________ Module Number __________ Page Number(s) __________

Description

____

____

____

(Optional) Correction

____

____

(Optional) Your Name and Address

____

____

Module 08105-06

Ladders and Scaffolds

NCCER STANDARDIZED CRAFT TRAINING PROGRAM

The National Center for Construction Education and Research (NCCER) provides a standardized national program of accredited craft training. Key features of the program include instructor certification, competency-based training, and performance testing. The program provides trainees, instructors, and companies with a standard form of recognition through the National Registry. The program is described in full in the *Guidelines for Accreditation*, published by NCCER. For more information on standardized craft training, contact NCCER by writing to P.O. Box 141104, Gainesville, FL 32614-1104; calling 352-334-0911; or e-mailing info@nccer.org. More information is available at www.nccer.org.

HOW TO USE THIS ANNOTATED INSTRUCTOR'S GUIDE

Each page presents two sections of information. The larger section displays each page exactly as it appears in the Trainee Module. The narrow column ties suggested trainee and instructor actions to each page and provides icons (detailed below) to call your attention to material, safety, audiovisual, or testing requirements. The bottom of each page includes space for your notes.

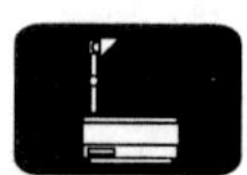

The **Audiovisual** icon indicates an appropriate time to show a transparency or other audiovisual aid.

The **Classroom** icon prompts you to define a term, stress a point, ask trainees to explain a concept, or give examples.

The **Demonstration** icon directs you to show trainees how to perform tasks.

The **Examination** icon tells you to administer the written module examination.

The **Homework** icon is placed where you may wish to assign reading for the next class, assign a project, or advise trainees to prepare for an examination.

The **Laboratory** icon is used when trainees are to practice performing tasks.

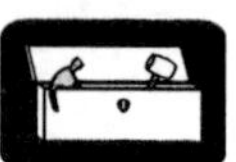

The **Materials** icon is a reminder for you to gather materials needed for classes, laboratories, and testing.

The **Performance Testing** icon tells you to administer a performance test or a portion thereof.

The **Safety** icon is used to emphasize safety issues. It is often keyed to *Caution* and *Warning!* statements in the Trainee Module.

The **Teaching Tip** icon indicated additional guidance is available, such as how to conduct an exercise, get the most educational value from a field trip, or encourage class participation. Teaching Tips may expand on a feature (*Think About It, Did You Know?*) or provide *Quick Quizzes* or similar exercises. You will be referred to the Teaching Tips section at the back of the module if there is additional material.

The **Combination** icon indicates that the laboratory listed corresponds with a performance task. If desired, you can note the proficiency of the trainees during the laboratory, and use it to satisfy performance testing requirements.

PREPARATION

Before teaching this module, you should review the Objectives, Performance Tasks, Materials and Equipment List, and Module Outline. Be sure to allow ample time to prepare your own training or lesson plan and gather all required materials and equipment.

MODULE OVERVIEW

This module covers hazards and general safety procedures governing the use of stepladders, straight and extension ladders, fixed scaffolds, and rolling scaffolds.

PREREQUISITES

Prior to training with this module, it is recommended that the trainee shall have successfully completed *Core Curriculum;* and *Pipefitting Level One*, Modules 08101-06 through 08104-06.

OBJECTIVES

Upon completion of this module, the trainee will be able to do the following:

1. Identify the different types of ladders and scaffolds used on a work site.
2. Describe how to safely use ladders and scaffolding.
3. Properly set up, inspect, and use stepladders, extension ladders, and scaffolding.

PERFORMANCE TASKS

Under the supervision of the instructor, the trainee should be able to do the following:

1. Select, inspect, and use stepladders.
2. Select, inspect, and use straight and extension ladders.
3. Erect, inspect, and disassemble tubular buck scaffolding.

MATERIALS AND EQUIPMENT LIST

Overhead projector and screen
Transparencies
Blank acetate sheets
Transparency pens
Whiteboard/chalkboard
Markers/chalk
Pencils and scratch paper
Appropriate personal protective equipment
Stepladder
Platform ladder
Straight ladder
Extension ladder
Personal fall arrest system
Pliers
Four base plates
Four caster wheels
Four leveling jacks
Four toe boards
Transporter
Twelve hinge pins
Two middle guard rails
Two scaffold cross braces
Two scaffold planks with safety
Two scaffold upper end frames
Two scaffold vertical supports
Two top guardrails
Company safety manual with procedures for fall protection and rescue after a fall
Scaffolding tags
OSHA requirements for scaffolds: *29 CFR 1926.450, Subpart L Scaffolds*
Television with VCR or DVD (optional)
Safety training video (optional)
Module Examinations*
Performance Profile Sheets*

* Located in the Test Booklet.

SAFETY CONSIDERATIONS

Ensure that the trainees are equipped with appropriate personal protective equipment and know how to use it properly. This module requires trainees to use ladders and scaffolding. Review fall hazards and personal fall arrest systems.

ADDITIONAL RESOURCES

This module is intended to present thorough resources for task training. The following reference work is suggested for both instructors and motivated trainees interested in further study. This is optional material for continued education rather than for task training.

Occupational Safety and Health Standards for the Construction Industry, Latest Edition. Occupational Safety and Health Administration. U.S. Department of Labor. Washington, DC: U.S. Government Printing Office.

TEACHING TIME FOR THIS MODULE

An outline for use in developing your lesson plan is presented below. Note that each Roman numeral in the outline equates to one session of instruction. Each session has a suggested time period of 2½ hours. This includes 10 minutes at the beginning of each session for administrative tasks and one 10-minute break during the session. Approximately 12.5 hours are suggested to cover *Ladders and Scaffolds*. You will need to adjust the time required for hands-on activity and testing based on your class size and resources. Because laboratories often correspond to Performance Tasks, the proficiency of the trainees may be noted during these exercises for Performance Testing purposes.

Topic	Planned Time
Sessions I and II. Introduction and Ladders	
A. Introduction	____________
B. Stepladders	____________
C. Laboratory – Trainees practice selecting, inspecting, and using a stepladder. This laboratory corresponds to Performance Task 1.	____________
D. Straight and Extension Ladders	____________
E. Laboratory – Trainees practice selecting, inspecting, and using straight and extension ladders. This laboratory corresponds to Performance Task 2.	____________
Sessions III and IV. Scaffolding	
A. Using and Caring for Tubular Buck Scaffolding	____________
B. Using and Caring for Pole Scaffolding	____________
C. Rolling Scaffolding	____________
D. Scaffolding Hazards	____________
E. Scaffolding Safety Guidelines	____________
F. Rescue After a Fall	____________
G. Laboratory – Trainees practice erecting, inspecting, and disassembling scaffolding. This laboratory corresponds to Performance Task 3.	____________
Session V. Review, Module Examination, and Performance Testing	
A. Review	____________
B. Module Examination	____________
1. Trainees must score 70 percent or higher to receive recognition from NCCER.	
2. Record the testing results on Craft Training Report Form 200, and submit the results to the Training Program Sponsor.	
C. Performance Testing	____________
1. Trainees must perform each task to the satisfaction of the instructor to receive recognition from NCCER. If applicable, proficiency noted during laboratory exercises can be used to satisfy the Performance Testing requirements.	
2. Record the testing results on Craft Training Report Form 200, and submit the results to the Training Program Sponsor.	

Pipefitting Level One

Assign reading of Module 08105-06.

08105-06

Ladders and Scaffolds

08105-06
Ladders and Scaffolds

Topics to be presented in this module include:

Overview

Ladders and scaffolding are some of the most important tools on a job site. They provide access to work areas above the ground level. However, when working above ground, carelessness can be fatal. Common accidents, like falling, being struck by falling objects, and electrocution, can be avoided if safety precautions are followed. Ladders and scaffolding have capacity limitations that cannot be exceeded.

Various ladders are designed for different uses. Select the one best suited for the job and inspect it before using it. Scaffolding is used to create a work platform above ground. In many cases, only a certified person can build scaffolding. In some cases, a fall arrest system must be used with scaffolding. Used properly, ladders and scaffolding make the pipefitter's job much easier.

Instructor's Notes:

Objectives

When you have completed this module, you will be able to do the following:

1. Identify the different types of ladders and scaffolding used on a work site.
2. Describe how to safely use ladders and scaffolding.
3. Properly set up, inspect, and use step ladders, extension ladders, and scaffolding.

Trade Terms

Base plates	Ladder
Casters	Leveling jack
Coupling pin	Scaffold
Cross braces	Scaffolding
Duty rating	Scaffold floor
Guardrails	Toeboard
Harness	Vertical upright
Hinge pin	

Required Trainee Materials

1. Pencil and paper
2. Appropriate personal protective equipment

Prerequisites

Before you begin this module, it is recommended that you successfully complete *Core Curriculum* and *Pipefitting Level One,* Modules 08101-06 through 08104-06.

This course map shows all of the modules in the first level of the Pipefitting curriculum. The suggested training order begins at the bottom and proceeds up. Skill levels increase as you advance on the course map. The local Training Program Sponsor may adjust the training order.

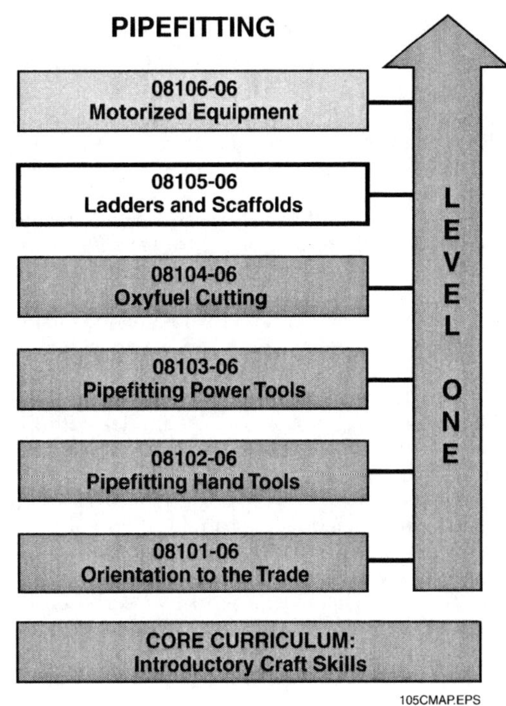

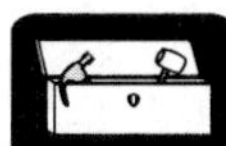

Ensure that you have everything required to teach the course. Check the Materials and Equipment list at the front of this module.

See the general Teaching Tip at the end of this module.

Explain that terms shown in bold are defined in the Glossary at the back of this module.

Review the modules covered in Level One and explain how this module fits in.

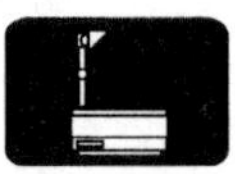

Show Transparency 1, Objectives, and Transparency 2, Performance Tasks. Review the goals of the module, and explain what will be expected of the trainee.

MODULE 08105-06 ◆ LADDERS AND SCAFFOLDS 5.1

Explain that pipefitters use ladders and scaffolding to work above ground level.

Discuss the safety hazards posed when using ladders and scaffolding.

Describe different types of ladders.

Provide different types of ladders for the trainees to examine.

1.0.0 ◆ INTRODUCTION

Ladders and **scaffolding** are common sights on construction jobs. They permit employees to work safely at elevated levels. Ladders can be used for short distances, while scaffolding can be used when the work site is many stories above the ground. See *Figure 1.*

Unsafe use of ladders and **scaffolds** has always been a safety problem in the workplace. Any time work is done above ground, there is a chance that an accident can happen. The most common accidents associated with ladders and scaffolding are falling, being struck by falling objects, and electrocution.

The risk of accidents can be greatly reduced if ladders and scaffolds are properly inspected before use. If the worker is also familiar with the safe operation of ladders and scaffolds, the chance of an accident can be reduced. Workers must always remember to wear a safety **harness** when working from a ladder or scaffold.

When working above ground, one careless mistake can mean the difference between life and death. In one instance, a worker died when he slipped from a fixed ladder attached to a water tower and fell 40 feet to the ground. He died because he was not using the proper fall protection equipment.

Safety is your top priority on a job. It is your responsibility to learn how to set up, use, and maintain equipment. Not only will this keep you safe, it could save the lives of your co-workers.

2.0.0 ◆ LADDERS

Ladders are some of the most important tools on a job site. They are designed to assist workers in reaching levels above the normal reaching point. The basic types of ladders are the stepladder, straight ladder, and extension ladder. Different types of ladders are used in different situations (*Figure 2*). Aluminum ladders are corrosion-resistant and can be used in situations where they might be exposed to the elements. They are also lightweight and can be used where they need to be frequently lifted and moved. Wooden ladders, which are heavier and sturdier than fiberglass or aluminum ladders, can be

105F01.EPS

Figure 1 ◆ Scaffolding can be many stories above the ground.

Figure 2 ◆ Types of ladders.

Instructor's Notes:

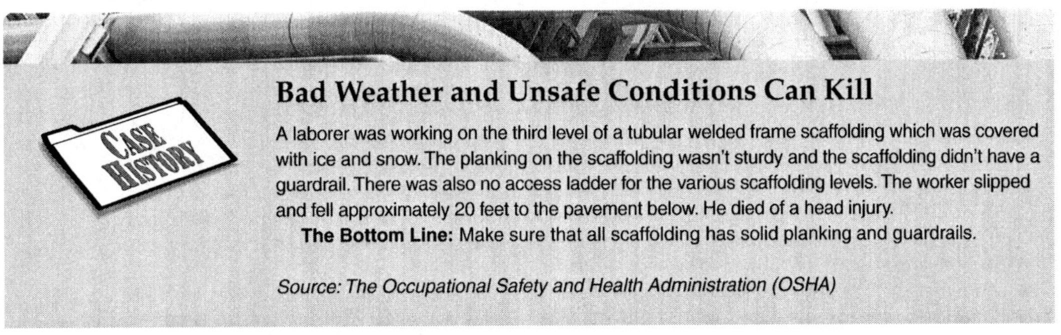

Bad Weather and Unsafe Conditions Can Kill

A laborer was working on the third level of a tubular welded frame scaffolding which was covered with ice and snow. The planking on the scaffolding wasn't sturdy and the scaffolding didn't have a guardrail. There was also no access ladder for the various scaffolding levels. The worker slipped and fell approximately 20 feet to the pavement below. He died of a head injury.

The Bottom Line: Make sure that all scaffolding has solid planking and guardrails.

Source: The Occupational Safety and Health Administration (OSHA)

used when heavy loads must be moved up and down. Fiberglass ladders are nonconductive and also very durable, so they are useful in situations involving electrical work or where some amount of rough treatment is unavoidable. Both fiberglass and aluminum are easier to clean than wood.

Selecting the right ladder for the job at hand is important to completing a job as safely and efficiently as possible. When selecting a ladder, consider its features and how it meets the needs of the job. Always consider the highest **duty rating** and weight limit needed, as well as the height requirements. A ladder that is too long or too short will not allow the work surface to be reached easily, safely, or comfortably.

The following safety guidelines must be followed when using any type of ladder:

- Inspect ladders carefully before each use. Test all working parts for proper attachment and operation.
- Never use a ladder with broken or missing rungs or steps, broken or split side rails, or other faulty construction.
- When defective ladders are discovered, first remove them from service and clearly mark them with *Do Not Use*. Then report the defect to the appropriate person on your work site or to your supervisor.
- Install and use ladders in compliance with safe practices and with all applicable governmental regulations, codes, and ordinances.
- Keep steps and rungs, as well as the soles of your shoes, free of grease, oil, paint, and other slippery substances.
- Place ladder feet on a firm, suitable surface, and keep the area around the bottom and the top of the ladder clear.
- Place the ladder so that it leaves 6 inches of clearance in back of the ladder and 30 inches of clearance in front of the ladder.
- Place the ladder so that it leans against a solid and immovable surface. Never place a ladder against a window, door, doorway, sash, loose or movable wall, or box.
- Do not place a ladder in a doorway, passageway, driveway, or other area where it is in the way of any other work unless it is protected by barricades or guards.
- Always use appropriate safety feet or nonslip bases. If the ladder has to be placed on a slippery surface, take additional precautions.
- Never use a ladder in the horizontal position as a platform, walkboard, or scaffold.
- Do not use metal, metal-reinforced, or wet ladders where direct contact with a live power source is possible. Use extreme caution when working near electrical wires, service, and equipment.
- Face the ladder when climbing up or down.
- Most ladders are intended to carry one person at a time. Do not overload.
- Double stepladders are designed with steps on the front and back so they can be climbed from either side. These ladders are usually identified with a sticker that indicates that both sides may be used to climb. When two people use the ladder, the maximum load limit must not be exceeded.

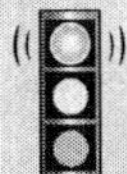

WARNING!

If you aren't sure about whether a stepladder can be climbed from either side, ask your supervisor for help before you use the ladder. You may prevent an accident.

- Climb or descend the ladder one rung at a time. Never run up or slide down a ladder.
- Do not use ladders during high winds. If you must use a ladder in windy conditions, make sure you securely lash the ladder to prevent slippage.
- Keep both hands free so that you can hold the ladder securely while climbing. Use a rope to raise and lower any tools and materials that you might need.

Discuss considerations for selecting the right ladder for a job. Define duty rating.

Discuss general safety considerations when using ladders.

Review the case history. Discuss the additional hazards posed when working with ladders in bad weather.

MODULE 08105-06 ◆ LADDERS AND SCAFFOLDS 5.3

Discuss the importance of securing ladders and other safety considerations.

Describe a stepladder and identify the components.

Explain how to inspect a stepladder.

Review the safety rules for using stepladders.

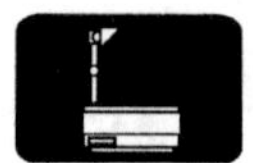

Show Transparency 3 (Figure 5). Discuss common ladder safety precautions.

- Never rest any tools or materials on the top of a ladder.
- Move the ladder in line with the work to be done. Never lean sideways away from the ladder in order to reach the work area.
- Store ladders in a cool, dry, ventilated place and provide easy access for inspection. Provide sufficient support to prevent sagging when laying ladders flat.
- Block, tie, or otherwise secure portable ladders in use to prevent them from being displaced.
- Use ladders only for short periods of elevated work. If you must work from a ladder for extended periods, use a lifeline fastened to a safety belt.
- Lay the ladder on the ground when you have finished using it, unless it is anchored securely at the top and bottom where it is being used.
- Never use makeshift substitutes for ladders.
- Never use stepladders for straight-ladder work.
- Do not paint wood ladders, because paint can hide defects and flaws.
- Do not change or alter any type of ladder.
- Do not stand or sit on the ladder top or the top step of the ladder.
- Use a stabilizer leg (*Figure 3*) if the ladder is on uneven ground. Do not use wood blocks or bricks.

105F03.EPS

Figure 3 ◆ Stabilizer leg.

2.1.0 Stepladders

Stepladders are self-supporting ladders made of two sections hinged at the top (*Figure 4*). The section of a stepladder used for climbing consists of rails and rungs like those on straight ladders. The other section consists of rails and braces. Spreaders are the hinged arms between the sections that keep the ladder stable and prevent it from folding while in use. A stepladder may have a pail shelf to hold paint or tools.

Inspect stepladders the way you inspect straight and extension ladders. Pay special attention to the hinges and spreaders to be sure they are in good repair. Be sure the rungs are clean. A stepladder's rungs are usually flat, so oil, grease, or dirt can easily build up on them and make them slippery.

Follow these rules when using stepladders:

- Be sure that all four feet are on a hard, even surface when you position a stepladder. If they're not, the ladder can rock from side to side or corner to corner when you climb it.
- Never stand on the top step or the top of a stepladder. Putting your weight this high will make the ladder unstable. The top of the ladder is made to support the hinges, not to be used as a step.
- Make sure the spreaders are locked in the fully open position when the ladder is in position.
- If the ladder is designed to be used from only one side, never use the braces for climbing even though they may look like rungs. They are not designed to support your weight.
- Some stepladders are designed to be used from either side, but others are not. Be sure you know if the ladder you are using can be climbed from either side before you use it.

Figure 5 shows some common ladder safety precautions.

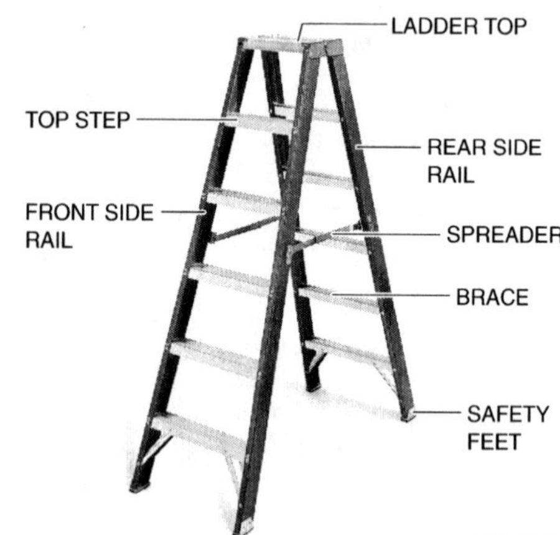

105F04.EPS

Figure 4 ◆ Typical stepladder.

Instructor's Notes:

5.4 PIPEFITTING ◆ LEVEL ONE

DOs

- Be sure your ladder has been properly set up and is used in accordance with safety instructions and warnings.
- Wear shoes with non-slip soles.

- Keep your body centered on the ladder. Hold the ladder with one hand while working with the other. Never let your belt buckle pass beyond either ladder rail.

- Move materials with extreme caution. Be careful pushing or pulling anything while on a ladder. You may lose your balance or tip the ladder.

- Get help with a ladder that is too heavy to handle alone. If possible, have another person hold the ladder when you are working on it.

- Climb facing the ladder. Center your body between the rails. Maintain a firm grip.
- Always move one step at a time, firmly setting one foot before moving the other.

- Haul materials up on a line rather than carry them up an extension ladder.
- Use extra caution when carrying anything on a ladder.

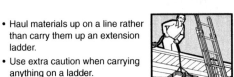

Read ladder labels for additional information.

DON'Ts

- DON'T stand above the highest safe standing level.
- DON'T stand above the second step from the top of a stepladder and the 4th rung from the top of an extension ladder. A person standing higher may lose their balance and fall.

- DON'T climb a closed stepladder. It may slip out from under you.
- DON'T climb on the back of a stepladder. It is not designed to hold a person.

- DON'T stand or sit on a stepladder top or pail shelf. They are not designed to carry your weight.
- DON'T climb a ladder if you are not physically and mentally up to the task.

- DON'T exceed the Duty Rating, which is the maximum load capacity of the ladder. Do not permit more than one person on a single-sided stepladder or on any extension ladder.

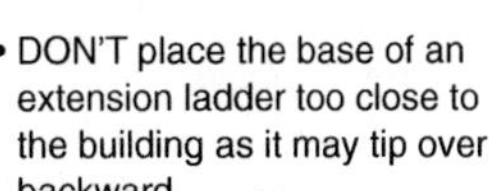

- DON'T place the base of an extension ladder too close to the building as it may tip over backward.
- DON'T place the base of an extension ladder too far away from the building, as it may slip out at the bottom. **Please refer to the 4 to 1 Ratio Box.**

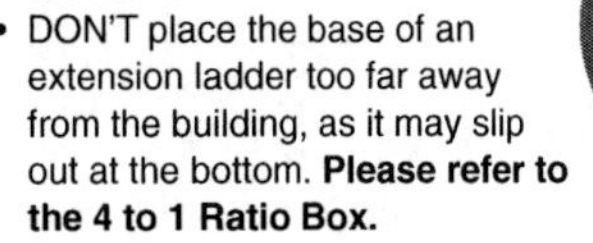

- DON'T over-reach, lean to one side, or try to move a ladder while on it. You could lose your balance or tip the ladder. **Climb down and then reposition the ladder closer to your work!**

4 TO 1 Ratio

Place an extension ladder at a 75-1/2° angle. The set-back ("S") needs to be 1 ft. for each 4 ft. of length ("L") to the upper support point.

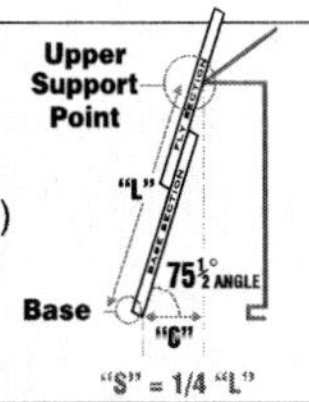

105F05.EPS

Figure 5 ◆ Ladder safety.

MODULE 08105-06 ◆ LADDERS AND SCAFFOLDS 5.5

Explain how to use and care for a stepladder.

Discuss how to carry a ladder safely. Explain the hazards of trying to move a ladder while on it. Explain how to determine if the ladder can be safely climbed from both sides. Stress the importance of descending safely.

Show trainees how to how to select, inspect, and use a stepladder.

Have trainees practice selecting, inspecting, and using a stepladder. Note the proficiency of each trainee. This laboratory corresponds to Performance Task 1.

Show Transparency 4 (Figure 6). Describe a straight ladder.

All the general safety guidelines for ladders apply to stepladders. Follow these steps to use and care for a stepladder:

Step 1 Determine the height of the work to be performed.

Step 2 Select the type of ladder best suited for the job.

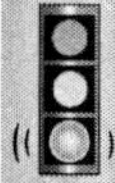

NOTE
According to OSHA standards, industrial stepladders are designed for heavy-duty work. They may not exceed 20 feet in height.

Step 3 Inspect the selected ladder for obvious damage, such as cracks or breaks, dirt, grease, and dry rot (on wooden ladders), and make sure that all steps are securely attached to the side rails.

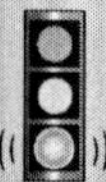

NOTE
Clean any dirt, grease, or rust off the ladder, especially the foot rungs. Replace the ladder if there is any damage.

Step 4 Carry the ladder to the work area.

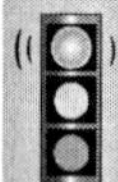

WARNING!
Be careful when carrying a ladder not to injure other people or damage other equipment with the ladder. If the ladder is too heavy or awkward, get help.

Step 5 Place the ladder into position for the job.

Step 6 Make sure that all four feet are on a firm, flat surface and that the spreader is locked in the fully open position.

Step 7 Climb up the ladder using both hands to hold on to the sides and rungs.

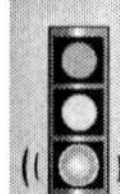

NOTE
Tools to be used while you are up on a ladder should be either attached to you or lifted by rope after you are in position.

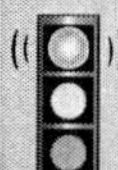

WARNING!
Some stepladders can be climbed from both sides, but many others can be safely climbed from one side only. Stepladders that can be safely climbed from only one side will have braces on their back side to stabilize them. Although the braces look like rungs, they will not support your weight. Always be sure of the type of ladder that you are using. When in doubt, ask your supervisor for help. Do not step on or above the top step of a stepladder. Maintain your balance at all times. Do not reach out to the sides beyond normal arm length or lean out.

Step 8 Perform the work from the ladder.

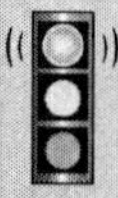

WARNING!
Do not try to move the ladder while on it. If it must be moved, climb down first, then move it. Be careful not to drop anything from a ladder. Remember to stay off the top step of the stepladder.

Step 9 Climb down the ladder using both hands to hold onto the sides and rungs.

WARNING!
Do not slide down the outer rails. Do not try to turn around to climb down.

Step 10 Lower and fold the ladder.

Step 11 Make sure the ladder is clean and in good condition before storing it.

Step 12 Return the ladder to the storage location.

2.2.0 Straight and Extension Ladders

Straight ladders consist of two rails, rungs between the rails, and safety feet on the bottom of the rails (*Figure 6*). The straight ladders used in construction are generally made of wood or fiberglass. Metal ladders conduct electricity and should not be used around electrical equipment.

Straight ladders are typically 8 to 12 feet in length and are similar to one section of an extension ladder. The most important rule to remember when using a straight or extension ladder is to maintain the correct ladder angle. Always place the ladder so that

Instructor's Notes:

Review the case history and discuss the importance of following safety instructions.

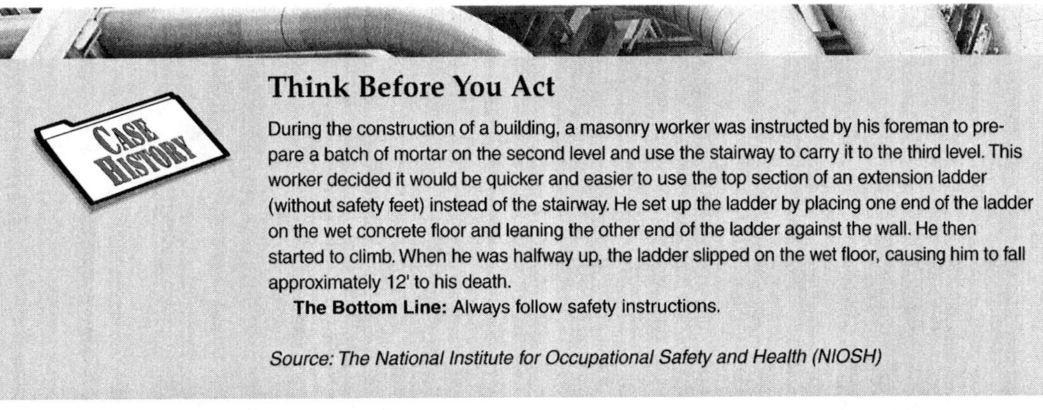

Think Before You Act

During the construction of a building, a masonry worker was instructed by his foreman to prepare a batch of mortar on the second level and use the stairway to carry it to the third level. This worker decided it would be quicker and easier to use the top section of an extension ladder (without safety feet) instead of the stairway. He set up the ladder by placing one end of the ladder on the wet concrete floor and leaning the other end of the ladder against the wall. He then started to climb. When he was halfway up, the ladder slipped on the wet floor, causing him to fall approximately 12' to his death.

The Bottom Line: Always follow safety instructions.

Source: The National Institute for Occupational Safety and Health (NIOSH)

Show Transparency 5 (Figure 7). Explain how to properly position a straight ladder.

the distance between the base of the ladder and the supporting wall is one-fourth the working distance of the ladder. The working distance of the ladder is the distance between the base of the ladder and the point where the ladder makes contact with the supporting wall. *Figure 7* shows the correct ladder angle. If the base of the ladder is too far from the wall, the ladder can slip out from under you. If the base of the ladder is too close to the wall, it is more likely to tip over backwards.

Explain how to inspect a straight ladder.

Ladders should always be inspected before use. Be sure to follow these guidelines when inspecting a ladder:

- Check the rails and rungs for cracks or other damage, including loose rungs. If you find any damage, do not use the ladder.
- Check the entire ladder for loose nails, screws, brackets, or other hardware. If you find any hardware problems, tighten the loose parts or have the ladder repaired before you use it.
- Make sure the feet are securely attached and that they are not damaged or worn down (Figure 8). Do not use a ladder if its safety feet are not in good working order.

Show Transparency 6 (Figure 8). Explain how to inspect the feet of a ladder.

Explain how to set up a straight ladder safely.

Setup is the next step after inspecting a ladder. Use these guidelines to make sure you are setting up the ladder safely:

- Place the straight ladder at the proper angle before using it. A ladder placed at an improper angle will be unstable and could cause you to fall.
- Straight ladders should be used only on stable and level surfaces unless they are secured at both the bottom and the top to prevent any accidental movement (*Figure 9*).
- The distance between the foot of a ladder and the base of the structure it is leaning against must be one-quarter the distance between the ground and the point where the ladder touches the structure.

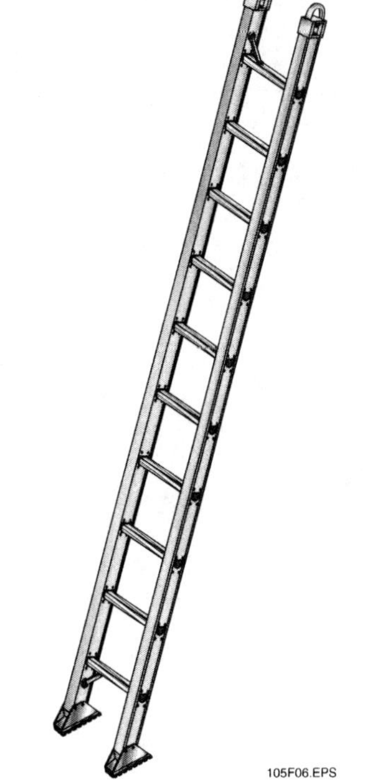

Figure 6 ◆ Straight ladder.

Show Transparency 7 (Figure 9). Explain how to secure a straight ladder on the top and bottom.

Once you've inspected and set up the ladder, you can begin working on it. Here are some safeguards to use while working on a ladder:

- Never try to move a ladder while someone is on it.
- If a ladder must be placed in front of a door that opens toward the ladder, the door must be locked or blocked open so that it cannot strike the ladder.
- Never use a ladder as a work platform by placing it horizontally.

MODULE 08105-06 ◆ LADDERS AND SCAFFOLDS 5.7

Discuss guidelines for working safely on a straight ladder.

Figure 7 ◆ Proper positioning.

Figure 8 ◆ Ladder safety feet.

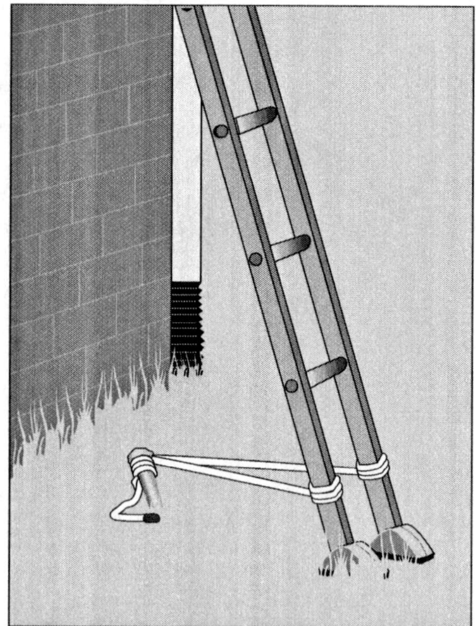

BOTTOM SECURED

TOP SECURED

105F09.EPS

Figure 9 ◆ Securing a ladder.

Instructor's Notes:

5.8 PIPEFITTING ◆ LEVEL ONE

- Make sure the ladder you are about to climb or descend is properly secured.
- Make sure the ladder's feet are solidly positioned on firm, level ground.
- Ensure that the top of the ladder is firmly positioned and in no danger of shifting once you begin your climb.
- Keep both hands on the rails when climbing a straight ladder.
- Always keep your body's weight in the center of the ladder between the rails.
- Never go up or down a ladder while facing away from it. Face the ladder at all times.
- Don't carry tools in your hands while you are climbing a ladder. Use a hand line or tagline and pull tools up once you've reached the place you will be working.

> **WARNING!**
> Remember that the addition of your own weight will affect the ladder's steadiness once you mount it. It is important to test the ladder first by applying some of your weight to it without actually beginning to climb. This way, you will be sure that the ladder remains steady as you ascend.

An extension ladder is actually two straight ladders. They are connected so you can adjust the overlap between them and change the length of the ladder as needed (*Figure 10*).

Extension ladders are positioned and secured following the same rules as straight ladders. There are, however, some safety rules that are unique to extension ladders:

- Inspect the rope used to raise the moveable section of an extension ladder before each use for frayed and worn spots. Replace damaged ropes.
- Inspect rung locks closely for damage. Do not use the ladder if the rung locks do not work correctly.
- Make sure the section locking mechanism is fully hooked over the desired rung.
- Make sure that all ropes used for raising and lowering the extension are clear and untangled.
- Make sure the extension ladder overlaps between the two sections (*Figure 11*). For ladders up to 36 feet long, the overlap must be at least 3 feet. For ladders 36 to 48 feet long, the overlap must be at least 4 feet. For ladders 48 to 60 feet long, the overlap must be at least 5 feet.

Figure 10 ◆ Extension ladders.

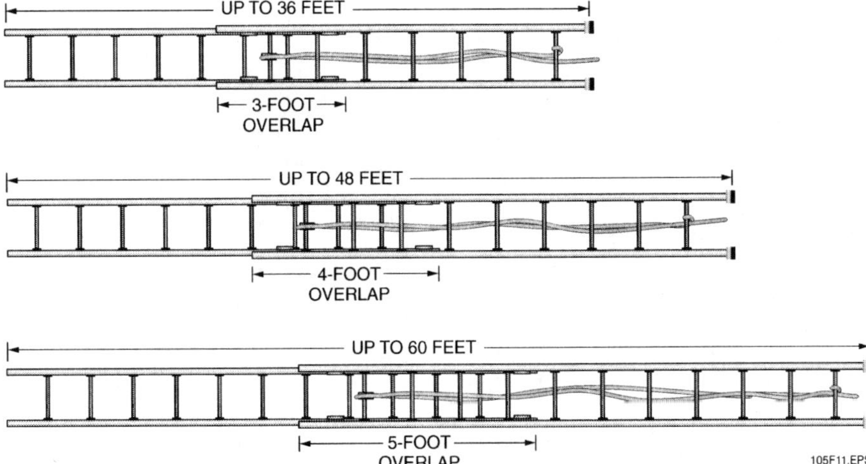

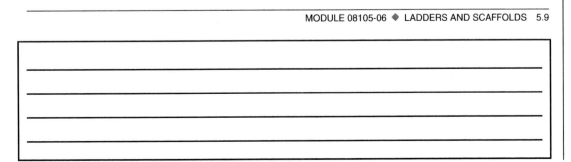

Figure 11 ◆ Overlap lengths for extension ladders.

Emphasize the importance of testing the ladder before putting your full weight on it.

Describe an extension ladder.

Explain that extension ladders are positioned and secured following the same rules as a straight ladder.

Show Transparency 8 (Figure 11). Discuss overlap lengths on an extension ladder.

Provide a straight ladder and an extension ladder for the trainees to examine.

MODULE 08105-06 ◆ LADDERS AND SCAFFOLDS 5.9

Discuss safety guidelines for working on extension ladders.

Explain how to use and care for an extension or straight ladder.

Discuss how to carry a ladder safely and the hazards posed when raising an extension ladder. Explain the hazards of trying to move a ladder while on it. Stress the importance of descending safely.

Show trainees how to select, inspect, and use straight and extension ladders.

Have trainees practice selecting, inspecting, and using straight and extension ladders. Note the proficiency of each trainee. This laboratory corresponds to Performance Task 2.

- Never stand above the highest safe standing level on a ladder. On an extension ladder, this is the fourth rung from the top. If you stand higher, you may lose your balance and fall. Some ladders have colored rungs to show where you should not stand.
- Always adjust the ladder from the bottom, never from the top. The rung locks cannot always be seen from the top.
- Rungs must be on 12-inch centers.
- Make sure that the rungs have been treated or constructed to prevent slipping.
- Two-section ladders must not exceed 48 feet, and three-section ladders must not exceed 60 feet.
- If you are going to step off the top of a ladder onto a roof or platform, the side rails of the ladder must extend over the top of the roof or platform by at least 36 inches. If this is not possible, grab rails must be installed on the top of the ladder to provide a secure grip for personnel.
- When using extension ladders, make sure that the narrow, smaller section is over the wider section.
- Do not splice ladders together to make longer ladders.

Follow these steps to use and care for an extension or straight ladder:

Step 1 Determine the height of the work to be performed.

Step 2 Select the type of ladder best suited for the job.

Step 3 Inspect the selected ladder for obvious damage, such as cracks or breaks, dirt, grease, and dry rot (on wooden ladders). Check that the ropes on an extension ladder are in good condition.

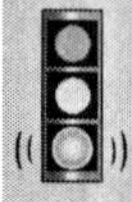

NOTE

Clean any dirt, grease, or rust off the ladder, especially from the foot rungs. Replace the ladder if there is any damage.

Step 4 Carry the ladder to the work area.

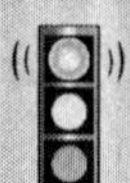

WARNING!

Be careful when carrying a ladder not to injure other people or damage other equipment with the ladder. If the ladder is too heavy or awkward, seek assistance.

Step 5 Place the ladder into position for the job.

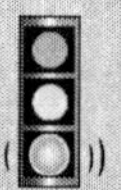

NOTE

When raising an extension ladder, pull the rope until the top section is at the desired position. Allow the rung hooks to slip over the desired rung and lock into place.

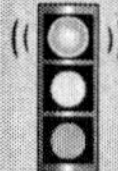

WARNING!

Be careful when raising a ladder. Make sure the ladder feet are securely set, that the ladder is balanced and properly spaced from the base of the wall. Once extended, secure the ladder at the bottom and top to prevent moving. Rope off the area around the base so that no one can run into the ladder and possibly knock it over.

Step 6 Climb up the ladder, using both hands to hold on to the sides and rungs.

WARNING!

Maintain your balance at all times. Do not reach out to the sides beyond normal arm length. Do not lean out or lean back.

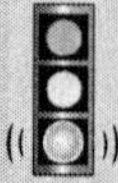

NOTE

Tools to be used while you are up on a ladder should be either attached to you or lifted by rope after you are in position.

Step 7 Perform the work from the ladder.

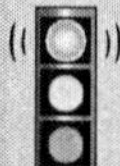

WARNING!

Do not try to move the ladder while on it. If it must be moved, climb down first, then move it. Be careful not to drop anything from a ladder.

Step 8 Climb down the ladder using both hands to hold on to the sides and rungs.

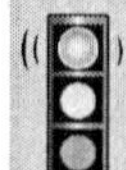

WARNING!

Do not slide down the outer rails. Do not try to turn around to climb down.

Instructor's Notes:

Step 9 Lower the ladder.

Step 10 Make sure the ladder is clean and in good condition before storing it.

Step 11 Return the ladder to the storage location.

3.0.0 ◆ SCAFFOLDING

Scaffolding consists of elevated working platforms that support workers and materials. See *Figure 12*. They are a very common sight on many construction jobs.

Scaffolding makes it easy to work on hard to reach areas, but it also increases the danger to workers because the work site is now above ground level. Be sure that the scaffolding you use has been tested using the American National Standards Institute (ANSI) standard. The ANSI standard for testing and rating scaffolding is *SC100*, which was written by the Scaffolding, Shoring, and Forming Institute (SSFI).

OSHA requires that the scaffolding supervisor be a competent person who has been specially trained to inspect scaffolding, oversee its assembly, and train workers in the safe and proper use of scaffolding. In addition, the scaffolding supervisor must inspect erected scaffolding at the start of each shift before it is used and whenever the integrity of the scaffold is in question, such as after an accident.

The main part of the scaffolding is the working platform. A working platform must have a guardrail system that includes a top rail, midrail, toeboard, and screening. To be safe and effective, the top rail should be approximately 42 inches high, the midrail should be located halfway between the toeboard and the top rail, and the toeboard should be a minimum of 4 inches high.

If people will be passing or working under the scaffolding, the area between the top rail and the toeboard must be screened. Finally, the platform planks must be laid close together. For safety purposes, the ends of the planks must overlap at least 6 inches and no more than 12 inches.

Scaffolds are built and designed according to high safety standards, but normal wear and tear and accidental overstress can weaken a scaffold and make it unsafe to use. Many different types of scaffolds are built to meet specific needs. Most job sites require scaffold builders to be certified according to company policies and procedures. Always check with your immediate supervisor before erecting a scaffold. A minimum of two people is recommended to safely erect a scaffold. The two types of scaffolds most used by pipefitters are tubular buck scaffolds and pole scaffolds. *Figure 13* shows these types of scaffolds.

DID YOU KNOW?

Scaffolding, Shoring, and Forming Institute

Manufacturers of scaffolding products take safety very seriously. Many of these companies voluntarily belong to the Scaffolding, Shoring, and Forming Institute (SSFI), which promotes the technical advancement of scaffolding and shoring products. You can learn more about SSFI by visiting its website (www.ssfi.org).

Figure 12 ◆ Typical scaffolding.

A scaffold is classified by the weight it is designed to support. The design capacity of a scaffold includes the total weight of all loads, which includes the working load and the weight of the scaffold. The three basic scaffold classes are the following:

- *Heavy duty* – Designed to carry a maximum working load of 75 pounds per square foot
- *Medium duty* – Designed to carry a maximum working load of 50 pounds per square foot
- *Light duty* – Designed to carry a maximum working load of 25 pounds per square foot

The following safety guidelines must be followed when using any type of scaffold:

- Follow all state, local, and government codes, ordinances, and regulations pertaining to scaffolding.

Have trainees review Sections 3.0.0–4.0.0.

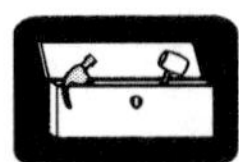

Ensure that you have everything required for teaching this session.

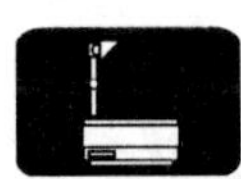

Show Transparency 9 (Figure 12). Describe typical scaffolding.

Explain that scaffolding is tested and rated against the *ANSI SC100 Standard*.

Describe OSHA requirements that scaffolding assembly must be supervised by a trained scaffolding inspector.

Describe the primary components of scaffolding.

Show Transparency 10 (Figure 13). Describe the types of scaffold commonly used by pipefitters.

Describe the classification of scaffolding based on design capacity.

Review the safety guidelines for using scaffolding.

Show Transparency 11 (Figure 15). Explain how leveling jacks are used.

Discuss the importance of using extreme caution when working on scaffolds.

- Ensure that the footing a scaffold is erected on will withstand the maximum intended load without settling or displacement.
- Do not use unstable objects, such as barrels, boxes, loose bricks, or concrete blocks, to support scaffolding.
- Inspect all equipment before using it. Do not use any damaged equipment. Inspect erected scaffolds regularly to make sure they are maintained in safe condition.
- Use **leveling jacks** (*Figure 14*) instead of blocking to erect scaffolds on uneven grades. *Figure 15* shows a scaffold using leveling jacks.

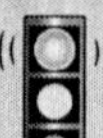

WARNING!

Use extreme caution when working on scaffolds. Do not stand on the handrails and do not jump from platform to platform.

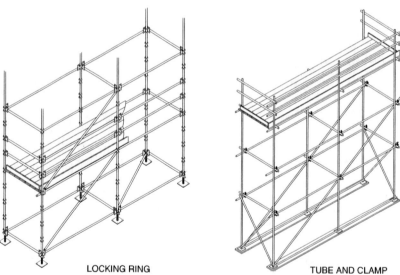

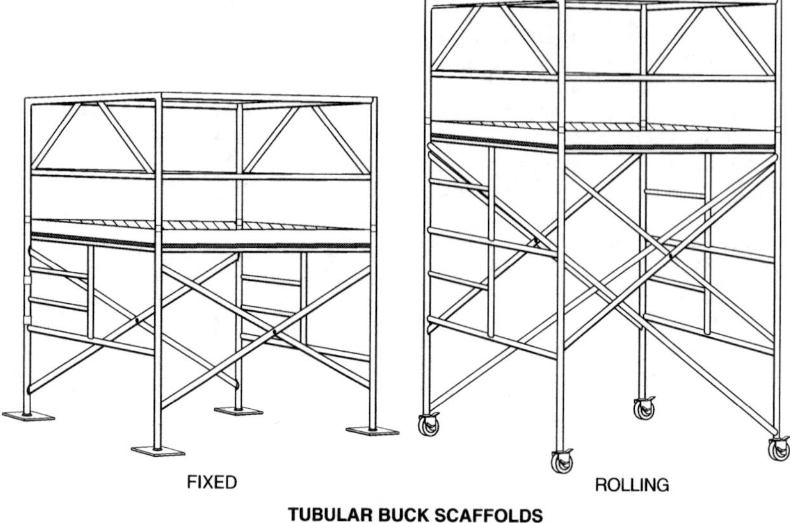

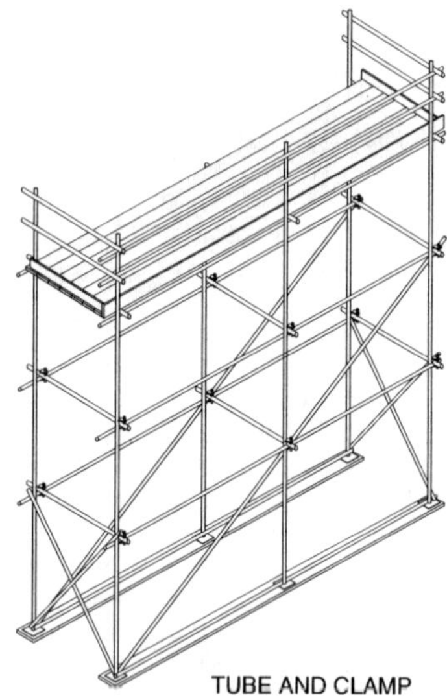

Figure 13 ◆ Scaffolds.

Instructor's Notes:

5.12 PIPEFITTING ◆ LEVEL ONE

3.1.0 Using and Caring for Tubular Buck Scaffolds

Erection methods for tubular buck scaffolds may differ by manufacturer, but most are erected in the same manner. The scaffold forms a rectangular framework made up of **vertical uprights**, or bucks, on the short sides of the scaffold and diagonal **cross braces** on the long sides of the scaffold. Access to the top of the scaffold is provided by ladders built into the vertical uprights or by a scaffold ladder attached to the scaffold. *Figure 16* shows basic tubular buck scaffold components.

3.1.1 Erecting Tubular Buck Scaffolds

A critical factor that must be considered before erecting a scaffold is the placement of the scaffold. You must fully examine the work area and consult with other workers in the area. Erect the scaffold so that it will not interfere with the work being done in the general area and will not jeopardize safety. Follow these steps to erect a tubular buck scaffold:

Step 1 Determine the height of the work to be performed.

Step 2 Identify the intended placement of the scaffold so that it will not interfere with your planned work activity or any other work being performed in the area.

Step 3 Calculate the number of vertical supports and cross braces needed to reach the work area.

105F14.EPS

Figure 14 ◆ Typical adjustable leveling jack.

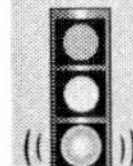

NOTE

The size of the vertical supports will determine the number of supports needed to reach the work area.

Step 4 Obtain all the required components needed to erect the scaffold from the storeroom or supply house.

105F15.EPS

Figure 15 ◆ Scaffold using leveling jacks.

Show Transparency 12 (Figure 16). Describe the components of tubular buck scaffolding.

Explain the steps to construct a tubular buck scaffold.

Emphasize the importance of using gloves when erecting scaffolding.

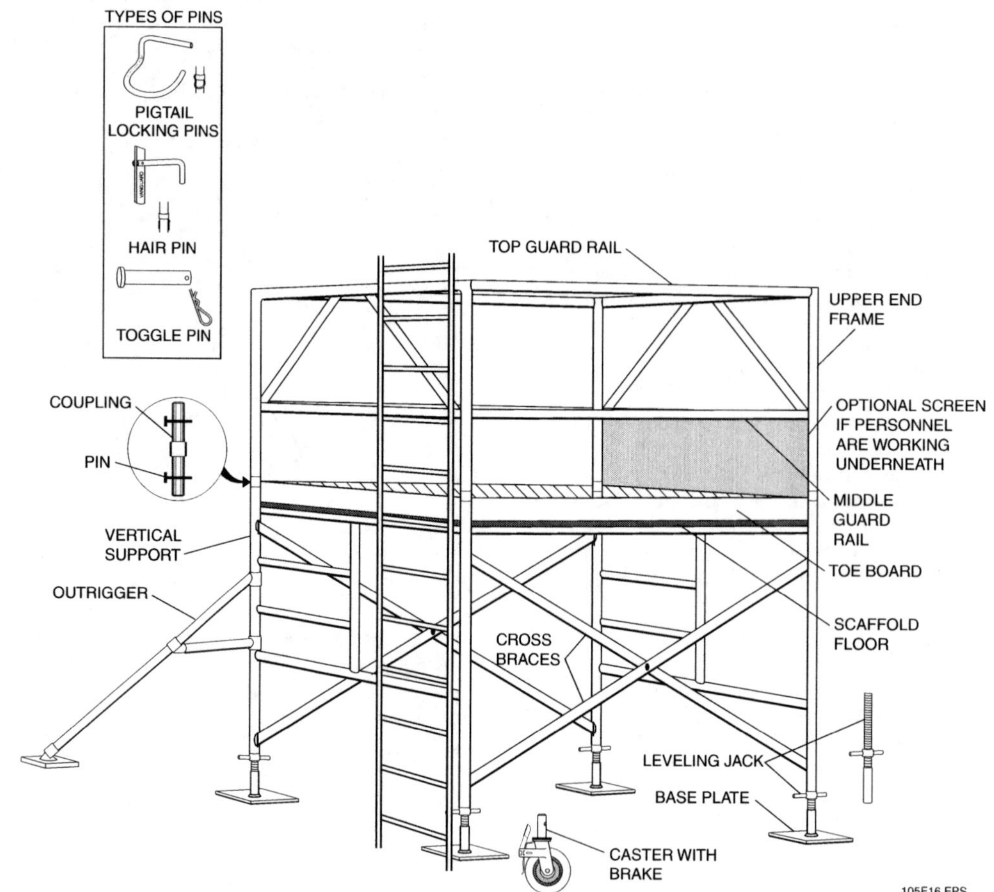

Figure 16 ◆ Basic tubular buck scaffold components.

Step 5 Inspect all parts for damage and excessive rust, dirt, or grease.

NOTE

If a part is damaged or has excessive rust, do not use it. If there is excessive dirt or grease on the parts, clean them before using.

Step 6 Assemble two cross braces to one vertical support and lock them into place.

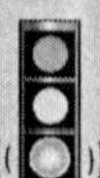

NOTE

For exterior installations, a mud sill must be placed under the scaffold.

Step 7 Assemble the cross braces into the second vertical support and lock them into place.

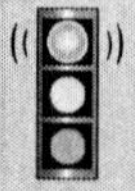

WARNING!

Wear gloves when assembling the scaffold to avoid getting pinched in the process.

Step 8 Using two people, lift one leg of the scaffold and insert a **base plate** into the bottom of the leg.

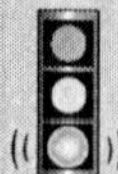

NOTE

If leveling jacks are needed, insert the leveling jacks between the legs and the base plates.

Instructor's Notes:

5.14 PIPEFITTING ◆ LEVEL ONE

Step 9 Insert a **hinge pin** through the scaffold leg (or bottom part of the leveling jack) to secure the base plate, and lock it into place.

Step 10 Repeat steps 8 and 9 to insert the base plates into the other scaffold legs.

Step 11 Ensure that the scaffold is plumb and level. Adjust leveling jacks as necessary.

Step 12 Place scaffold planks between the vertical supports on the top rails to form the floor of the scaffold.

Step 13 Insert **coupling pins** into the tops of each of the vertical support legs, and lock them into place.

Step 14 Place all items to complete the **guardrails** and the toeboards on the scaffold platform.

Step 15 Climb the ladder to the scaffold platform.

NOTE

If the scaffold must be built higher, attach vertical supports over the coupling pins in the lower vertical supports, lock in the cross braces, and move the scaffold planks to this level. Then proceed with step 16.

Step 16 Insert one upper end frame on the coupling pins of the lower vertical support, and lock it into place.

Step 17 Set the other upper end frame over the other vertical support and lock it into place.

Step 18 Secure the top guardrails to the scaffold.

Step 19 Secure the middle guardrails to the scaffold.

Step 20 Set the toeboards on all four sides of the scaffold.

3.1.2 Inspecting Tubular Buck Scaffolds

After erecting a tubular buck scaffold, it must be inspected before use. Follow these steps to inspect a tubular buck scaffold:

Step 1 Check the base plates to ensure they are securely locked into place.

Step 2 Check the cross braces to ensure they are securely locked into the vertical supports.

Step 3 Check the top and middle guardrails to ensure that they are securely mounted.

Step 4 Check the toeboards to ensure they are properly installed.

Step 5 Check the coupling pins to ensure they are securely locked into place and cannot slip out.

Step 6 Check the **scaffold floor** to ensure it is securely locked into place and cannot slip out.

Step 7 Ensure that the scaffold is level and standing on a firm surface that is capable of supporting it.

3.1.3 Disassembling Tubular Buck Scaffolds

The scaffold is disassembled from the top down. Follow these steps to disassemble a tubular buck scaffold:

Step 1 Remove the toeboards from all four sides of the scaffold.

Step 2 Remove the middle guardrails from the scaffold.

Step 3 Remove the top guardrails from the scaffold.

Step 4 Remove the two upper end frames from the scaffold.

Step 5 Remove the four coupling rings from the tops of the two vertical supports.

Step 6 Remove the scaffold platform planks from the tops of the vertical supports.

Step 7 Remove the hinge pin from one of the base plates.

Step 8 Lift the leg of the scaffold and remove the base plate from that leg.

Step 9 Repeat steps 7 and 8 to remove the other base plates.

Step 10 Disconnect the cross braces from one vertical support, and remove that vertical support.

Step 11 Disconnect the cross braces from the remaining vertical support.

Step 12 Store all scaffold parts in their proper places.

Describe procedures for inspecting and disassembling tubular buck scaffolding.

Show trainees how to erect, inspect, and disassemble tubular buck scaffolding.

Have trainees practice erecting, inspecting, and disassembling tubular buck scaffolding. Note the proficiency of each trainee. This laboratory corresponds to Performance Task 3.

Describe pole scaffolds.

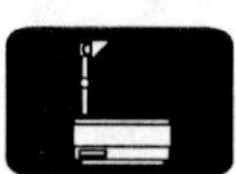

Show Transparencies 13 and 14 (Figures 17 and 18). Describe locking ring pole scaffolds and explain how to assemble and disassemble them.

Show Transparency 15 (Figure 20). Explain how to set up tube and clamp scaffolds.

Describe rolling scaffolds and discuss safety considerations specific to rolling scaffolds.

3.2.0 Using and Caring for Pole Scaffolds

Erection methods for pole scaffolding also differ by manufacturer, but most are erected in basically the same manner. The pole scaffold consists of separate poles that serve as vertical posts, diagonal cross braces, and horizontal bearers or runners. The leveling jacks, guardrails, walk boards, and toeboards used for pole scaffolding are basically the same as those used for tubular buck scaffolding. All the safety rules that apply to tubular buck scaffolding also apply to pole scaffolds. The maximum interval for horizontal runners on a pole scaffold is 7 feet. Two types of pole scaffolds are locking ring scaffolds and tube and clamp scaffolds.

3.2.1 Locking Ring Scaffolds

Locking ring scaffolds are joined by locking rings and end connectors. They are easily assembled and disassembled without the use of bolts and nuts or loose pins. *Figure 17* shows the components of a locking ring pole scaffold.

The vertical posts contain locking ring sets every 21 inches. The locking ring sets each contain two rings spaced 3½ inches apart. The end connectors of the horizontal runners and vertical cross braces lock into these rings. To assemble the end connectors to the locking rings, hook the connector onto the rings, and hammer the wedge underneath the bottom ring. To disassemble, pry the wedge out from underneath the bottom ring and unhook the end connector. The horizontal runners can be attached to the locking rings at any angle around the vertical post to allow for a wide variety of scaffolding configurations. *Figure 18* shows assembly and disassembly of the end connectors to the locking rings.

3.2.2 Tube and Clamp Scaffolds

Like the locking ring scaffold, the tube and clamp scaffold is versatile and can be set up in many different configurations to suit different jobs. The vertical posts, horizontal runners, and diagonal cross braces are joined together with right-angle and swivel couplers (*Figure 19*). These couplers are secured to the poles with nuts and bolts. *Figure 20* shows the components of a tube and clamp pole scaffold.

3.3.0 Rolling Scaffolds

All the safety rules that apply to fixed scaffolds also apply to rolling scaffolds. Other safety rules that are specific to rolling scaffolds include the following:

- Do not use a working platform with a height that exceeds four times the smallest base dimension unless outriggers are used or the scaffold is guyed or braced against tipping.
- The scaffold as well as the **caster** wheels must be designed to support four times the intended maximum load.
- Use caster wheels with locking brakes to hold the scaffold in position.
- Brace rolling scaffolds properly with cross bracing and horizontal supports.
- Ensure that platforms are tightly planked for the full width of the scaffold and secured in place on each end.
- When moving a rolling scaffold, apply force as near to the base of the scaffold as possible to prevent tipping the scaffold over.
- Do not ride on a rolling scaffold.
- Lock the caster brakes at all times when work is being performed from the scaffold.
- Apply stationary scaffold rules for the guardrails and toeboards of rolling scaffolds.
- Do not extend the leveling jacks on a rolling scaffold more than 12 inches.

The procedures for assembling, inspecting, and disassembling rolling scaffolds are similar to those for the fixed scaffolds. Instead of installing base plates, you install caster wheels and lock them into place. As with the fixed scaffolds, actual assembly and securing procedures will vary slightly depending on the scaffold manufacturer.

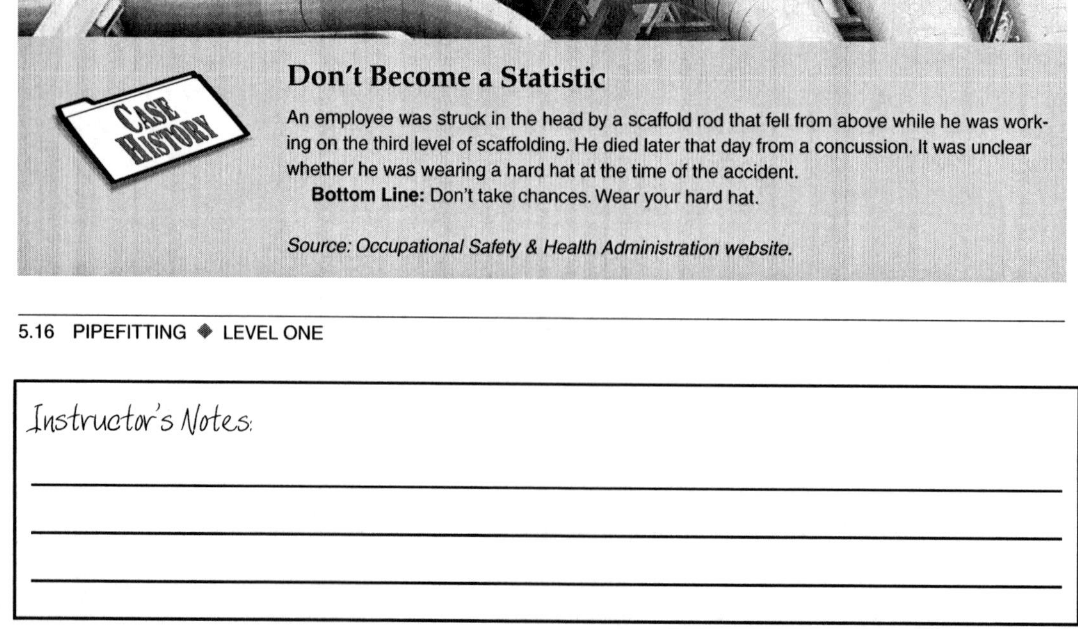

Don't Become a Statistic

An employee was struck in the head by a scaffold rod that fell from above while he was working on the third level of scaffolding. He died later that day from a concussion. It was unclear whether he was wearing a hard hat at the time of the accident.

Bottom Line: Don't take chances. Wear your hard hat.

Source: Occupational Safety & Health Administration website.

Instructor's Notes:

5.16 PIPEFITTING ◆ LEVEL ONE

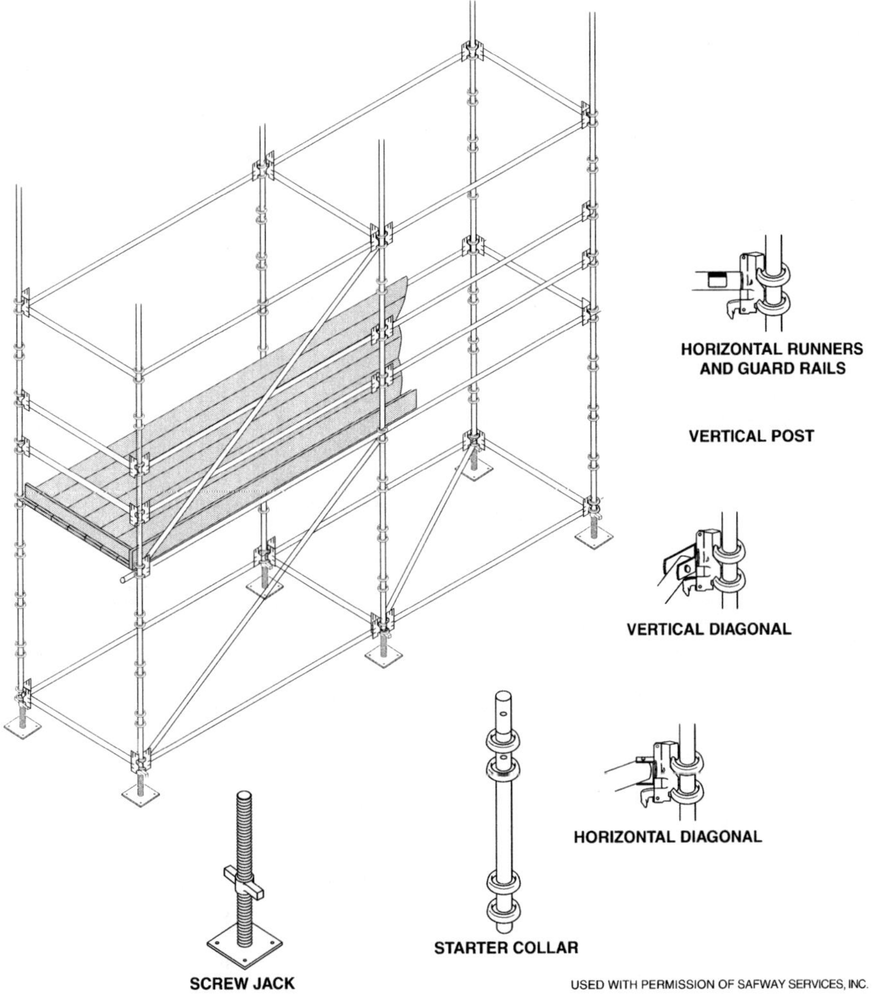

Figure 17 ◆ Components of locking ring pole scaffold.

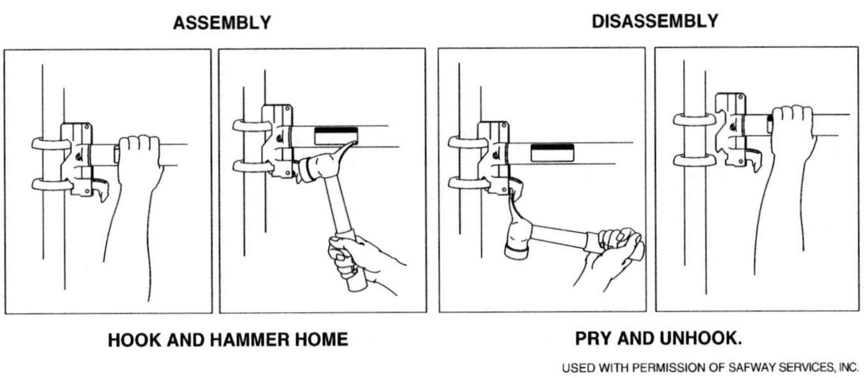

Figure 18 ◆ Assembly and disassembly of end connectors to locking rings.

MODULE 08105-06 ◆ LADDERS AND SCAFFOLDS 5.17

See the Teaching Tip for Section 3.4.0 at the end of this module.

Explain that improper or careless use of scaffolding can result in serious injury or death.

Identify the primary hazards associated with scaffold use.

Discuss safety measures to minimize these hazards, including fall protection, guardrails, toeboards, and debris nets.

Stress the importance of exiting the scaffolding if it shifts or begins to collapse.

Review the Case History and discuss the hazards from falls from a short distance.

Discuss electrical hazards and how they can be minimized.

3.4.0 Scaffolding Hazards

Improper or careless use of scaffolding can result in accidents, injury, and/or death. Those who work on scaffolding can minimize their risks by being aware of the hazards involved and following the proper safety procedures and guidelines to minimize those hazards.

The main hazards involved with the use of scaffolding are:

- Falls
- Workers being struck by falling objects
- Electric shock

Falls can happen because fall protection has not been provided, is not used, or is installed or used improperly. Poorly planked scaffolding causes many falls. Working on scaffolding when conditions are dangerous, such as in high winds, ice, rain, and lightning, also leads to accidents. Falls also happen when scaffolding collapses because of improper construction.

Fall protection is required on any scaffolding 6 feet or more above a lower level. Fall-protection devices consist of guardrail systems, personal fall-arrest systems, and/or safety nets. A guardrail system is adequate fall protection for most scaffolding.

Guardrail systems must extend around all open sides of the scaffolding. The side facing the work surface need not have a guardrail if it is located less than 14 inches away from the work surface. Any opening on a scaffolding platform must be protected by a guardrail system, including the access opening(s) and platforms that do not extend across the entire width of the scaffolding.

People who work or pass under scaffolding may be hit by falling objects. Tools, materials, debris, and scaffolding parts may fall to the surface below. Those working on scaffolding may also be injured if there are others working above them, or if the structure or workpiece extends above the work level of the scaffolding. Any worker who is exposed to the danger of falling objects is required to wear a hard hat. Depending on the situation, additional protection such as debris nets, screens or mesh, canopy structures, and toeboards may be needed. Barricades that prevent access under the scaffolding can also be used to protect workers and others.

105F19.EPS

Figure 19 ◆ Pole scaffolding coupler.

WARNING!

If the scaffolding you are working on shifts or begins to collapse, stop what you are doing and safely exit the scaffolding.

Because most scaffolding is made of metal, the chance of electric shock is always a hazard. Never assume that you can work around high-voltage wires just by avoiding contact. High voltages can arc through the air and cause electrocution without direct contact. When scaffolding must be erected close to power lines, the utility company must be called in to de-energize, move, and/or cover the lines with insulating protective barriers.

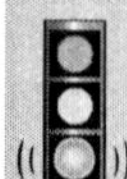

NOTE

Always refer to the competent person on site if you have any questions about the safety of scaffolding.

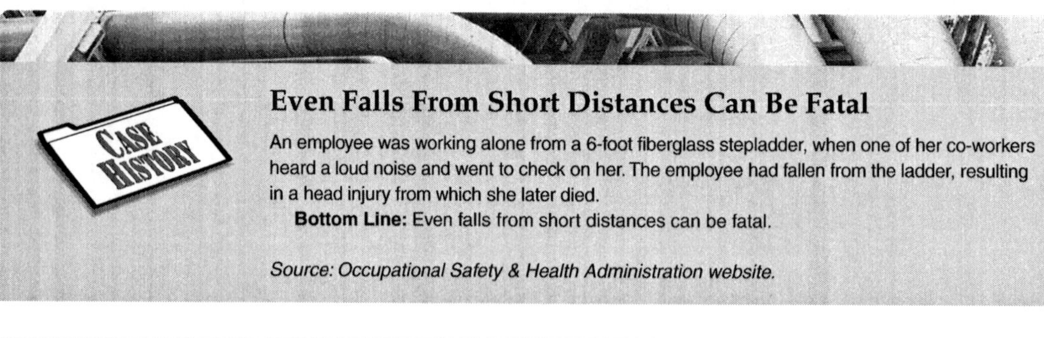

Even Falls From Short Distances Can Be Fatal

An employee was working alone from a 6-foot fiberglass stepladder, when one of her co-workers heard a loud noise and went to check on her. The employee had fallen from the ladder, resulting in a head injury from which she later died.

Bottom Line: Even falls from short distances can be fatal.

Source: Occupational Safety & Health Administration website.

Instructor's Notes:

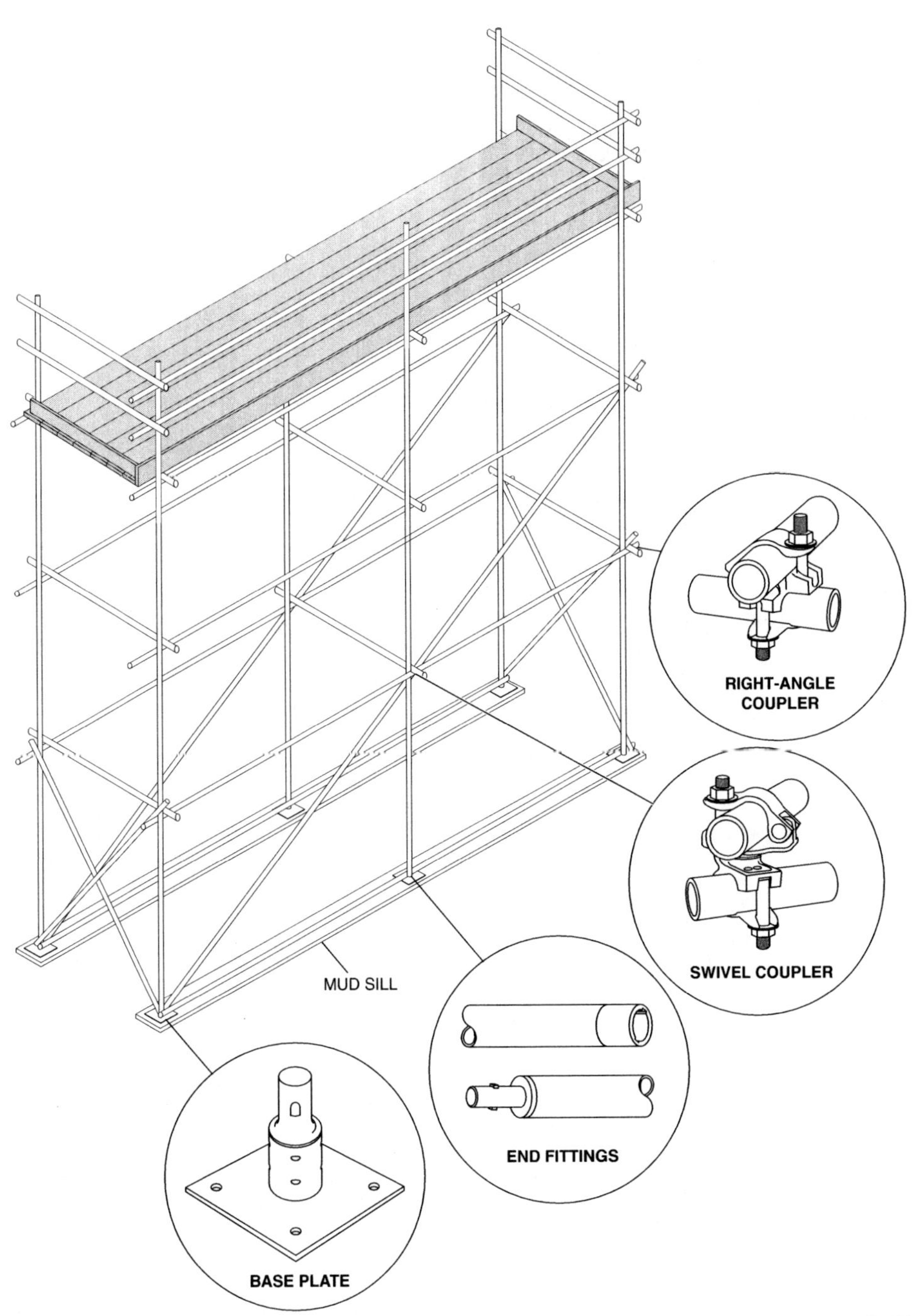

Figure 20 ◆ Components of tube and clamp pole scaffold.

MODULE 08105-06 ◆ LADDERS AND SCAFFOLDS 5.19

Provide a copy of *29 CFR 1926.450 Subpart L Scaffolds* for the trainees to examine.

Discuss OSHA safety guidelines for erecting and using scaffolding.

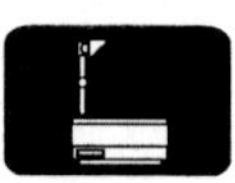

Show Transparency 16 (Figure 21). Explain how to use scaffolding tags.

Provide scaffolding tags for the trainees to examine.

3.5.0 Scaffolding Safety Guidelines

Safety begins by getting training in the proper use of scaffolding. OSHA requires that the scaffolding supervisor be a competent person who has been specially trained to inspect scaffolding, oversee its assembly, and train workers in the safe and proper use of scaffolding. But you need to protect yourself. Always use the right safety equipment, including a hard hat and personal fall-protection systems.

Never work on scaffolding if you:

- Are subject to seizures
- Become dizzy or lightheaded when working at an elevation
- Take medication that might affect your stability and/or performance
- Are under the influence of drugs and/or alcohol

When working on scaffolding, always follow these guidelines:

- Erect and use scaffolding according to the manufacturer's instructions. They must also be erected and used in accordance with all local, state, and federal/OSHA requirements.
- Check all scaffolds for complete or incomplete inspection tags before mounting the scaffold. Do not mount a scaffold tagged incomplete.
- Install guardrails and toeboards on all open sides and ends of platforms that are more than 6 feet high.
- OSHA requires that the scaffolding supervisor approve alterations to scaffolding plans.
- A professional engineer must design tube and coupler scaffolding that exceeds 125 feet in height and pole scaffolding that exceeds 60 feet in height.
- The front edge of scaffolding must not be more than 14 inches from the face of the work, unless guardrail systems are erected along the front edge and/or personal fall arrest systems are used to protect employees from falling. When outriggers are used, the front edge of the scaffolding must not be more than 3 inches from the face of the work.
- Outriggers or some other means of securing the scaffolding, such as guys, ties, and braces, must be used according to the scaffold manufacturer's recommendations or when the vertical height of the scaffolding is four times the minimum base dimension. Securing devices are set both vertically and horizontally. Vertical devices are set at no more than 20-foot intervals for scaffolds 3 feet (or less) wide, and at no more than 26-foot intervals for scaffolds greater than 3 feet wide. On completed scaffolds, the last securing device must be placed no greater than four times the minimum base dimension from the top of the scaffolding. The horizontal devices must be set not greater than 30 feet apart (measured from one end towards the other).
- Guardrails must be constructed from 2 by 4 inch lumber or OSHA-approved equivalent metal and must be supported every 8 feet.
- The distance from the upper edge of the toprail to the scaffold platform must be 38 to 45 inches on units placed in service after January 1, 2000, while the distance must be 36 to 45 inches for scaffolds placed in service before January 1, 2000.
- Toeboards must be solid or with openings not over one inch at any point. They must be fastened at the outer edge of the platform with not more than ¼ inch clearance above the platform. Toeboards must be able to withstand at all points a force of at least 50 pounds in a horizontal or vertical direction without failure. There must be at least three and one-half inches from the top edge of the toeboard to the platform support.
- The space between the platform and the scaffolding upright support must not be more than 1 inch wide and there may be no more than 2 inches between planks.
- Ladders are not normally permitted on scaffolding except under the most stringent circumstances. You as a pipefitter should never make a decision to place a ladder on scaffolding except under the direction of the scaffolding supervisor.
- Install a screen between the toeboard and the guardrail when persons must work or pass underneath the scaffold. This screen must be 18-gauge U.S. Standard 1-inch mesh or the equivalent.
- Scaffolds must be capable of supporting at least four times their maximum intended load weight.
- Provide an access ladder or equivalent safe access. Do not climb cross braces.
- Plumb and level all scaffolds as the erection proceeds. Do not force braces to fit, but level the scaffold until the proper fit can be made easily.
- Wooden scaffold boards must be No. 1 grade lumber and be equipped with cleats on each end. They must be secured to the scaffold with OSHA-certified 9-gauge wire.
- Aluminum scaffold boards must be equipped with a turn key latch to secure the board to the scaffolding.
- Never interchange parts of a scaffolding system made by different manufacturers.
- Attach a green, red, or yellow tag (*Figure 21*), as applicable, to any scaffolding that is assembled and erected to alert users of its current mechanical and/or safety status. Do not rely solely on

Instructor's Notes:

5.20 PIPEFITTING ◆ LEVEL ONE

Remind trainees to use extreme caution when working on scaffolds.

Discuss safety guidelines for erecting and using tubular built-up scaffolding.

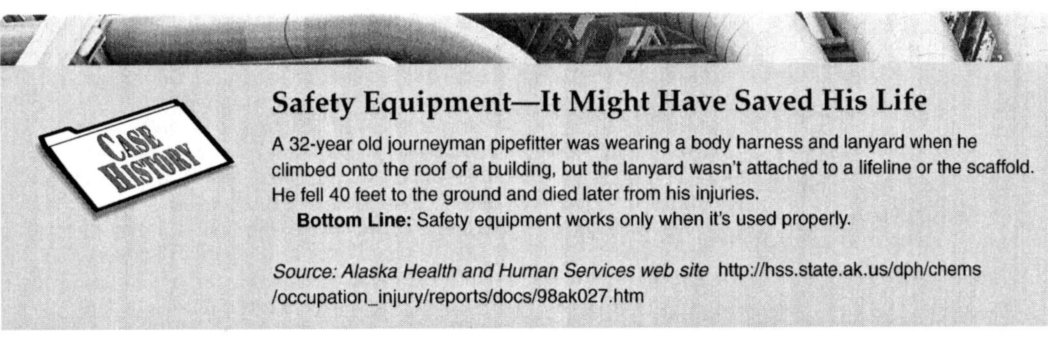

Safety Equipment—It Might Have Saved His Life

A 32-year old journeyman pipefitter was wearing a body harness and lanyard when he climbed onto the roof of a building, but the lanyard wasn't attached to a lifeline or the scaffold. He fell 40 feet to the ground and died later from his injuries.

Bottom Line: Safety equipment works only when it's used properly.

Source: Alaska Health and Human Services web site http://hss.state.ak.us/dph/chems/occupation_injury/reports/docs/98ak027.htm

the tag. Inspect all parts of scaffolding before each use.

- If the scaffolding shifts, exit the scaffolding immediately.

CAUTION

Take extreme caution when working on scaffolds. Do not stand on the handrails or jump from platform to platform.

3.5.1 Safety Guidelines for Built-Up Scaffolding

Built-up scaffolding is built from the ground up at a job site. Use the following guidelines when erecting and using tubular built-up scaffolding:

- Inspect all scaffolding parts before assembly.
- Never use parts that are broken, damaged, or deteriorated. Be cautious of rusted materials.
- Follow the manufacturer's recommendations for the proper methods of erecting and using scaffolding.
- Do not interchange parts from different manufacturers.
- Do not force braces or other parts to fit. Adjust the level of the scaffolding until the connections can be made easily.
- Provide adequate sills or underpinnings for all scaffolding built on filled or soft ground. Compensate for uneven ground by using adjusting screws or leveling jacks.
- Do not use boxes, concrete blocks, bricks, or other similar objects to support scaffolding.
- Keep scaffolding free of clutter and slippery material.
- Be sure scaffolding is plumb and level at all times. Follow the prescribed spacing and positioning requirements for the parts of the scaffolding. Anchor or tie-in scaffolding to the building at prescribed intervals.

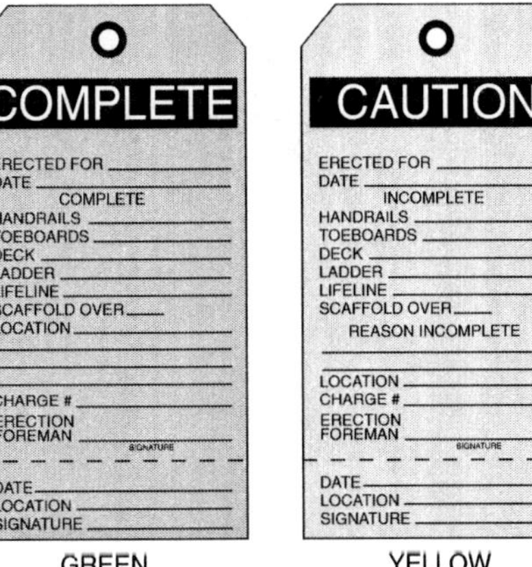

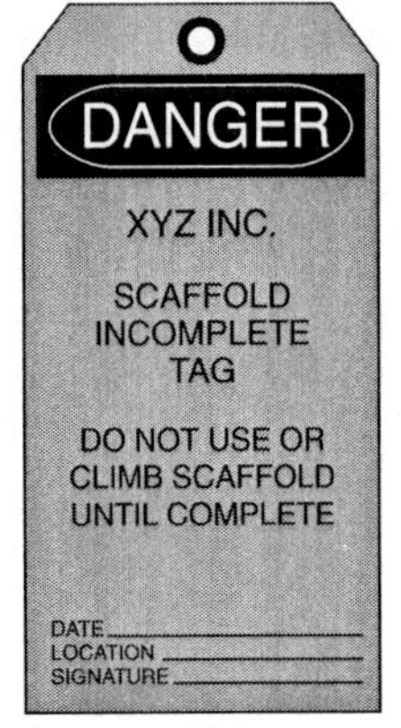

Figure 21 ◆ Typical scaffolding tags.

MODULE 08105-06 ◆ LADDERS AND SCAFFOLDS 5.21

See the Teaching Tip for Section 3.5.2 at the end of this module.

Provide a copy of a company safety manual with procedures for fall protection and rescue after a fall for the trainees to examine.

- Use ladders rather than cross braces to climb the scaffolding. Position ladders with caution to prevent the scaffolding tower from tipping.
- Do not work on scaffolding that is more than 6 feet high without guardrails, midrails, and toeboards on open sides and ends.
- Lock the casters of mobile scaffolding when it is positioned for use.
- Do not ride on mobile scaffolding.
- Avoid building scaffolding near power lines.

3.5.2 Personal Fall-Arrest Systems

Describe a personal fall arrest system, specialized equipment needed, and how to don it.

Personal fall-arrest systems catch workers after they have fallen. They are designed and rigged to prevent a worker from free-falling a distance of more than 6 feet and hitting the ground or a lower work area.

Personal fall-arrest systems use specialized equipment, which includes the following:

- Body harnesses (*Figure 22*)
- Lanyards (*Figure 23*)
- Deceleration devices (*Figure 24*)
- Lifelines (*Figures 25* and *26*)
- Anchoring devices and equipment connectors

Explain that the expansion of a shock-absorbing lanyard must be included when calculating free-fall distance from an anchor point.

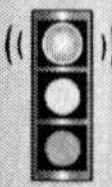

NOTE

In the past, body belts were often used instead of a full-body harness as part of a fall-arrest system. As of January 1, 1998, however, they were banned from such use. This is because body belts concentrate all of the arresting force in the abdominal area, causing the worker to hang in an uncomfortable and potentially dangerous position while awaiting rescue.

WARNING!

When activated during the fall-arresting process, a shock-absorbing lanyard stretches in order to reduce the arresting force. This potential increase in length must always be taken into consideration when determining the total free-fall distance from an anchor point.

Show trainees how to properly don a personal fall arrest system.

4.0.0 ◆ RESCUE AFTER A FALL

Have trainees practice properly donning a personal fall arrest system.

Every elevated job site should have an established rescue and retrieval plan. Planning is especially important in remote areas without ready access to a telephone. Before beginning work, make sure that you know what your employer's rescue plan calls for you to do in the event of a fall. Find out what rescue equipment is available and where it is located. Learn how to use equipment for self-rescue and the rescue of others.

If a fall occurs, any employee hanging from the fall-arrest system must be rescued safely and quickly. Your employer should have previously determined the method of rescue for fall victims, which may include equipment that lets the victim rescue himself or herself, a system of rescue by co-workers, or a way to alert a trained rescue squad. If a rescue depends on calling for outside help such as the fire department or rescue squad, all the needed phone numbers must be posted in plain view at the work site. In the event a co-worker falls, follow your employer's rescue plan. Call any special rescue service needed. Communicate with the victim and monitor him or her constantly during the rescue.

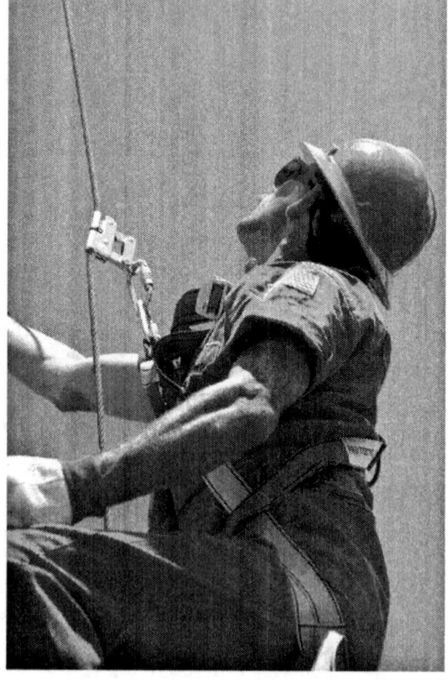

105F22.EPS

Figure 22 ◆ Harness.

105F23.EPS

Figure 23 ◆ Lanyard.

Instructor's Notes:

5.22 PIPEFITTING ◆ LEVEL ONE

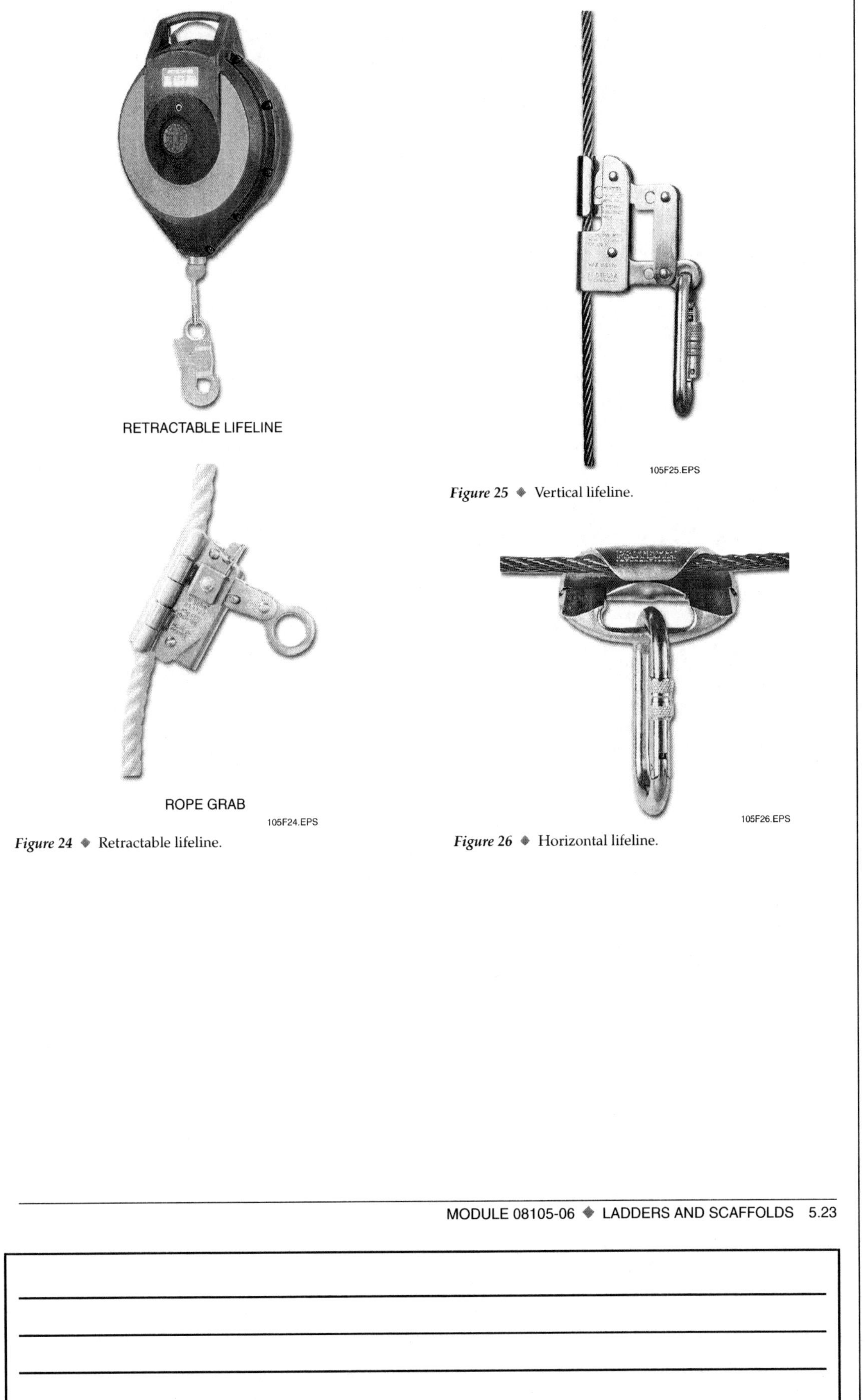

Figure 24 ◆ Retractable lifeline.

Figure 25 ◆ Vertical lifeline.

Figure 26 ◆ Horizontal lifeline.

MODULE 08105-06 ◆ LADDERS AND SCAFFOLDS 5.23

Have the trainees complete the Review Questions, and go over the answers prior to administering the Module Examination.

Review Questions

1. When you are working on ladders or scaffolding, your first responsibility is to _____.
   a. get the job done rapidly and efficiently
   b. work carefully whenever it is possible
   c. keep yourself and your co-workers safe
   d. complete your part of the job correctly

For Questions 2 through 4, match the type of ladder to the corresponding description.

2. _____ Aluminum
3. _____ Fiberglass
4. _____ Wooden
   a. Corrosion-resistant and can be used in situations where it might be exposed to the elements
   b. Heavier than other ladders and can be used when heavy loads must be moved up and down
   c. Durable, non-conductive, and useful in situations where some amount of rough treatment is unavoidable

5. Before climbing a ladder, you must check all of the following *except*:
   a. the ladder's rungs to be sure they are clean and in good repair
   b. that when placed against a window the ladder is securely tied down
   c. the soles of your shoes are free from any slippery substances
   d. that the ladder is placed on stable ground where it will not slip

6. If you lean a straight ladder against the top of a 16-foot wall, the base of the ladder should be _____ feet from the base of the wall.
   a. 1½
   b. 3
   c. 4
   d. 6

7. It is safe to stand on the top step of a stepladder as long as you are holding onto a solid component of the building.
   a. True
   b. False

8. Straight ladders can be used on unstable surfaces if safety feet are used.
   a. True
   b. False

9. What is the best way to access tools when you are working from a straight ladder?
   a. Carry the tools you need in your hand and be extra careful when climbing the ladder.
   b. Set the tools on a rung before you climb the ladder and move them up as you climb.
   c. Plan to use a rope to lift tools that you will need after you have climbed the ladder.
   d. Using tools on a ladder is not safe. Use scaffolding if tools are needed above ground.

10. At most job sites, scaffold builders _____.
    a. are certified according to company policy
    b. learn to build scaffolds by trial and error
    c. work alone to reduce the odds of a mishap
    d. are used to erect only complex scaffolds

11. Before you erect any scaffolding, you must _____.
    a. examine the area and place the scaffolding so it will not hamper other work
    b. ask all other workers to leave the area to decrease the chances of an accident
    c. reschedule tasks of your co-workers so you can quickly build the scaffolding
    d. ask the local OSHA agent to inspect the area you have selected for the scaffold

Instructor's Notes:

5.24 PIPEFITTING ◆ LEVEL ONE

Review Questions

12. Fall protection is required on any scaffolding that is _____ feet or more above a lower level.
    a. 6
    b. 10
    c. 12
    d. 14

13. Electric shock is one of the hazards of working on scaffolding.
    a. True
    b. False

14. Scaffolds must be capable of supporting at least _____ their maximum intended load weight.
    a. twice
    b. three times
    c. four times
    d. five times

15. When you are assigned to work on scaffolding you need to _____.
    a. take the least amount of equipment onto the scaffold with you including safety equipment
    b. notify your supervisor when you are taking prescription medication that makes you dizzy
    c. ask any non-essential workers to leave the work area to reduce the potential of accidents
    d. climb the scaffolding quickly since that is the time that you are at greatest risk of falling

MODULE 08105-06 ◆ LADDERS AND SCAFFOLDS 5.25

Summarize the major concepts presented in the module.

Summary

Ladders and scaffolding are used in a variety of different construction jobs. The type of ladders or scaffolding will vary with each job, but the dangers won't. There is always a risk of falling and being struck by falling objects, each of which can result in serious injury or death. It is your responsibility to be aware of and follow the safety procedures associated with ladders and scaffolding. If you are involved in assembly of scaffolding, you must make absolutely sure it is properly assembled and leveled, and equipped with guardrails, toeboards, and other safety features.

Notes

Instructor's Notes:

5.26 PIPEFITTING ◆ LEVEL ONE

Trade Terms Quiz

Fill in the blank with the correct trade term that you learned from your study of this module.

1. Discs or rectangles used to distribute the weight of a scaffold are _________.
2. A(n) _________ is the end section of a scaffold, also known as a buck.
3. Some scaffolds have wheels, or _________, attached to the bottoms of the scaffold legs.
4. To keep tools from falling off the scaffold, attach a(n) _________ around the scaffold floor.
5. The vertical uprights of a scaffold are connected and supported by _________.
6. A temporary built-up framework to support workers, materials, and equipment is called _________.
7. A pin used to hold a base plate or castor wheel onto a leg is called a(n) _________.
8. Part of a personal fall arrest system, the _________ straps around your body and is connected to a lifeline.
9. Ladders are assigned a _________ based on the intended use and maximum load limit of the ladder.
10. Sections of the scaffold are lined up and joined with a(n) _________.
11. A(n) _________ is an elevated work platform.
12. To level a scaffold on an uneven surface, use _________.
13. The top _________ should be 42 inches above the scaffold floor.
14. Use a _________ to get onto the scaffold, rather than climbing the cross braces.
15. The platform on top of the scaffolding is the _________.

Trade Terms

Base plates
Casters
Coupling pin
Cross braces
Duty rating
Guardrail
Harness
Ladder
Leveling jacks
Scaffold
Scaffolding
Scaffold floor
Toeboard
Vertical upright

Have the trainees complete the Trade Terms Quiz, and go over the answers prior to administering the Module Examination.

Administer the Module Examination. Record the results on Craft Training Report Form 200, and submit the results to the Training Program Sponsor.

Administer the Performance Test, and fill out Performance Profile Sheets for each trainee. If desired, trainee proficiency noted during laboratory sessions may be used to complete the Performance Test. Record the results on Craft Training Report Form 200, and submit the results to the Training Program Sponsor.

MODULE 08105-06 ◆ LADDERS AND SCAFFOLDS 5.27

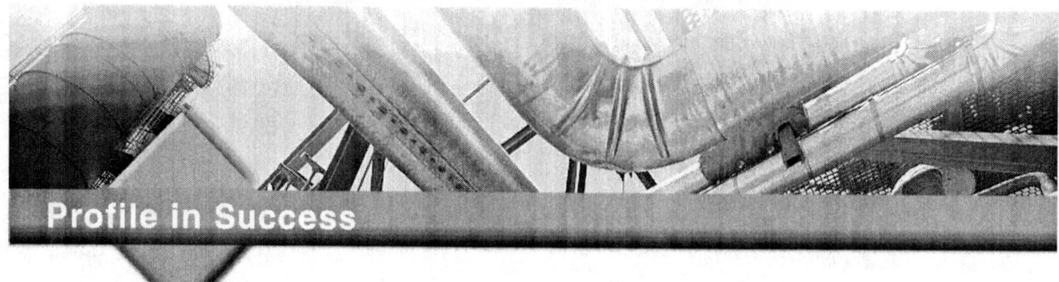

Ned Bush

Lee College, Baytown, TX
Pipefitting Lead Instructor

Pipefitting runs in the family for Ned Bush. He started as a helper when he was in high school working with his father. After working and progressing in the trade for 26 years, Ned became an instructor at the community college. He actively participates in the Associated Builders and Contractors annual National Craft Championship competition. He has been the Technical Advisor for the past 12 years. He drafted the written exam and developed the hands-on skills competition projects for the pipefitting program.

How did you choose a career in the pipefitting field?
Pipefitting is a family tradition. The men in my family were all pipefitters.

What types of training have you been through?
Five years apprentice training
AAS degree in Welding Technology
AAS degree in Drafting Technology

What kinds of work have you done in your career?
Pipefitting has always been my career. When I was in high school, I worked during the summer as a pipefitter's helper. After I graduated, I went through five years of apprenticeship training and have continued in the trade for another 21 years.

Currently, I am the lead instructor for Pipefitting at the community college level. I have taught welding, drafting, blueprint reading, and pipefitting classes for Lee College over the past 17 years. Our classes are offered both for college credit and industrial training documentation.

Tell us about your present job.
I am the lead instructor for the Pipefitting Technology Program at Lee College in Baytown, Texas. This is a very rewarding position. In the pipefitting program we offer an AAS in Pipefitting Technology, a Certificate of Completion in Pipefitting Technology and a Certificate of Completion in Pipefitter Helper. The program offers students both hands-on experience and book knowledge.

It gives me great satisfaction to see my students excel in their craft training and put those skills to use. Many of my students return to Lee College to tell me of their successes and accomplishments.

What factors have contributed most to your success?
Pride in what I do and the desire to be the best.

What advice would you give to those new to pipefitting field?
I believe that a formal training program along with on the job work experience is the best preparation for a pipefitting career.

Instructor's Notes:

5.28 PIPEFITTING ◆ LEVEL ONE

Trade Terms Introduced in This Module

Base plates: Flat discs or rectangles under scaffold legs that are used to evenly distribute the weight of the scaffold. Base plates come in different sizes for different scaffold heights.

Casters: Wheels that are attached to the bottoms of scaffold legs instead of base plates. Most casters come with brakes.

Coupling pin: A steel pin used to line up and join sections of a scaffold.

Cross braces: Steel pieces used to connect and support the vertical uprights of a scaffold.

Duty rating: American National Standards Institute (ANSI) rating assigned to ladders. It indicates the type of use the ladder is designed for (industrial, commercial, or household) and the maximum working load limit (weight capacity) of the ladder. The working load limit is the maximum combined weight of the user, tools, and any materials bearing down on the rungs of a ladder.

Guardrails: Protective rails attached to a scaffold. The top rail is 42 inches above the scaffold floor, and the middle rail is halfway between the top rail and the toeboard.

Harness: A device that straps securely around the body and is connected to a lifeline. It is part of a personal fall arrest system.

Hinge pin: A small pin inserted through a leg and base plate or caster wheel and used to hold these parts together.

Ladder: A wood, metal, or fiberglass framework consisting of two parallel side pieces (rails) connected by rungs on which a person steps when climbing up or down. Ladders may either be of a fixed length that is permanently attached to a building or structure, or portable. Portable ladders have either fixed or adjustable lengths and are either self-supporting or not self-supporting.

Leveling jack: A threaded, adjustable screw located between the legs and the base plates or caster wheels of a scaffold. It is used to raise or lower parts of a scaffold to level it on uneven surfaces.

Scaffold: An elevated work platform for workers and materials.

Scaffolding: A temporary built-up framework or suspended platform or work area designed to support workers, materials, and equipment at elevated or otherwise inaccessible job sites.

Scaffold floor: A work area platform made of metal or wood.

Toeboard: A 4-inch railing attached around the scaffold floor to prevent tools and materials from falling off the scaffold.

Vertical upright: The end section of a scaffold, also known as a buck, made of welded steel. It supports other vertical uprights or upper end frames and the scaffold floor.

MODULE 08105-06 ◆ LADDERS AND SCAFFOLDS 5.29

Additional Resources

This module is intended to be a thorough resource for task training. The following reference work is suggested for further study. This is optional materials for continued education rather than for task training.

Occupational Safety and Health Standards for the Construction Industry, Latest Edition. Occupational Safety and Health Administration. U.S. Department of Labor. Washington, DC: U.S. Government Printing Office.

Figure Credits

Topaz Publications, Inc., 105F01, 105F14, 105F19

Louisville Ladder Group, 105F02

Levelok Corporation, 105F03

Ridge Tool Company (RIDGID®), 105F04, 105F10

Werner Company, 105F05

Safway Services, Inc., 105F17, 105F18

DBI/SALA & Protecta, 105F22–105F26

Zachary Construction Corporation, Title Page

Instructor's Notes:

5.30 PIPEFITTING ◆ LEVEL ONE

MODULE 08105-06 — TEACHING TIPS

The following are suggested activities or instructional methods to help you teach the material in this AIG.

General

When you call on someone to answer a question, the rest of the class relaxes or even tunes out because they expect that the question and answer will take place only between you and the trainee you called on. Instead, use this technique to involve more trainees in answering questions and to keep them on their toes.

1. Ask the trainees to define a term or explain a concept.
2. After one trainee has answered, ask a trainee seated nearby if the answer is right. Then ask whether a trainee in the back of the room agrees.
3. Ask the trainees to explain why they think an answer is right or wrong.
4. Use the session to clear up incorrect ideas and encourage the trainees to learn from their mistakes.

Section 3.4.0 *Scaffolding Safety*

Pipefitters often use scaffolding to perform their job. This exercise will familiarize trainees with various aspects of scaffolding safety. Trainees will need appropriate personal protective equipment, pencils, and paper. Arrange for an opportunity to observe a construction site or have a safety professional or OSHA representative give a presentation on scaffolding safety. If possible have the site safety manager give a tour of the area. Some areas to include in the presentation are different types of ladders and scaffolding, setting up ladders and scaffolding, and proper climbing techniques. Allow 20 to 40 minutes for this exercise.

1. Introduce the topic and the speaker. Ask the site safety manager to give a tour of the site and point out various safety features or ask the safety professional to give a presentation.
2. Have the trainees take notes and write down questions during the tour or presentation.
3. Ask the presenter to spend some time after the tour to answer any questions the trainees may have.

Section 3.5.2 *Fall Protection*

Pipefitters frequently work at elevation and must understand how to use fall protection equipment. Trainees will need pencils and paper. Obtain a safety video or arrange for a safety professional to give a presentation on how to use fall protection equipment. Allow 20 to 30 minutes for this exercise.

Obtain one of the following or another safety training video:

Construction Fall Protection: Get Arrested! 11 minutes. Coastal Training Technologies Corp. www.coastal.com. $195.

What's the Fall Distance? 22 minutes. U.S. Department of Labor, Occupational Safety and Health Administration. www.osha.gov. Free.

Fall Protection Training. 12 minutes. All About OSHA. www.allaboutosha.com. $99.95

1. Introduce the video or speaker who will give a presentation on fall protection equipment.
2. Have the trainees take notes and write down questions during the video or presentation.
3. After the video or presentation answer any questions the trainees may have.

MODULE 08105-06 — ANSWERS TO REVIEW QUESTIONS

Answer	Section
1. c	1.0.0
2. a	2.0.0
3. c	2.0.0
4. b	2.0.0
5. b	2.0.0
6. c	2.0.0
7. b	2.1.0
8. b	2.2.0
9. c	2.2.0
10. a	3.0.0
11. a	3.1.1
12. a	3.4.0
13. a	3.4.0
14. c	3.5.0
15. b	3.5.0

Answers to Trade Terms Quiz

1. Base plates
2. Vertical upright
3. Casters
4. Toeboard
5. Cross braces
6. Scaffolding
7. Hinge pin
8. Harness
9. Duty rating
10. Coupling pin
11. Scaffold
12. Leveling jacks
13. Guardrail
14. Ladder
15. Scaffold floor

CONTREN® LEARNING SERIES — USER UPDATE

NCCER makes every effort to keep these textbooks up-to-date and free of technical errors. We appreciate your help in this process. If you have an idea for improving this textbook, or if you find an error, a typographical mistake, or an inaccuracy in NCCER's Contren® textbooks, please write us, using this form or a photocopy. Be sure to include the exact module number, page number, a detailed description, and the correction, if applicable. Your input will be brought to the attention of the Technical Review Committee. Thank you for your assistance.

Instructors – If you found that additional materials were necessary in order to teach this module effectively, please let us know so that we may include them in the Equipment/Materials list in the Annotated Instructor's Guide.

Write: Product Development and Revision
National Center for Construction Education and Research
P.O. Box 141104, Gainesville, FL 32614-1104

Fax: 352-334-0932

E-mail: curriculum@nccer.org

Craft ______________________ Module Name ______________________

Copyright Date ____________ Module Number ____________ Page Number(s) ____________

Description

____

____

____

(Optional) Correction

____

____

(Optional) Your Name and Address

____

____

Module 08106-06

Motorized Equipment

NCCER STANDARDIZED CRAFT TRAINING PROGRAM

The National Center for Construction Education and Research (NCCER) provides a standardized national program of accredited craft training. Key features of the program include instructor certification, competency-based training, and performance testing. The program provides trainees, instructors, and companies with a standard form of recognition through the National Registry. The program is described in full in the *Guidelines for Accreditation*, published by NCCER. For more information on standardized craft training, contact NCCER by writing to P.O. Box 141104, Gainesville, FL 32614-1104; calling 352-334-0911; or e-mailing info@nccer.org. More information is available at www.nccer.org.

HOW TO USE THIS ANNOTATED INSTRUCTOR'S GUIDE

Each page presents two sections of information. The larger section displays each page exactly as it appears in the Trainee Module. The narrow column ties suggested trainee and instructor actions to each page and provides icons (detailed below) to call your attention to material, safety, audiovisual, or testing requirements. The bottom of each page includes space for your notes.

The **Audiovisual** icon indicates an appropriate time to show a transparency or other audiovisual aid.

The **Classroom** icon prompts you to define a term, stress a point, ask trainees to explain a concept, or give examples.

The **Demonstration** icon directs you to show trainees how to perform tasks.

The **Examination** icon tells you to administer the written module examination.

The **Homework** icon is placed where you may wish to assign reading for the next class, assign a project, or advise trainees to prepare for an examination.

The **Laboratory** icon is used when trainees are to practice performing tasks.

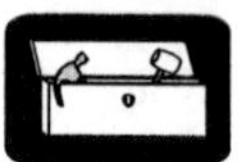

The **Materials** icon is a reminder for you to gather materials needed for classes, laboratories, and testing.

The **Performance Testing** icon tells you to administer a performance test or a portion thereof.

The **Safety** icon is used to emphasize safety issues. It is often keyed to *Caution* and *Warning!* statements in the Trainee Module.

The **Teaching Tip** icon indicated additional guidance is available, such as how to conduct an exercise, get the most educational value from a field trip, or encourage class participation. Teaching Tips may expand on a feature (*Think About It, Did You Know?*) or provide *Quick Quizzes* or similar exercises. You will be referred to the Teaching Tips section at the back of the module if there is additional material.

The **Combination** icon indicates that the laboratory listed corresponds with a performance task. If desired, you can note the proficiency of the trainees during the laboratory, and use it to satisfy performance testing requirements.

PREPARATION

Before teaching this module, you should review the Objectives, Performance Tasks, Materials and Equipment List, and Module Outline. Be sure to allow ample time to prepare your own training or lesson plan and gather all required materials and equipment.

MODULE OVERVIEW

This module explains the applications, proper use, and safety considerations for using engine-driven generators, welding machines, air compressors, pumps, forklift trucks, and hydraulic cranes.

PREREQUISITES

Prior to training with this module, it is recommended that the trainee shall have successfully completed *Core Curriculum;* and *Pipefitting Level One*, Modules 08101-06 through 08105-06.

OBJECTIVES

Upon completion of this module, the trainee will be able to do the following:

1. State the safety precautions associated with the use of motor-driven equipment on job sites.
2. Identify and explain the operation and use of the following motor-driven equipment.
   - Welding machines
   - Portable generators
   - Air compressors
   - Portable pumps
   - Aerial lifts
   - Forklifts
   - Compaction equipment
   - Trenching equipment
   - Backhoe loaders
   - Mobile cranes
3. Perform prestart checks and operate the following equipment:
   - Portable generators
   - Welding machines
   - Portable pumps
   - Air compressors

PERFORMANCE TASKS

Under the supervision of the instructor, the trainee should be able to do the following:

1. Perform all prestart checks for engine-driven generators.
2. Set up and operate engine-driven welding machines.
3. Operate engine-driven generators.
4. Perform all prestart checks for portable air compressors.
5. Operate portable air compressors.
6. Identify portable pumps to use for specific applications.
7. Identify forklift trucks and recognize safety hazards involved in working around them.
8. Identify types of hydraulic cranes and recognize safety hazards involved in working around them.

MATERIALS AND EQUIPMENT LIST

Overhead projector and screen
Transparencies
Blank acetate sheets
Transparency pens
Whiteboard/chalkboard
Markers/chalk
Pencils and scratch paper
Appropriate personal protective equipment
Portable generators and accessories
Portable generator operator's manual
Welding machine and accessories
Welding machine operator's manual
Portable air compressor and accessories
Portable air compressor operator's manual
Portable pumps and accessories
Portable pump operator's manual
29 CFR 1926.453
Aerial lift operator's manual
Compactor operator's manual
Forklift operator's manual
Backhoe operator's manual
Trencher operator's manual
Module Examinations*
Performance Profile Sheets*

* Located in the Test Booklet.

SAFETY CONSIDERATIONS

Ensure that the trainees are equipped with appropriate personal protective equipment and know how to use it properly. This module requires trainees to use various types of motorized equipment. Review hazards associated with each type of equipment and general precautions needed when operating motorized equipment.

ADDITIONAL RESOURCES

This module is intended to present thorough resources for task training. The following reference works are suggested for both instructors and motivated trainees interested in further study. These are optional materials for continued education rather than for task training.

Construction Equipment Guide, Latest Edition. New York, NY: John Wiley & Sons.

Machinery Handbook, Latest Edition. Erik Oberg, Franklin D. Jones, Holbrook L Horton, and Henry H Ryffel. New York, NY: Industrial Press, Inc.

TEACHING TIME FOR THIS MODULE

An outline for use in developing your lesson plan is presented below. Note that each Roman numeral in the outline equates to one session of instruction. Each session has a suggested time period of 2½ hours. This includes 10 minutes at the beginning of each session for administrative tasks and one 10-minute break during the session. Approximately 10 hours are suggested to cover *Motorized Equipment*. You will need to adjust the time required for hands-on activity and testing based on your class size and resources. Because laboratories often correspond to Performance Tasks, the proficiency of the trainees may be noted during these exercises for Performance Testing purposes.

Topic	Planned Time
Session I. Introduction, Safety, Generators, and Welding Machines	
A. Introduction	____________
B. Safety Precautions	____________
C. Generators	____________
D. Laboratory – Trainees practice performing prestart checks and operating generators. This laboratory corresponds to Performance Tasks 1 and 3.	____________
E. Welding Machines	____________
F. Laboratory – Trainees practice setting up and operating welding machines. This laboratory corresponds to Performance Task 2.	____________
Session II. Air Compressors, Portable Pumps, Aerial Lifts, and Compaction Equipment	
A. Air Compressors	____________
B. Laboratory – Trainees practice performing prestart checks and operating air compressors. This laboratory corresponds to Performance Tasks 4 and 5.	____________
C. Portable Pumps	____________
D. Laboratory – Trainees identifying portable pumps to use for specific applications. This laboratory corresponds to Performance Task 6.	____________
E. Aerial Lifts	____________
F. Compaction Equipment	____________
Session III. Forklifts, Backhoes, Trenchers, and Cranes	
A. Forklifts	____________
B. Laboratory – Trainees practice identifying forklifts and recognizing hazards associated with them. This laboratory corresponds to Performance Task 7.	____________
C. Backhoes	____________
D. Trenchers	____________
E. Cranes	____________
F. Laboratory – Trainees practice identifying cranes and recognizing hazards associated with them. This laboratory corresponds to Performance Task 8.	____________
Session IV. Review, Module Examination, and Performance Testing	
A. Review	____________
B. Module Examination	____________
1. Trainees must score 70 percent or higher to receive recognition from NCCER.	
2. Record the testing results on Craft Training Report Form 200, and submit the results to the Training Program Sponsor.	
C. Performance Testing	____________
1. Trainees must perform each task to the satisfaction of the instructor to receive recognition from NCCER. If applicable, proficiency noted during laboratory exercises can be used to satisfy the Performance Testing requirements.	
2. Record the testing results on Craft Training Report Form 200, and submit the results to the Training Program Sponsor.	

Assign reading of Module 08106-06.

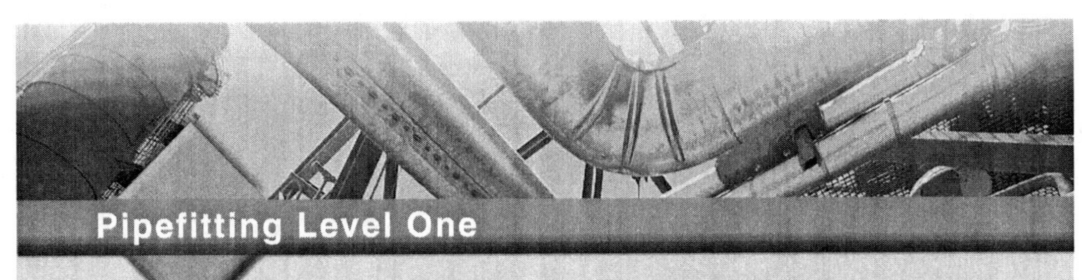

08106-06

Motorized Equipment

08106-06
Motorized Equipment

Topics to be presented in this module include:

Overview

Pipefitters work with various types of motorized equipment. Motorized equipment, like power tools, allows the user to be more productive. However, increased power also increases the hazards. Anyone operating motorized equipment must be properly trained, and in some cases certified, to operate the equipment. Always follow the recommended safety precautions, the manufacturer's instructions, and maintenance schedule.

Pipefitters often use smaller equipment such as generators, compressors, and portable pumps. Generators and compressors provide power for tools on job sites. Pipefitters may also operate or work around larger equipment such as forklifts, backhoe loaders, trenchers, and mobile cranes. A pipefitter will be able to work safely and efficiently by understanding the capacities and limitations of the equipment.

Instructor's Notes:

Objectives

When you have completed this module, you will be able to do the following:

1. State the safety precautions associated with the use of motor-driven equipment used on job sites.
2. Identify and explain the operation and uses of the following motor-driven equipment:
   - Welding machines
   - Portable generators
   - Air compressors
   - Portable pumps
   - Aerial lifts
   - Compaction equipment
   - Forklifts
   - Trenching equipment
   - Backhoe loaders
   - Mobile cranes
3. Perform prestart checks and operate the following equipment:
   - Portable generators
   - Welding machines
   - Portable pumps
   - Air compressors

Trade Terms

Aerial lift
Ampere (amp)
Centrifugal force
Centrifugal pump
Cherry picker
Circuit breaker
Compactor
Compressor
Diaphragm
Diaphragm pump
Forklift
Fuse holder
Generator
Governor
Impeller
Mobiling
Pneumatic
Powered industrial truck
Proportional control
Straight blade duplex connector
Trencher
Twist-lock connector
Velocity
Volt
Watt

Required Trainee Materials

1. Pencil and paper
2. Appropriate personal protective equipment

Prerequisites

Before you begin this module, it is recommended that you successfully complete *Core Curriculum; Pipefitting Level One*, Modules 08101-06 through 08105-06.

This course map shows all of the modules in the first level of the *Pipefitting* curriculum. The suggested training order begins at the bottom and proceeds up. Skill levels increase as you advance on the course map. The local Training Program Sponsor may adjust the training order.

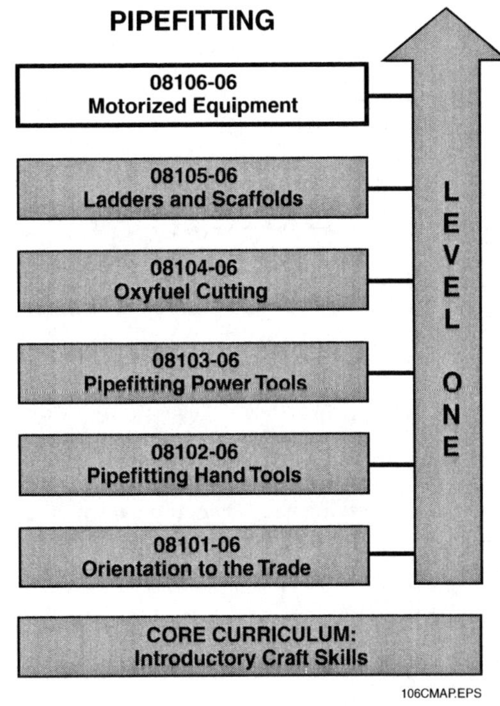

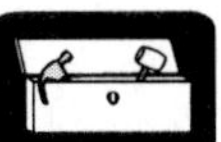

Ensure that you have everything required to teach the course. Check the Materials and Equipment list at the front of this module.

See the general Teaching Tip at the end of this module.

Explain that terms shown in bold are defined in the Glossary at the back of this module.

Review the modules covered in Level One and explain how this module fits in.

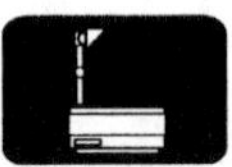

Show Transparency 1, Objectives, and Transparency 2, Performance Tasks. Review the goals of the module, and explain what will be expected of the trainee.

Explain that pipefitters use various types of motorized equipment.

Explain that safety is the primary concern when operating any type of motorized equipment. Emphasize the importance of reading the operator's manual and following the manufacturer's safety precautions.

Discuss the importance of following basic safety rules when operating motorized equipment.

Describe safety interlocks that prevent unsafe operation.

1.0.0 ◆ INTRODUCTION

Pipefitters work with and around various types of motorized equipment throughout their careers. Motorized equipment includes portable equipment such as **generators**, **compressors**, and pumps, as well as larger equipment, such as trucks, mobile cranes, and personnel lifts. A piece of motorized equipment is considered portable if it can be transported from job site to job site or to different areas of the same job site. Portable equipment can either be moved or rolled to a different location by hand or transported on a trailer behind a truck or tractor. A piece of equipment is also considered to be portable if it can provide its own power source in the field. This module explains the use, care, and preventive maintenance of portable motorized equipment. It also covers **forklifts**, backhoe loaders, and mobile cranes that pipefitters will be working with and around throughout their careers.

2.0.0 ◆ SAFETY PRECAUTIONS

Safety is the primary concern no matter what type of equipment you may be operating. Most of the equipment contained in this module share common safety precautions. These safety precautions are discussed in this section. As each piece of light equipment is discussed, remember this safety discussion and think about how these precautions apply to the equipment being covered. Additional safety precautions that are unique to particular equipment are discussed later. Appropriate personal protective equipment must be worn while operating or working near any equipment, and all use must be performed under the direct supervision of your instructor or other qualified personnel.

Hydraulic Fluid Leaks Can be Dangerous

Hydraulic fluid escaping through pinholes in hoses and hose fittings can be almost invisible. Escaping hydraulic fluid under pressure can penetrate the skin or eyes, causing serious injury. When inspecting light construction equipment, hydraulic hoses, and hose couplings for fluid leaks, always wear protective clothing and eye protection. Do not trace for leaks along hydraulic hoses and hydraulic system components with bare hands; instead, use a piece of cardboard or wood. In the event that hydraulic fluid penetrates the skin, seek immediate medical attention from a doctor familiar with this type of injury.

WARNING!

Anyone operating power equipment must be properly trained in its use. Part of this training includes understanding the associated safety precautions. Some localities may require the operator to be trained and qualified in the use of certain equipment. The primary points to remember are:

- Read and fully understand the operator's manual, and follow the instructions and safety precautions.
- Know the capacity and operating characteristics of the equipment being operated.
- Inspect the equipment before each use to make sure everything is in proper working order. Have any defects repaired before using the equipment. Never modify or remove any part of the equipment unless authorized to do so by the manufacturer.
- Check for hazards above, below, and all around the job site. Be sure to maintain a safe distance from electrical power lines and other electrical hazards.
- Learn as much about the work area as you can before beginning work.
- Fasten seat belt or operator restraints before starting.
- Set up warning barriers and keep others away from the equipment and job site.
- Know where to get assistance.
- Know how to use a first aid kit and fire extinguisher.

The manufacturer provides safety precautions for each piece of light equipment. These safety precautions should be observed whenever you are using any equipment. All of the equipment discussed can cause serious injury or death if the safety precautions are not followed.

2.1.0 Interlocking Systems

Some equipment comes with safety interlocks to prevent unsafe operation. Keep the following in mind when using this type of equipment:

- Interlock systems must not be modified. Interlock systems help prevent incorrect control operation.
- If an interlock system fails, contact the manufacturer's repair representative immediately.
- Runaway equipment is always possible; learn how to use all the controls and emergency procedures.

2.2.0 Transporting Equipment

When transporting equipment between jobs, remember these safety guidelines:

Instructor's Notes:

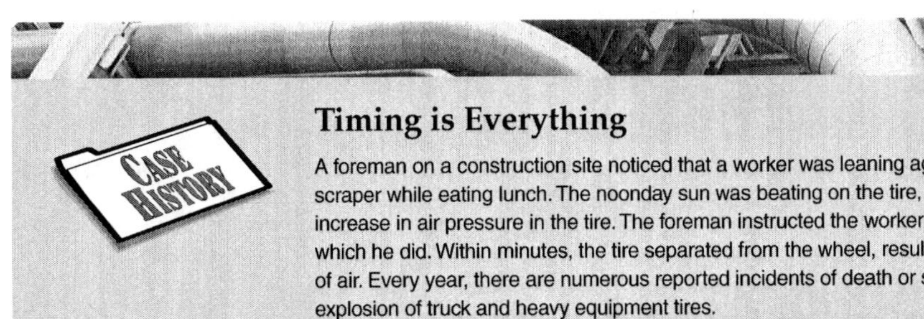

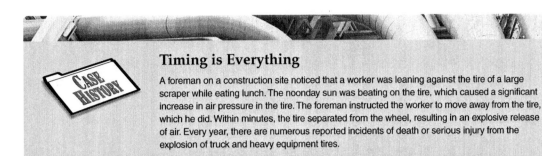

Timing is Everything

A foreman on a construction site noticed that a worker was leaning against the tire of a large scraper while eating lunch. The noonday sun was beating on the tire, which caused a significant increase in air pressure in the tire. The foreman instructed the worker to move away from the tire, which he did. Within minutes, the tire separated from the wheel, resulting in an explosive release of air. Every year, there are numerous reported incidents of death or serious injury from the explosion of truck and heavy equipment tires.

- Park, unload, and load the equipment from a trailer on level ground.
- To prevent tipping, connect the trailer to the tow vehicle before loading or unloading.
- Follow the manufacturer's requirements for towing and transporting the equipment.
- Inspect the tires and wheels before towing. Do not tow with high or low tire pressures, cuts, excessive wear, bubbles, damaged rims, or missing lug bolts or nuts.
- Do not inflate tires with flammable gases or from systems using an alcohol injector.

WARNING!

Large tires under pressure can explode, causing injury or death. Always maintain the correct tire pressure.

2.3.0 Hydraulic Systems

When operating equipment that uses hydraulics, always remember the following:

- Before disconnecting any hydraulic lines, relieve system pressure by cycling controls.
- Before pressurizing the system, be sure all connections are tight and the lines are undamaged.
- Do not perform any work on the equipment unless you are authorized and qualified to do so.
- Check for leaks using a piece of cardboard or wood; never use bare hands.

2.4.0 Fueling Safety

Fueling safety precautions are applicable to all light equipment using gasoline. Make sure you follow all the manufacturer's instructions before fueling the equipment. The following are general safety precautions associated with fueling light equipment:

- Never fill the fuel tank with the motor running, while smoking, or near an open flame.
- Be careful not to overfill the tank or spill fuel. If fuel is spilled, clean it up immediately.
- Use clean fuel only.
- Do not operate the equipment if fuel has been spilled inside or near the unit.
- Ground the fuel funnel or nozzle against the filler neck to prevent sparks.
- Replace the fuel tank cap after refueling.

2.5.0 Battery Safety

Battery safety precautions are applicable to all light equipment equipped with a battery for starting or operating electric components.

- Chargers can ignite flammable materials and vapors. They should not be used near fuels, grain dust, solvents, or other flammables.
- To reduce the possibility of electric shock, a charger should only be connected to a properly grounded single-phase outlet. Do not use an extension cord longer than 25 feet.
- If the battery is frozen, do not charge or attempt to jump-start the equipment, as the battery may explode.
- Battery and fuel fumes could ignite and cause explosions and burns. Keep batteries away from flames or sparks. Do not smoke.

3.0.0 ◆ GENERATORS

Generators are used to provide electrical power and/or lighting at the job site. Generators are available in many different sizes and configurations (*Figures 1* and *2*). They range from small portable machines to large installed backup power systems for emergency power generation. Construction site requirements for power will vary depending on the scale of the work being performed. This section describes a typical tow-behind generator used to provide electrical power to a job site.

Discuss safety requirements for properly transporting equipment.

Discuss the hazards posed from large tires under pressure and safety precautions needed.

Explain safety considerations when using hydraulic equipment.

Explain precautions necessary when fueling.

Discuss the hazards and precautions needed when working with batteries.

Describe a generator.

Provide a generator and operator's manual for the trainees to examine.

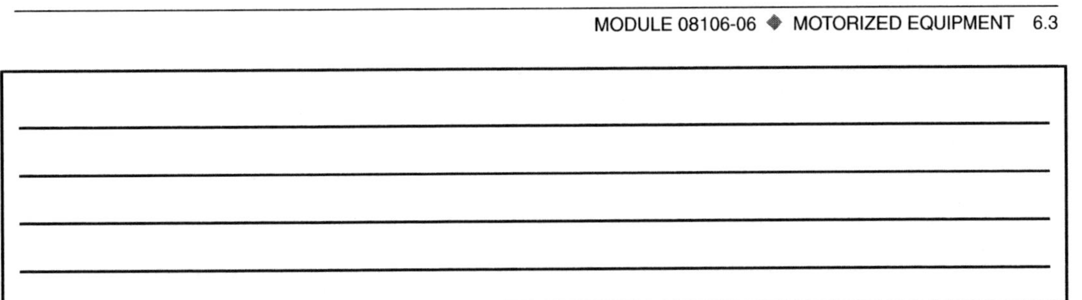

MODULE 08106-06 ◆ MOTORIZED EQUIPMENT 6.3

Describe the training necessary to operate a generator.

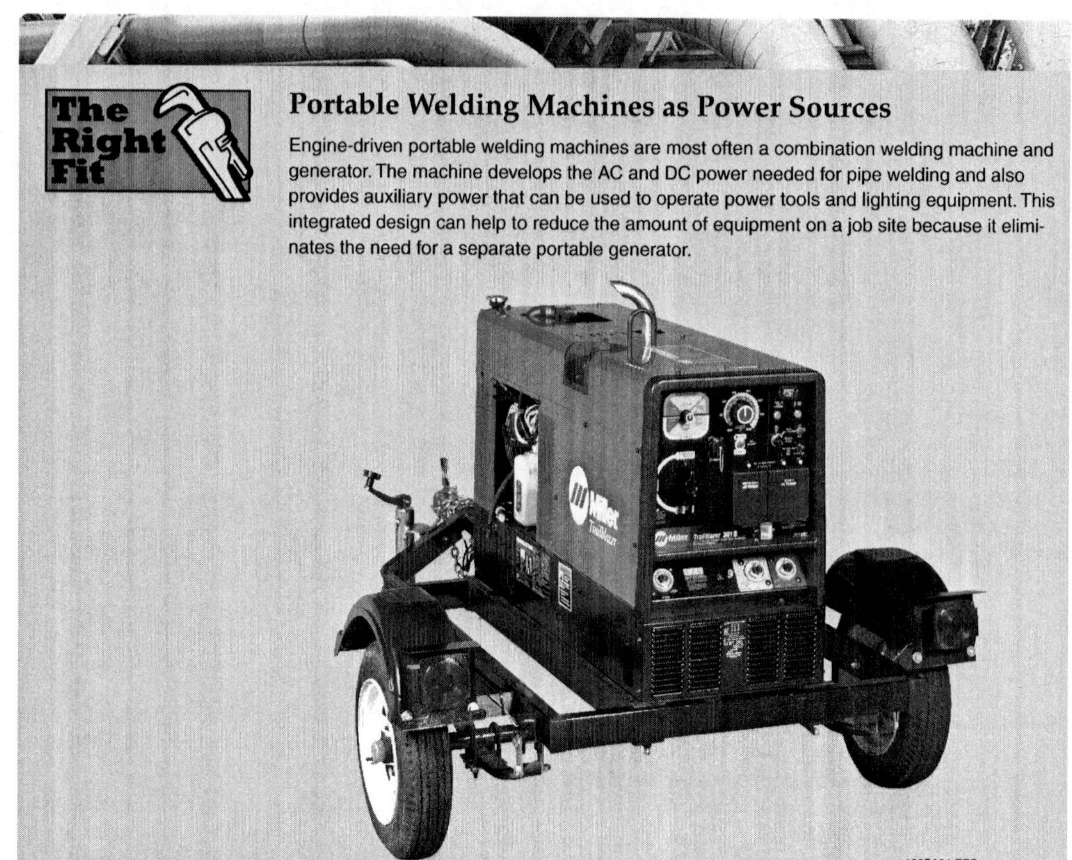

The Right Fit

Portable Welding Machines as Power Sources

Engine-driven portable welding machines are most often a combination welding machine and generator. The machine develops the AC and DC power needed for pipe welding and also provides auxiliary power that can be used to operate power tools and lighting equipment. This integrated design can help to reduce the amount of equipment on a job site because it eliminates the need for a separate portable generator.

106SA01.EPS

TOW-BEHIND GENERATOR

PORTABLE GENERATOR

106F01.EPS

Figure 1 ◆ Generators.

106F02.EPS

Figure 2 ◆ Tow-behind generator for lighting.

3.1.0 Generator Operator Qualifications

Only trained, authorized persons are permitted to operate a generator. Personnel who have not been trained in generator operation may only operate them for the purpose of training. The training shall be conducted under the direct supervision of a qualified trainer.

Instructor's Notes:

6.4 PIPEFITTING ◆ LEVEL ONE

3.2.0 Typical Generator Controls

Tow-behind generators have operator controls for the engine and an electrical control panel to control and monitor the power produced by the generator. *Figure 3A* shows a typical engine control panel. This panel provides the operator with the controls required to start and stop the motor and monitor critical motor functions.

Controls and indicators include:

- *Engine over-speed indicator* – This indicator monitors engine speed. If speed exceeds the manufacturer's set limit, the engine will shut down automatically.
- *High engine temperature indicator* – This indicates when the engine coolant temperature has exceeded the manufacturer's set limit. The engine will shut down automatically.
- *Low engine oil pressure indicator* – This indicates when the engine oil pressure has fallen below the manufacturer's preset limit. The engine will shut down automatically.
- *Alternator not charging* – This indicates the engine alternator is not outputting enough energy to charge the unit's battery.
- *Ignition switch* – Place this switch in the on position to run; place it in the off position to stop.

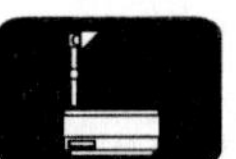

Show Transparency 3 (Figure 3). Describe the engine controls on the control panel on a typical tow-behind generator.

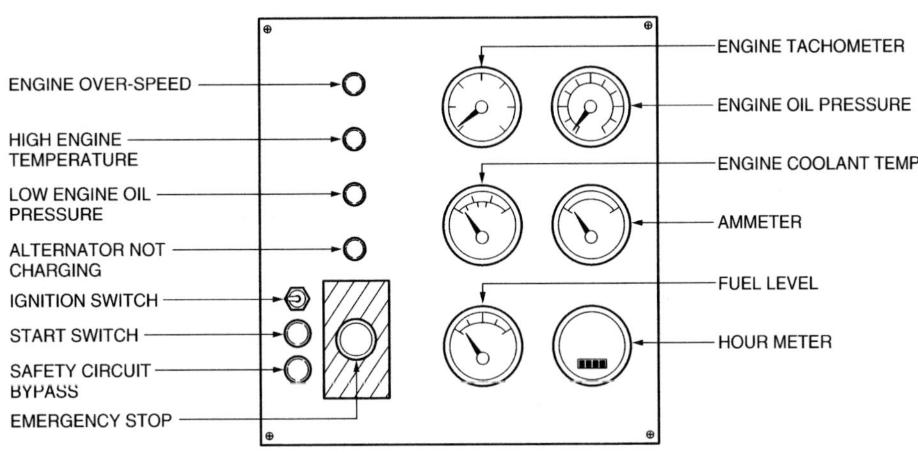

(A) ENGINE CONTROL PANEL

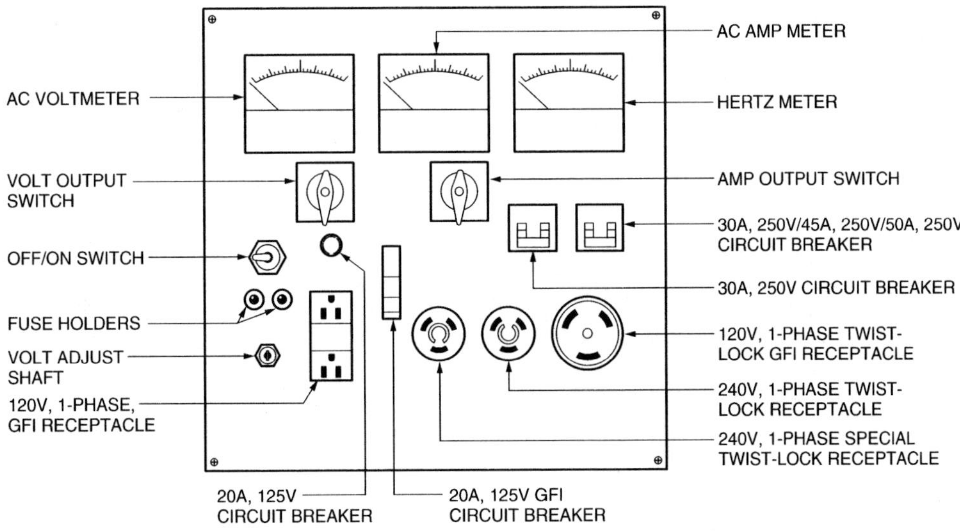

Figure 3 ◆ Generator control panels.

MODULE 08106-06 ◆ MOTORIZED EQUIPMENT 6.5

Keep Transparency 3 (Figure 3) showing. Describe the generator controls on the control panel on a typical tow-behind generator.

Emphasize the importance of following safety precautions when operating a generator.

Discuss the hazards of high voltage and electrical shock.

Review safety precautions necessary when operating a generator.

Describe grounding requirements for operating a generator.

- *Start switch* – Pressing this switch activates the engine starting motor.
- *Safety circuit bypass* – Pressing this switch bypasses automatic shutdowns when starting the engine.
- *Emergency stop* – Pressing this switch causes the engine to shut down with no other operator action required.
- *Engine tachometer gauge* – This gauge indicates engine rpm.
- *Engine oil pressure gauge* – This gauge displays engine oil pressure.
- *Engine coolant temperature* – This gauge indicates the temperature of the engine coolant.
- *Ammeter gauge* – This gauge indicates the charge rate of the engine alternator.
- *Fuel level gauge* – This gauge indicates the level of fuel in the fuel tank(s).
- *Hour meter* – This meter records total engine operating hours. It is used in determining the maintenance schedule on the unit.

The second control panel in a tow-behind generator is the generator control panel. This panel contains meters, monitor switches, a voltage regulator, **circuit breakers**, and receptacles to control and monitor the output of the generator. *Figure 3B* shows a typical generator control panel.

- *AC voltmeter* – This meter indicates the generator output voltage level.
- *AC amperes meter* – This meter indicates the generator output load in **amperes (amp)**.
- *Hertz meter* – This meter indicates the frequency of the generator output.
- *Voltage output monitor switch* – This switch selects the reference for the generator voltage displayed on the AC voltmeter.
- *Amperage output monitor switch* – This switch selects the line-to-line (phase) amperage displayed on the AC ammeter.
- *Voltage regulator on/off switch* – In the off position, this switch removes excitation to the generator field, stopping the generation of power.
- *Voltage regulator fuse holders* – The **fuse holders** house the fuses for the voltage regulator.
- *Voltage regulator voltage adjust shaft* – This shaft is turned to adjust generator output voltage.
- *Circuit breakers* – Various styles of circuit breakers are used, including push-to-reset, ground fault circuit interrupter (GFCI), and flip-to-reset styles.
- *Receptacles* – Various types of receptacles are provided, depending on the size and style of generator used. They will include GFCI **straight blade duplex connectors** and **twist-lock connectors**.

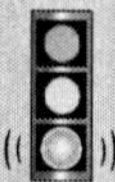

NOTE

Generator selection is based on the total wattage of the tools and devices to be connected to the generator. The selection must be done by a qualified electrician.

3.3.0 Generator Safety Precautions

The safe operation of a generator is the operator's responsibility. The other safety precautions, already discussed in this module, also apply.

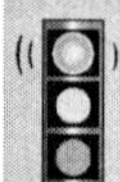

WARNING!

High voltages are present when the generator is operating. Exercise caution to avoid electric shock.

The following are general safety precautions that must be observed when operating a generator:

- Do not operate electrical equipment when standing in water or on wet ground, or with wet hands or shoes.
- Grounding should be performed in compliance with local electrical codes and in accordance with the manufacturer's manual.
- Use a generator as an alternate power supply only after the main feed at the service entrance panel has been opened and locked.
- Do not change voltage selection while the engine is running. Voltage selection, adjustment, and electrical connections may only be performed by qualified personnel.
- Do not exceed the generator power (**watt**) rating during operation.
- If welding is required on the unit, follow the manufacturer's instructions to prevent damage to the circuitry.
- Never make electrical connections with the unit running.

3.4.0 Generator Operation

Before operating the generator, local requirements for grounding must be investigated and followed. The generator set can produce high voltages, which can cause severe injury or death to personnel and damage to equipment. The generator should have proper internal and external grounds when required by the *National Electrical Code®* (*NEC®*). A qualified, licensed electrical contractor, knowledgeable in local codes, should be consulted. Follow all

Instructor's Notes:

6.6 PIPEFITTING ◆ LEVEL ONE

Grounding Portable and Vehicle-Mounted Generators

The grounding requirements for portable and vehicle-mounted generators are listed in ***NEC Section 250.34***. ***NEC Section 250.20*** governs the grounding of portable generators for applications that supply fixed wiring systems.

manufacturer's requirements to ensure that the generator is connected properly before operation.

The following discussion covers generic operation of a generator. Each manufacturer provides detailed operating instructions with their equipment. Refer to the operator's manual provided with the generator for specific instructions. The general guidelines given here are intended to help you understand the operator controls. The operator should perform a complete inspection of the generator and correct any deficiencies before beginning operation.

3.4.1 Generator Setup and Preoperational Checks

Many sites have a prestart checklist which must be completed and signed prior to starting or operating an engine-driven generator. Check with your supervisor, and if your site has such a checklist, complete and sign it. A typical procedure for setting up a generator for use at the job site is as follows:

Step 1 Place the unit as level as possible. Follow the manufacturer's directions concerning equipment placement and any special considerations for equipment location.

Step 2 Disconnect the generator from the towing vehicle.

Step 3 Chock the wheels of the generator.

Step 4 Unlock the jack and lower it to the service position.

Step 5 Lock the jack in the service position.

Step 6 Disconnect the safety chains and crank the jack to raise the coupling off the hitch.

Before putting the generator in operation, check it for evidence of arcing on or around the control panel. If any arcing is noted, the problem must be located and repaired before beginning operation. Check for loose wiring or loose routing clamps within the housing.

If your site does not have a prestart checklist, perform the following checks before starting the engine.

- Check the oil, using the engine oil dipstick. If the oil is low, add the appropriate grade oil for the time of year.
- Check the coolant level in the radiator if the engine is liquid-cooled. If the coolant level is low, add coolant.

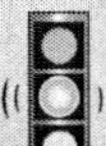

CAUTION

Do not add plain water to radiators that contain antifreeze. Antifreeze not only protects radiators from freezing in cold weather, it also has rust inhibitors and additives to aid in cooling. If the antifreeze is diluted, it will not function properly. If the weather turns cold, the system may freeze, causing damage to the radiator, engine block, and water pump.

- Check the fuel. The unit may have a fuel gauge or a dipstick. If the fuel is low, add the correct fuel (diesel or gasoline) to the fuel tank. The type of fuel required should be marked on the fuel tank. If it is not marked, contact your supervisor to verify the fuel required and have the tank marked.

WARNING!

Never add fuel to an engine-driven generator while the motor is running. Always make sure that no one is using the generator, and shut down the generator to refuel it.

CAUTION

Adding gasoline to a diesel engine or diesel to a gasoline engine causes severe engine problems. It can also cause a fire hazard. Always be sure to add the correct fuel to the fuel tank.

- Check the battery water level unless the battery is sealed. Add room-temperature water if the battery water level is low. Never add cold water to a battery.
- If it is a welding machine, check the electrode holder to be sure it is not grounded. If the electrode holder is grounded, it will arc and overheat

Emphasize the importance of reading the operator's manual and inspecting a generator before operating it.

Describe the procedure for setting up a generator.

Describe the prestart checks that should be performed before operating a generator.

Discuss the hazards of adding fuel to a generator that is running.

Emphasize the importance of wearing safety glasses when using compressed air to clean equipment.

Explain how to start a generator.

Show Transparency 4 (Figure 4). Describe routine maintenance that must be performed on a generator.

Show trainees how to how to perform all prestart checks and operate an engine-driven generator.

Have trainees practice performing all prestart checks and operating an engine-driven generator. Note the proficiency of each trainee. This laboratory corresponds to Performance Tasks 1 and 3.

Describe an engine-driven welding machine.

the welding system when the welding machine is started. This is a fire hazard and can cause damage to the equipment.

- Open the fuel shutoff valve if the equipment has one. The fuel shutoff valve is located in the fuel line between the fuel tank and the carburetor.
- Record the hours from the hour meter if the equipment has one. An hour meter records the total number of hours the engine runs. This information is used to determine when the engine needs to be serviced. The hours are displayed on a gauge similar to an odometer.
- Clean the unit. Use a compressed air hose to blow off the engine and generator or alternator. Use a rag to remove heavier deposits that cannot be removed with the compressed air.

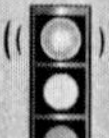

WARNING!

Whenever using compressed air to clean equipment, you must wear safety glasses with side shields or mono-goggles to protect your eyes from flying debris. You should also warn all others in the immediate area to ensure their safety.

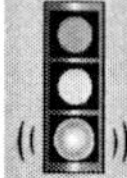

NOTE

Cleaning may not be required on a daily basis. Clean the unit as required.

3.4.2 Starting the Engine

Read and follow the manufacturer's starting procedures found in the operator's manual. Most engines have an on/off ignition switch and a starter. They may be combined into a key switch similar to the ignition on a car. To start the engine, turn on the ignition switch and press the starter. Release the starter when the engine starts. Larger diesel engines have glow plugs that must be warmed up before starting. The engine speed is controlled by the governor. If the governor switch is set for idle, the engine will slow to an idle after a few seconds. Small engine-driven generators may have an on/off switch and a pull cord to start the engine. Engine-driven generators should be started 5 to 10 minutes before they are needed to allow the engine to warm up before a load is placed on it.

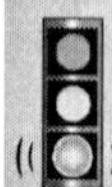

NOTE

Some engines are equipped with a choke, which provides extra combustion air during startup. The choke is closed once the engine is running.

If no power is required for 30 minutes or more, stop the generator by turning off the ignition switch. Never shut down an engine-driven generator before making sure that all workers using the generator have stopped.

3.5.0 Generator Operator's Maintenance Responsibility

Figure 4 shows a typical preventive maintenance schedule for a generator. The schedule is laid out with the performance period across the top and the items to be checked listed down the side of the page. For example, under the Daily column, the following items need to be checked:

- Evidence of arcing around electrical terminals
- Loose wire routing clamps
- Engine oil and coolant levels
- Grounding circuit
- Instruments
- Fan belts, hoses, and wiring insulation
- Air vents
- Fuel/water separator
- Service air indicator

Proper maintenance will extend the life and performance of the equipment.

3.6.0 Welding Machines

To produce welding current, the engine must turn the generator at a required number of revolutions per minute (rpm). The engines used to power generators have governors that control the engine speed. Most governors have a welding speed switch that can be set to idle the engine speed when no welding is taking place. When the welding electrode is touched to the base metal, the governor automatically increases the speed of the engine to the required rpms for welding. After about 15 seconds of no welding, the engine automatically returns to an idle.

Engine-driven welding machines often have an auxiliary power unit to produce 110-**volt** alternating current for lighting, power tools, and other electrical equipment. All power tools powered from a welding machine must have AC and DC capabilities. When 110-volt power is required, the engine-driven generator must run continuously at the welding speed.

Engine-driven welding machines have both engine controls and welding machine controls. The engine controls vary with the size and type of machine but are essentially the same as the generator engine controls previously discussed. The welding machine controls usually include an amperage control switch and gauge, a current-control switch

Instructor's Notes:

6.8 PIPEFITTING ◆ LEVEL ONE

Provide a engine-driven welding machine and operator's manual for the trainees to examine.

The Right Fit

Extension Cords

It is important that the extension cords used to supply power from a portable generator to electric power tools or other devices have an adequate current-carrying capacity for the job. If undersized extension cords are used, excessive voltage drops will result. This causes excessive heating of the extension cords and the portable tools, as well as additional generator loading.

PREVENTIVE MAINTENANCE SCHEDULE

	DAILY	WEEKLY	MONTHLY/ 150 HRS.	3 MONTHS/ 250 HRS.	6 MONTHS/ 500 HRS.	YEARLY 1000 HRS.
Evidence of Arcing Around Electrical Terminals	✓					
Loose Wire Routing Clamps	✓					
Engine Oil and Coolant Levels	✓					
Proper Grounding Circuit	✓					
Instruments	✓					
Frayed/Loose Fan Belts, Hoses, and Wiring Insulation	✓					
Obstructions in Air Vents	✓					
Fuel/Water Separator (drain)	✓					
Service Air Indicator	✓					
Precleaner Dumps		✓				
Tires		✓				
Battery Connections		✓				
Engine Radiator (exterior)			✓			
Air Intake Hoses and Flexible Hoses			✓			
Fasteners (tighten)			✓			
Emergency Stop Switch Operation			✓			
Engine Protection Shutdown System			✓			
Diagnostic Lamps			✓			
Voltage Selector and Direct Hook-up Interlock Switches				✓		
Air Cleaner Housing				✓		
Control Compartment (interior)					✓	
Fuel Tank (fill at end of each day)					Drain	
Fuel/Water Separator Element					Replace	
Wheel Bearings and Grease Seals					Repack	
Engine Shutdown System Switches (settings)						✓
Exterior Finish				(as needed)		
Engine			Refer to Engine Operator			
Decals		Replace decals if removed, damaged, or missing.				

✓ = Check or Clean (and Adjust or Replace, if necessary)

106F04.EPS

MODULE 08106-06 ◆ MOTORIZED EQUIPMENT 6.9

Show Transparency 5 (Figure 5). Describe the controls on a typical welding machine.

Explain how welding machines are sized. Discuss advantages and disadvantages of engine-driven welding machines.

Review the safety procedures for using a welding machine.

Show trainees how to how to set up and operate an engine-driven welding machine.

Have trainees practice setting up and operating an engine-driven welding machine. Note the proficiency of each trainee. This laboratory corresponds to Performance Task 2.

Have trainees review Sections 4.0.0–7.6.0.

Ensure that you have everything required for teaching this session.

and gauge, and a polarity switch. If the welding machine is not equipped with a polarity switch, manually change the lead cables at the welding current terminals to change the polarity. *Figure 5* shows a typical welding machine control panel.

The size of a welding machine is determined by the amperage output of the machine at a given duty cycle. The duty cycle of a welding machine is based on a 10-minute period, which is the percentage of 10 minutes that the machine can continuously produce its rated amperage without overheating. For example, a machine with a rated output of 300 amps at 60 percent duty cycle can deliver 300 amps of welding current for 6 minutes out of every 10 without overheating. The duty cycle of a welding machine will be 10, 20, 30, 40, 60, or 100 percent. A welding machine having a duty cycle of 10 to 40 percent is considered a light- to medium-duty machine. Most industrial, heavy-duty machines for manual welding are 60 percent duty cycle. With the exception of 100 percent duty cycle machines, the maximum amperage that a welding machine produces is always higher than its rated capacity. A welding machine rated 300 amps at 60 percent duty cycle generally puts out a maximum of 375 to 400 amps.

The advantage of using engine-driven welding machines is that they are portable and can be used in the field where electricity is not available for other types of welding machines. The disadvantage of engine-driven welding machines is that they are expensive to purchase, operate, and maintain.

Welding machine safety procedures include the following:

- Ensure that welding machines and all motorized equipment are shut down in the event of a plant evacuation.
- Operate arc welding machines and equipment only in clean, dry locations.
- Make sure that exhausts of gasoline- or diesel-powered welding generators are sufficiently ventilated when used indoors.
- Ensure that the arc welding machine is properly grounded to either your workpiece or to a ground rod driven into the ground, according to company policies.
- Check all connections to the machine to make sure they are secure before beginning to weld.
- Do not attempt to install or repair welding equipment. Call a qualified electrician.
- Protect arc welding machines in the field from weather conditions as much as possible.
- All welding machines must be equipped with a ground fault circuit interrupter (GFCI). This serves as an automatic shutoff to protect the user from injury due to a short circuit.

4.0.0 ◆ AIR COMPRESSORS

Compressors are used to provide compressed air for pneumatic tools at the job site. A jackhammer is one example of a pneumatic tool. Compressors are

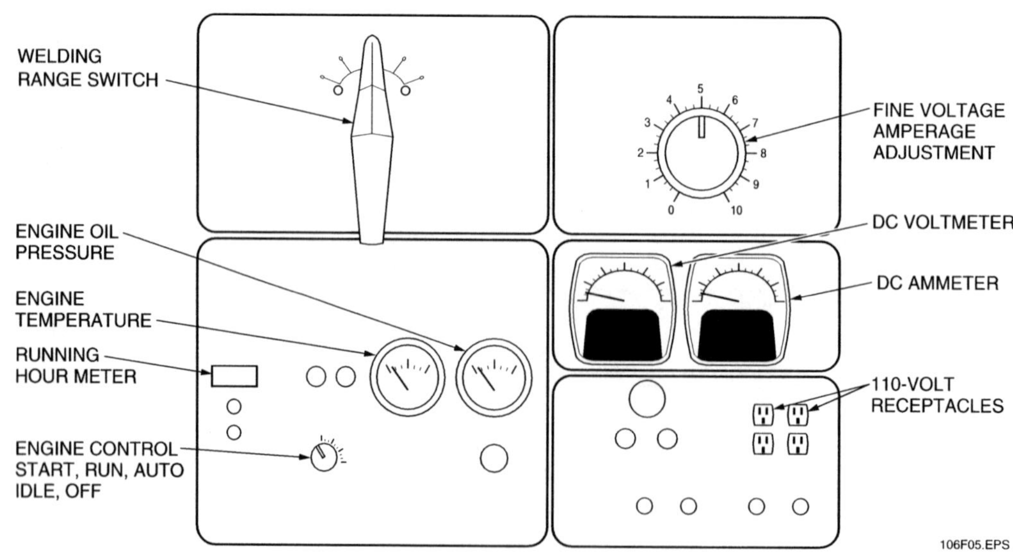

Figure 5 ◆ Typical welding machine control panel.

Instructor's Notes:

available in many different sizes and configurations (see *Figure 6*). They range from portable home-use machines to large installed units for industrial applications. Construction site requirements for compressed air vary depending on the scale of the work being performed. This section describes a typical tow-behind compressor.

The primary ratings of air compressor capacity are pounds per square inch (psi) and cubic feet per minute (cfm). The pressure in psi will depend on the pressure ratings of the tools or equipment to be operated by the compressor. Construction compressors typically range from 50 to 125 psi, but can go up to more than 390 psi. Most handheld pneumatic tools run on 90 psi. The term cfm refers to the amount of air delivered at the required pressure. The more tools and equipment, the more air delivery is required. A knowledgeable person must determine the size and type of compressor based on the tools and equipment to be used at a site.

4.1.0 Compressor Assemblies

Figure 7 shows a tow-behind compressor. Its major components include:

- Towing assembly and frame
- Protective cover and doors
- Engine and compressor assembly
- Operator control panel, located behind the access door

4.2.0 Compressor Operator Qualifications

Only trained, authorized persons are permitted to operate a compressor. Personnel who have not been trained in the operation of compressors may only operate them for the purpose of training. The training must be conducted under the direct supervision of a qualified trainer.

4.3.0 Typical Compressor Controls

Compressors normally have operator controls for the engine and compressor. This section discusses controls associated with a tow-behind compressor. *Figure 8* shows a typical compressor control panel. This panel provides the operator with the controls required to start and stop the motor and monitor critical motor and compressor functions.

Controls and indicators include the following:

- *Engine over-speed indicator* – This indicator monitors engine speed. If the speed exceeds the manufacturer's preset limit, the engine will shut down automatically.

TOW-BEHIND COMPRESSOR

SMALL PORTABLE GASOLINE ENGINE-DRIVEN COMPRESSOR

106F06.EPS

Figure 6 ◆ Types of compressors.

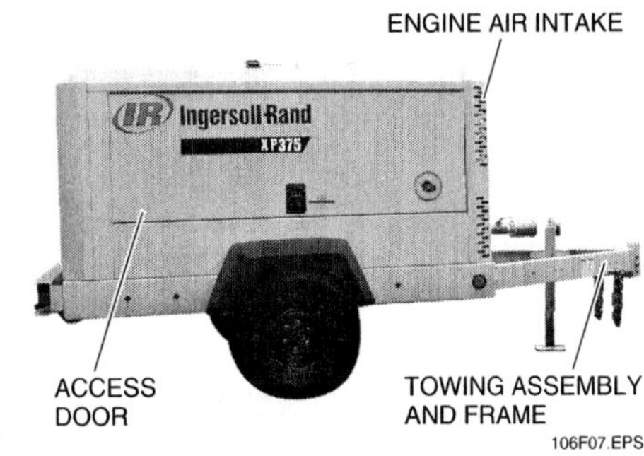

106F07.EPS

Figure 7 ◆ Large tow-behind compressor.

- *High engine temperature indicator* – This indicates that the engine coolant temperature has exceeded the manufacturer's preset limit; the engine will shut down automatically.
- *Low engine oil pressure indicator* – This indicates that the engine oil pressure has fallen below the manufacturer's preset limit; the engine will shut down automatically.

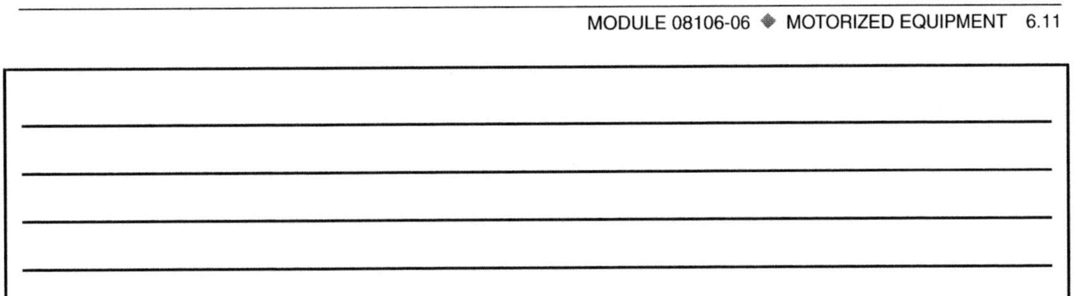

Describe an air compressor.

Explain how compressors are rated.

Describe the typical components of a tow-behind compressor.

Explain that only trained, authorized persons may operate a compressor.

Show Transparency 6 (Figure 8). Describe the controls on a typical compressor.

Provide an operator's manual for a compressor for the trainees to examine.

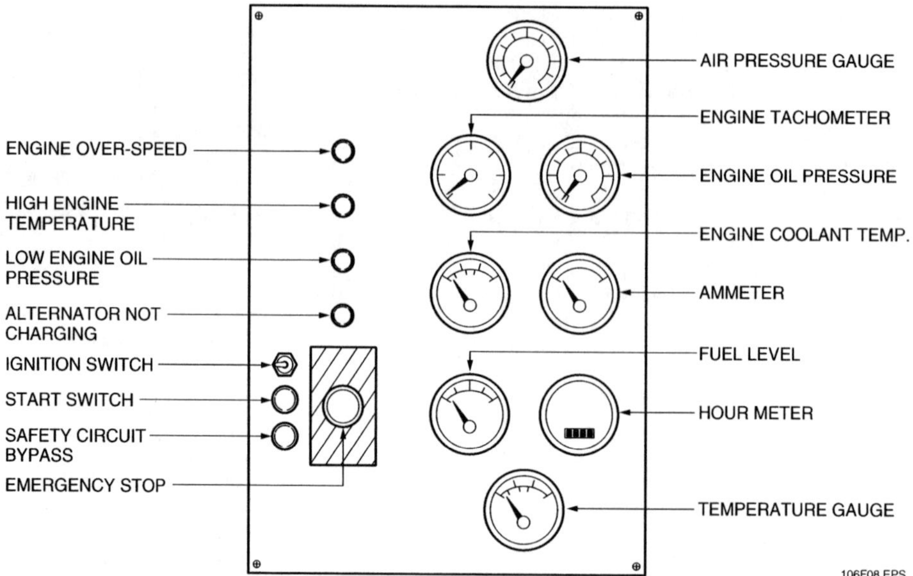

Figure 8 ◆ Compressor control panel.

The Right Fit

Sizing Compressors

Many people size compressors based on a horsepower rating. This is incorrect. Selection of a larger horsepower compressor motor/engine allows the compressor to run at faster speeds or larger displacement to produce greater air flow at a rated pressure. The size of a compressor should be based on the total air requirements of the various tools or equipment to be used with the compressor. These are the total air flow in cubic feet per minute (cfm) and the required air pressure in pounds per square inch (psi). Pneumatic tool manufacturers provide this information for each of their tools. For example, a typical framing nailer gun operates at a pressure of between 70 and 100 psi and consumes an average air flow of 9.6 cfm.

A rule of thumb for sizing a compressor is to determine the total cfm required, then multiply it by 1.5. For example, if the total air flow required is 20 cfm, the compressor should be sized to produce 30 cfm (20 × 1.5 = 30). This means that the compressor must produce a minimum of 30 cfm at maximum rated pressure to operate this tool properly.

The compressor rated pressure in psi is determined by the pressure requirements of the tools and/or equipment that the compressor is supplying. It is important that the compressor maintain pressures that are higher than what is required. For example, if the tools/equipment being used require a pressure of 120 psi, then a compressor capable of producing and maintaining a pressure higher than 120 psi is required.

The operating duty cycle of the tools being used with the compressor is another factor to be considered. Tools such as nail guns tend to be operated intermittently so that they only consume small amounts of air over a long time interval. However, tools like sanders and grinders tend to be operated continuously, causing them to consume large amounts of air over a long time interval. If the application involves tools or equipment that will be used continuously, select a compressor capable of producing a higher cfm output.

Instructor's Notes:

- *Alternator not charging* – This indicates that the engine alternator is not outputting enough energy to charge the unit's battery.
- *Ignition switch* – Place this switch in the on position to run; place to the off position to stop.
- *Start switch* – Pressing the start switch activates the engine starting motor.
- *Safety circuit bypass* – Pressing this switch bypasses automatic shutdowns when starting the engine.
- *Emergency stop* – Pressing this switch causes the engine to shut down with no other operator action required.
- *Air pressure gauge* – This gauge indicates the output air pressure level.
- *Engine tachometer gauge* – This gauge indicates engine rpm.
- *Engine oil pressure gauge* – This gauge displays engine oil pressure.
- *Engine coolant temperature* – This gauge indicates the temperature of the engine coolant.
- *Ammeter gauge* – This gauge indicates the charge rate of the engine alternator.
- *Fuel level gauge* – This gauge indicates the level of fuel in the fuel tank(s).
- *Hour meter* – This meter records total engine operating hours. It is used for scheduling maintenance on the unit.
- *Temperature gauge* – This gauge indicates the temperature of the compressed air.

4.4.0 Compressor Safety Precautions

The safe operation of the air compressor is the operator's responsibility. The following are general safety precautions that are applicable to air compressors. Remember that the other safety precautions, already discussed in this module, also apply. Refer to the operator's manual for the particular machine being used for specific safety precautions.

- Compressed air venting to the atmosphere can cause hearing damage. Make sure proper hearing protection is worn by everyone in the area.
- Ether is highly flammable. Do not inject ether into a hot engine or an engine equipped with a glow-plug type of preheater.
- Do not inject ether into the compressor air filter or a common air filter for the engine and compressor.
- Do not attempt to move the compressor or lift the drawbar without adequate personnel or equipment to handle the weight.
- Do not exceed the machine's air pressure rating.
- Do not use compressed air for breathing.
- Never direct compressed air at anyone.
- Be sure no pressure is in the system before removing the compressor filler cap.
- Do not use tools that are rated for a pressure lower than that provided by the compressor.
- Be sure all connections are securely made and hoses under pressure are secured to prevent whipping. Use appropriate safety devices to secure hoses.
- Do not weld or perform any modifications on the air compressor receiver tank.
- Use a safe, nonflammable solvent when cleaning parts.

4.5.0 Compressor Operation

This section covers the generic operation of a compressor. Each manufacturer provides detailed operating instructions with their equipment. Refer to the operator's manual provided with the compressor for specific instructions. The general outline provided here is intended to help you understand the operator controls. The operator should perform a complete inspection of the compressor and correct any deficiencies before beginning operation.

The following is a discussion of the type of inspection the operator performs before beginning work. Each manufacturer provides specific inspection procedure(s) in the operator's manual.

4.5.1 Compressor Setup and Preoperational Checks

Before beginning operation, the operator must perform a setup and preoperational checklist. Each manufacturer will provide a checklist with the equipment. This section covers the setup and preoperational checks for compressors.

- Place the unit in as level a position as possible. Follow the manufacturer's directions concerning equipment placement and any special considerations for equipment location.
- Disconnect the compressor from the towing vehicle.
- Chock the wheels of the compressor.
- Unlock the jack and lower it to the service position. Lock the jack in the service position.
- Disconnect the safety chains and crank the jack to raise the coupling off the hitch.

Vent all internal pressure before opening any air line, fitting, hose, valve, drain plug, connector, oil filler, or filler caps, and before refilling anti-ice systems. Be sure all hoses, shutoff valves, flow-limiting valves, and other attachments are connected according to the manufacturer's instructions. If the

Discuss safety precautions necessary when operating an air compressor.

Emphasize the importance of reading the operator's manual and inspecting a compressor before operating it.

Describe the procedure for setting up a compressor.

Describe the prestart checks that should be performed before operating a compressor.

Emphasize the importance of adding the correct fluids, including antifreeze and fuel, to a compressor.

Explain how to start a compressor.

Discuss maintenance requirements for a compressor.

Show trainees how to how to perform all prestart checks and operate an engine-driven compressor.

Have trainees practice performing all prestart checks and operating an engine-driven compressor. Note the proficiency of each trainee. This laboratory corresponds to Performance Tasks 4 and 5.

air compressor is to be used along with other sources of air, be sure there is a check valve installed at the service valve.

The following checks should be performed before starting the engine:

- Check the oil, using the engine oil dipstick. If the oil is low, add the appropriate grade oil for the time of year.
- Check the coolant level in the radiator if the engine is liquid-cooled. If the coolant level is low, add coolant.

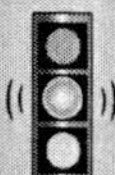

CAUTION

Do not add plain water to radiators that contain antifreeze. Antifreeze not only protects radiators from freezing in cold weather, it also has rust inhibitors and additives to aid in cooling. If the antifreeze is diluted, it will not function properly. If the weather turns cold, the system may freeze, causing damage to the radiator, engine block, and water pump.

- Check the fuel. The unit may have a fuel gauge or a dipstick. If the fuel is low, add the correct fuel (diesel or gasoline) to the fuel tank. The type of fuel required should be marked on the fuel tank. If it is not marked, contact your supervisor to verify the fuel required and have the tank marked.

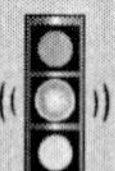

CAUTION

Adding gasoline to a diesel engine or diesel to a gasoline engine causes severe engine problems. It can also cause a fire hazard. Always be sure to add the correct fuel to the fuel tank.

- Check the battery water level unless the battery is sealed. Add room-temperature water if the battery water level is low. Do not add cold water to a battery.
- Open the fuel shutoff valve if the equipment has one. The fuel shutoff valve is located in the fuel line between the fuel tank and the carburetor.
- Record the hours from the hour meter if the equipment has one. An hour meter records the total number of hours the engine runs. This information is used to determine when the engine needs to be serviced. The hours are displayed on a gauge similar to an odometer.
- Check the tension and condition of the belts. If the belts are loose, tighten them. If they are worn, replace them.
- Check the safety valve setting to make sure it does not stick and is working properly.
- Make sure that the regulating valve between the compressor and the outlet valves is closed.

4.5.2 Starting the Air Compressor

Most engines have an on/off ignition switch and a starter, which may be combined into a key switch similar to the ignition on a car. To start the engine, turn on the ignition switch and press the starter. Release the starter when the engine starts. On diesel-engine generators, hold the glow plug button to warm up the glow plugs before starting the compressor engine. Small air compressors may have an on/off switch and a pull cord to start the engine. Engine-driven compressors should be allowed to build up to the operating pressure before they are used. While the compressor is building up pressure, connect all outlet lines and tools to the compressor. As soon as the operating pressure is reached, open the regulating valve to the outlet valves.

CAUTION

Before disconnecting any tools from the air lines, close the regulating valve to that particular line and bleed the air from the line. It is then safe to disconnect a tool from that line and connect another tool. Open the regulating valve before trying to use the tool.

When shutting down the air compressor, close the regulating valve at the outlet ports and then bleed the air from the air lines. Turn off the ignition switch and disconnect all tools and air lines from the compressor.

4.6.0 Air Compressor Operator's Maintenance Responsibility

The manufacturer's manual provides a schedule and procedures to be followed when performing maintenance on the air compressor. Before performing any maintenance, the operator must be trained, authorized, and have the proper tools to perform any procedures. The following are examples of the types of maintenance that could be performed on an air compressor:

- Changing the engine oil and filter
- Changing the engine and compressor air filters
- Changing the engine coolant
- Performing battery maintenance
- Lubricating parts
- Inspecting all guards and safety devices

Instructor's Notes:

5.0.0 ◆ PORTABLE PUMPS

Portable pumps are widely used by pipefitters for various applications. The three most common types of portable pumps are trash pumps, mud pumps, and submersible pumps. These pumps can be **pneumatic**, electric, or gasoline engine-driven pumps.

5.1.0 Trash Pumps

Most trash pumps are **centrifugal pumps**. Centrifugal pumps are designed primarily for pumping clear water. They can, however, be used where small quantities of up to 10 percent volume of mud, sand, or silt are encountered. They can also be used to control heavier seepage in excavations, trenches, pipelines, manholes, and foundations. Trash pumps are centrifugal pumps specifically designed for use with water containing solids, such as sticks, stones, sand, gravel, and other foreign materials that would clog a standard centrifugal pump. The trash pump has a compartment that collects the solids to prevent them from damaging the pump impeller. Trash pumps are adequate for pumping water from a maximum depth of 20 feet. To remove water from a depth greater than 20 feet, a submersible pump should be used. Any time pumps are to be left outside during cold weather, remove all water from the pump to prevent the water from freezing and cracking the pump housing. *Figure 9* shows portable trash pumps.

Trash pumps are continuous flow pumps that use **centrifugal force** generated by rotation. The liquid enters the **impeller** at the center, or the eye, of the impeller; and the rotation of the impeller blades causes a rotary motion of the liquid. Centrifugal force moves the liquid away from the center. As the liquid is moved away from the center of the impeller, its **velocity** increases until it is finally discharged out the discharge outlet.

PORTABLE TRASH PUMP IN USE

CENTRIFUGAL TRASH PUMP

106F09.EPS

Figure 9 ◆ Portable trash pumps.

5.2.0 Mud Pumps

Most mud pumps are **diaphragm pumps** that are designed for use where low, continuous flow is required for highly viscous fluids, such as thick mud, air mixed with water, and water containing a considerable amount of solids or abrasive materials. *Figure 10* shows a mud pump.

Most diaphragm pumps use hydraulic pressure delivered by a piston to actuate the flexible diaphragm. The side of the **diaphragm** that produces the pumping action is called the power side, and the side doing the pumping is called the fluid side. A suction check valve and a discharge check valve open and close to move fluids through the

106F10.EPS

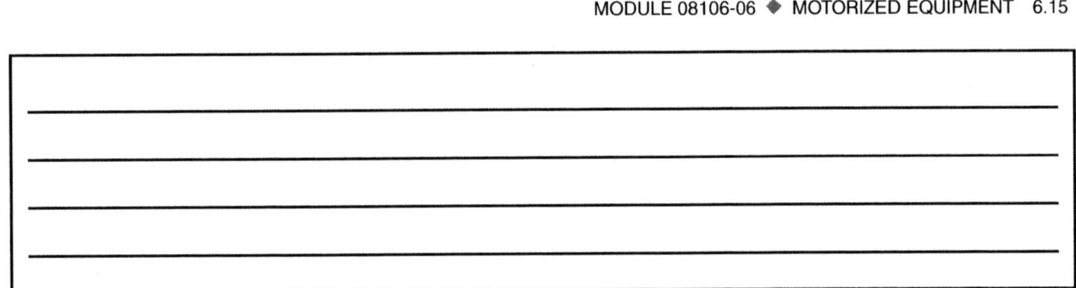

Identify the three most common types of portable pumps.

Explain that a trash pump is a modified centrifugal pump. Describe how a trash pump works.

Explain that most mud pumps are diaphragm pumps.

Show Transparency 7 (Figure 11). Explain how a diaphragm pump works.

Provide several portable pumps and the operator's manuals for the trainees to examine.

MODULE 08106-06 ◆ MOTORIZED EQUIPMENT 6.15

Describe a submersible pump.

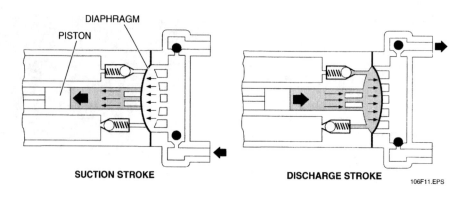

Show Transparency 8 (Figure 13). Explain how a submersible pump works.

Show trainees how to identify portable pumps to use for specific operations.

Have trainees practice identifying portable pumps to use for specific operations. Note the proficiency of each trainee. This laboratory corresponds to Performance Task 6.

Describe different types of aerial lifts.

Show Transparency 9 (Figure 15). Describe the components on an aerial lift.

Provide an operator's manual for an aerial lift for the trainees to examine.

See the Teaching Tip for Section 6.0.0 at the end of this module.

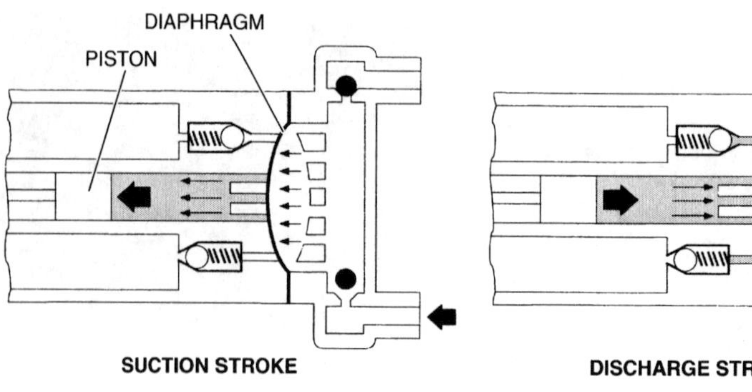

pump. *Figure 11* shows the operation of a diaphragm pump.

5.3.0 Submersible Pumps

A submersible pump (*Figure 12*) is encased with its motor in a protective housing that allows the entire unit to operate under water. Submersible pumps are used in many residential, commercial, and industrial wells. The operation of these pumps can be controlled by a float switch located inside the well that contains the pump or by manual or automatic control switches located outside the well. The float switch is a mercury-to-mercury contact switch that is controlled by the position of an attached float. As long as the water level is at an adequate height to remove water from the well, the pump is on. When the water level falls to a level too low for pumping, the float also drops and turns the pump off. As the water level builds back up, the float rises and turns the pump on again. *Figure 13* shows a submersible pump controlled by a float switch.

Figure 12 ◆ Submersible pump.

6.0.0 ◆ AERIAL LIFTS

Aerial lifts are used to raise and lower workers to and from elevated job sites. There are two main types of lifts: boom lifts and scissor lifts. Both types are made in various models. Some are transported on a vehicle to a job site where they are unloaded. Others are trailer-mounted and towed to the job site by a vehicle, and some are permanently mounted on a vehicle. Depending on their design, they can be used for indoor work, outdoor work, or both. *Figure 14* shows two commonly used types of aerial lifts.

Boom lifts are designed for both indoor and outdoor use. Boom lifts have a single arm that extends a work platform/enclosure capable of holding one or two workers. Some models have a jointed (articulated) arm that allows the work platform to be positioned both horizontally and vertically. Scissor lifts raise a work enclosure vertically by means of crisscrossed supports. Scissor lifts can also be used indoors and outdoors.

Most models of aerial lifts are self-propelled, allowing workers to move the platform as work is performed. The power to move these lifts is provided by several means, including electric motors, gasoline or diesel engines, and hydraulic motors.

6.1.0 Aerial Lift Assemblies

Aerial lifts normally consist of three major assemblies: the platform, a lifting mechanism, and the base. *Figure 15* shows these components for a scissor lift.

The platform of an aerial lift is constructed of a tubular steel frame with a skid-resistant deck sur-

Instructor's Notes:

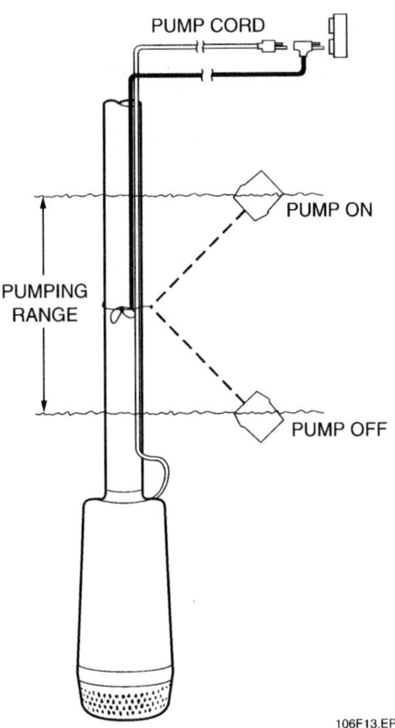

Figure 13 ♦ Submersible pump controlled by float switch.

face, railings, toe board, and midrails. Entry to the platform is normally from the rear. The entry opening is closed either with a chain or a spring-returned gate with a latch. The work platform may also be equipped with a retractable extension platform. The lifting mechanism is raised and lowered either by electric motors and gears or by one or more single-acting hydraulic lift cylinder(s). A pump, driven by either an AC or DC motor, provides hydraulic power to the cylinder(s). The base provides a housing for the electrical and hydraulic components of the lift. These components are normally mounted in swingout access trays. This allows easy access when performing maintenance or repairs to the unit. The base also contains the axles and wheels for moving the assembly. In the case of a self-propelled platform, electrical or hydraulic motors will drive two or more of the wheels to allow movement of the lift from one location to another. Brakes will be incorporated on one or more of the wheels to prevent inadvertent movement of the lift.

6.2.0 Aerial Lift Operator Qualifications

Only trained and authorized workers may use an aerial lift. Safe operation requires the operator to understand all limitations and warnings, operating procedures, and operator requirements for maintenance of the aerial lift. The following is a list of requirements the operator must meet:

- Understand and be familiar with the associated operator's manual for the lift being used.
- Understand all procedures and warnings within the operator's manual and those posted on decals on the aerial lift.
- Be familiar with employer's work rules and all related government (OSHA) safety regulations.
- Demonstrate this understanding and operate the associated model of aerial lift during training in the presence of a qualified trainer.

6.3.0 Aerial Lift Controls

Figure 16 shows an example of an electrical panel and its associated controls. This is only an example: other models of lifts may have different control configurations. This electrical panel contains the following switches and controls:

- *Up/down toggle switch* – By holding this switch in the up position, the platform can be raised to the desired level. The platform will stop moving when the switch is returned to the center position. By holding this switch in the down position, the platform can be lowered.
- *Buzzer alarm* – An audible alarm sounds when the platform is being lowered. On some models, this alarm may sound when any control function is being performed.
- *Hour meter* – This meter records the number of hours the platform has been operating. The meter will only register when the electric or hydraulic motor associated with operating the aerial lift is running.

One other control that may be available to the operator is an emergency battery disconnect switch. When this switch is placed in the off position, it will disconnect power to all control circuits.

The basic controls for an aerial lift with a hydraulic system generally include:

- *Emergency lowering valve* – This valve allows for platform lowering in the event of an electrical/hydraulic system failure.
- *Free wheeling valve* – Opening this valve allows hydraulic fluid to flow through the wheel motors. This allows the aerial lift to be pushed by hand and prevents damage to the motors when the aerial lift is moved between job site locations. There are usually strict limits to how fast the aerial lift may be moved without causing damage to hydraulic system components.

Other types of controls that may be available on aerial lifts include:

Describe the qualifications needed to operate an aerial lift.

Show Transparency 10 (Figure 16). Describe switches and controls on the electrical panel on an aerial lift.

Describe the controls used to operate an aerial lift.

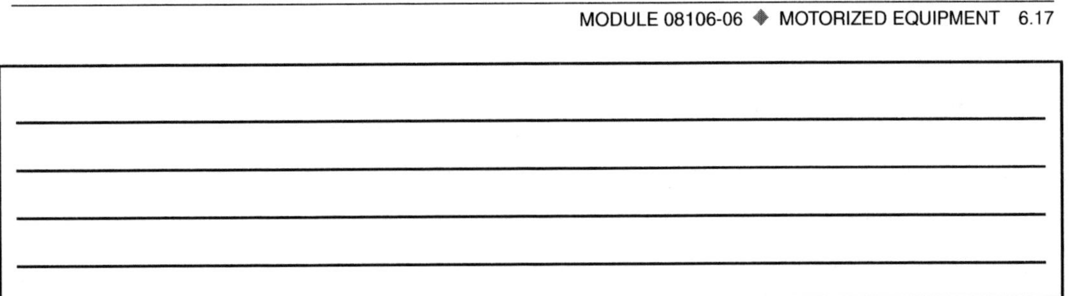

Figure 14 ◆ Aerial lifts.

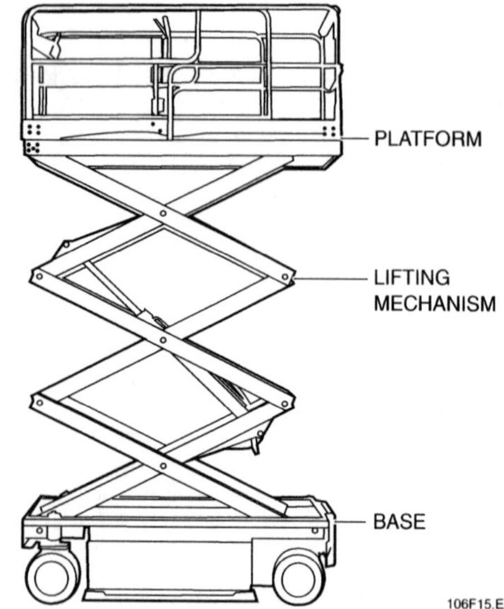

Figure 15 ◆ Aerial lift components.

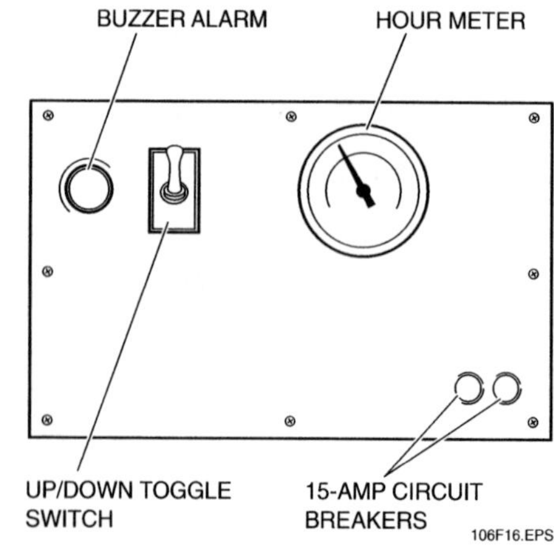

Figure 16 ◆ Aerial lift electrical panel.

Instructor's Notes:

6.18 PIPEFITTING ◆ LEVEL ONE

- *Parking brake manual release* – This control allows manual release of the parking brake. This control should only be used when the aerial lift is located on a level surface.
- *Safety bar* – The safety bar is used to support the platform lifting hardware in a raised position during maintenance or repair.
- *Up/down selector switch* – This switch is used to raise and lower the platform.
- *Emergency stop button* – This button used to cut off power to both the platform and base control boxes.

Stabilizer hardware is required on some aerial lifts. At a minimum, this hardware includes a stabilizer leg, stabilizer lock pin and cotter key, and a stabilizer jack at each corner of the aerial lift.

Most aerial lifts are equipped with a base control box. The base control box includes a switch to select whether operation will be controlled from the platform or the base control box. The base control overrides the platform control.

The worker on the platform designated as the operator uses the platform controls. In some aerial lifts, operation from the platform is limited to raising and lowering the platform. On others, the aerial platform is designed to be driven by the operator using controls provided on the platform. In either case, the operator will have an operator control box located on the platform assembly. *Figure 17* shows a typical platform control box for a self-propelled aerial lift.

Controls available for the operator include (but are not limited to) the following:

- *Drive/steer controller* – This is a one-handed toggle (joystick) lever for controlling speed and steering of the aerial platform. This is usually a deadman switch that returns to neutral and locks when released. The handle is moved forward to drive the aerial platform forward. The platform speed is determined by how far forward the handle is moved. The handle is moved backward to drive the aerial platform backward; speed is selected as before. Releasing the stick will stop the motion of the aerial lift. Steering is performed by depressing a rocker switch on the top of the stick in the desired direction of travel, either right or left.
- *Up/down selector switch* – Placing the switch in the up position raises the platform; placing the switch in the down position lowers the platform. When released, the switch returns to the middle position and stops the movement of the platform.
- *Lift/off/drive selector switch* – The lift position energizes the lift circuit; the off position removes power to the control box; the drive position energizes the drive and steering controls.
- *Emergency stop pushbutton* – When pushed, this button disconnects power to the platform control circuit. In an emergency, push the button in. To restore power, pull the button out.
- *Lift enable button* – When pushed and held down, this button enables the lift circuit. The button must be held down when raising or lowering the platform. Releasing the button stops the motion of the platform.

Different aerial lifts will have different controls. It is important that the operator fully understand all controls and their functions before operating the aerial lift. The operator must read and fully understand all control and safety information provided by the manufacturer before attempting to operate the equipment.

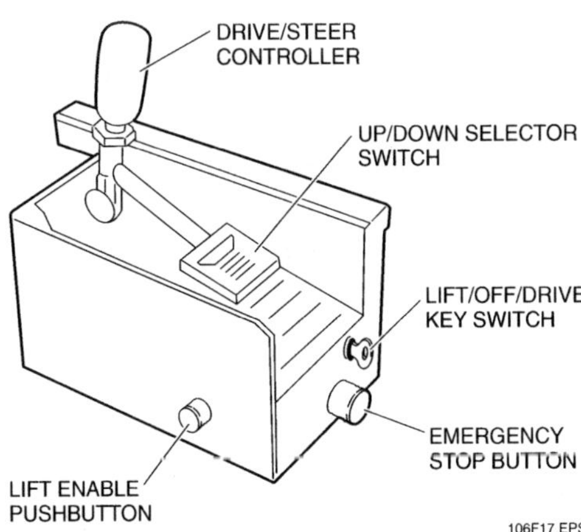

Figure 17 ◆ Aerial lift platform control box.

6.4.0 Aerial Lift Safety Precautions

Safety precautions unique to aerial lifts are listed here. Remember that the other safety precautions discussed in this module apply. Each manufacturer will provide specific safety precautions in the operator's manual provided with their equipment. Specifically, *OSHA Standard 1926.453* defines and governs the use of aerial lifts.

- Avoid using the lift outdoors in stormy weather or in strong winds.
- Prevent people from walking beneath the work area of the platform.
- Use personal fall arrest equipment (body harness and lanyard) as required for the type of lift being used. Use approved anchorage points.
- Lower the lift and lock it into place before moving the equipment. Also, lower the lift, shut off the engine, set the parking brake, and remove the key before leaving it unattended.

Explain the function of stabilizer hardware and the base control box.

Explain that on some machines the platform controls are limited.

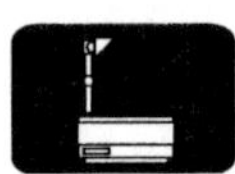

Show Transparency 11 (Figure 17). Describe the operator controls on an aerial lift.

Explain that the operator controls vary on different machines. Emphasize the importance of reading the operator's manual.

Explain that the use of aerial lifts is governed by OSHA. Review the safety precautions required when operating an aerial lift.

Provide a copy of an aerial lift operator's manual and *29 CFR 1926.453* for the trainees to examine.

Describe the operation of a scissor-type self-propelled aerial lift. Describe proportional controls.

Explain the pre-operation procedures.

Show Transparency 12 (Figure 18). Describe the operator's inspection checklist.

Emphasize the importance of maintenance procedures.

Show Transparency 13 (Figure 19). Describe the maintenance schedule for an aerial lift.

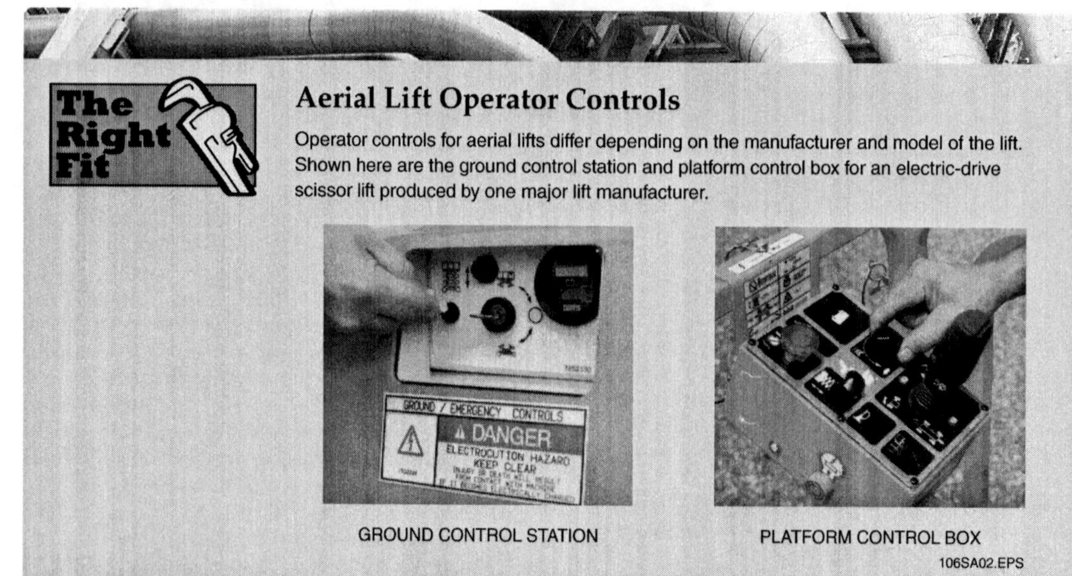

Aerial Lift Operator Controls

Operator controls for aerial lifts differ depending on the manufacturer and model of the lift. Shown here are the ground control station and platform control box for an electric-drive scissor lift produced by one major lift manufacturer.

GROUND CONTROL STATION

PLATFORM CONTROL BOX

106SA02.EPS

- Stand firmly on the floor of the basket or platform. Do not lean over the guardrails of the platform, and never stand on the guardrails. Do not sit or climb on the edge of the basket or use planks, ladders, or other devices to attain additional height.

6.5.0 Aerial Lift Operation

This section describes operations for a scissor-type, self-propelled aerial lift. Remember that operating procedures will vary from model to model and the operator must become familiar with all operating procedures, controls, safety features, and associated safety precautions before operating any type of aerial lift.

A typical **proportional control** procedure involves the operation of a lever or foot pedal to cause the aerial lift to move. The further the control is moved, the more power is applied to the motor and the faster the aerial lift will move.

The following tasks must be performed before operating the aerial lift:

- Carefully read and fully understand the operating procedures in the operator's manual and all warnings and instruction decals on the work platform.
- Check for obstacles around the work platform and in the path of travel such as holes, dropoffs, debris, ditches, and soft fill. Operate the aerial lift only on firm surfaces.
- Check overhead clearances. Make sure to stay at least 10' away from overhead power lines.
- Make sure batteries are fully charged (if applicable).
- Make sure all guardrails are in place and locked in position.
- Perform an operator's checklist.
- Never make unauthorized modifications to the components of an aerial lift.

Figure 18 shows an example of an operator's checklist.

The following is an example of the operator controls used to drive the aerial lift forward. The operator selects the drive position with the lift drive select switch. The operator then lifts the handle lock ring and moves the drive/steer controller forward. The speed can be adjusted by continuing to move the controller forward until the desired speed is reached. By releasing the controller, the forward motion of the lift is stopped. To drive in reverse, the lever is moved in the opposite direction.

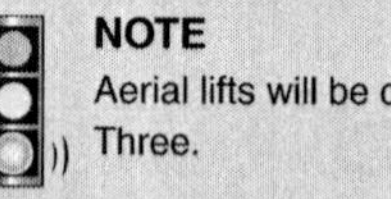

NOTE

Aerial lifts will be covered in more detail in Level Three.

6.6.0 Aerial Lift Operator's Maintenance Responsibilities

Death or injury to workers and damage to equipment can result if the aerial platform is not main-

Instructor's Notes:

6.20 PIPEFITTING ◆ LEVEL ONE

OPERATOR'S CHECKLIST

INSPECT AND/OR TEST THE FOLLOWING DAILY OR AT BEGINNING OF EACH SHIFT

___ 1. OPERATING AND EMERGENCY CONTROLS

___ 2. SAFETY DEVICES

___ 3. PERSONNEL PROTECTIVE DEVICES

___ 4. TIRES AND WHEELS

___ 5. OUTRIGGERS (IF EQUIPPED) AND OTHER STRUCTURES

___ 6. AIR, HYDRAULIC, AND FUEL SYSTEM(S) FOR LEAKS

___ 7. LOOSE OR MISSING PARTS

___ 8. CABLES AND WIRING HARNESSES

___ 9. DECALS, WARNINGS, CONTROL MARKINGS, AND OPERATING MANUALS

___ 10. GUARDRAIL SYSTEM

___ 11. ENGINE OIL LEVEL (IF SO EQUIPPED)

___ 12. BATTERY FLUID LEVEL

___ 13. HYDRAULIC RESERVOIR LEVEL

___ 14. COOLANT LEVEL (IF SO EQUIPPED)

106F18.EPS

Figure 18 ◆ Example of an aerial lift operator's checklist.

tained in good working condition. Inspection and maintenance should be performed by personnel who are authorized for such procedures. The operator must be assured that the work platform has been properly maintained before using it.

As a minimum, if the operator is not responsible for the maintenance of the aerial lift, the operator should perform the daily checks. *Figure 19* shows an example of a maintenance and inspection schedule for a typical aerial lift. The daily checks should be performed at the beginning of each shift or at the beginning of the day if only one shift is worked. The aerial lift must never be used until these checks are completed satisfactorily. Any deficiencies found during the daily check must be corrected before using the aerial lift.

Figure 19 is broken into four major categories. Each category is further broken down into the components that should be checked daily, weekly, monthly, every three months, every six months, and yearly. Footnotes (the numbers in parentheses) following each component tell the operator what type of inspection is required. For example, under the Electrical category, battery fluid level should be checked daily or at the beginning of each shift. The (1) refers the operator to the Notes portion of the schedule. Note 1 tells the operator to perform a visual inspection of the battery fluid level.

Different models of aerial lifts will have different maintenance requirements. Always refer to the operator's manual to determine exactly what checks are required before operation. General maintenance rules applicable to any aerial lift include the following:

- Disconnect the battery ground negative (–) lead before performing any maintenance.
- Properly position safety devices before performing maintenance with the work platform in the raised position.

Preventive maintenance is easier and less expensive than corrective maintenance.

Keep Transparency 13 (Figure 19) showing. Discuss the different sections of the maintenance schedule for an aerial lift.

Discuss the safety hazards posed from operating an aerial lift.

Explain that settlement can damage structures. Explain that compactors come in a variety of sizes.

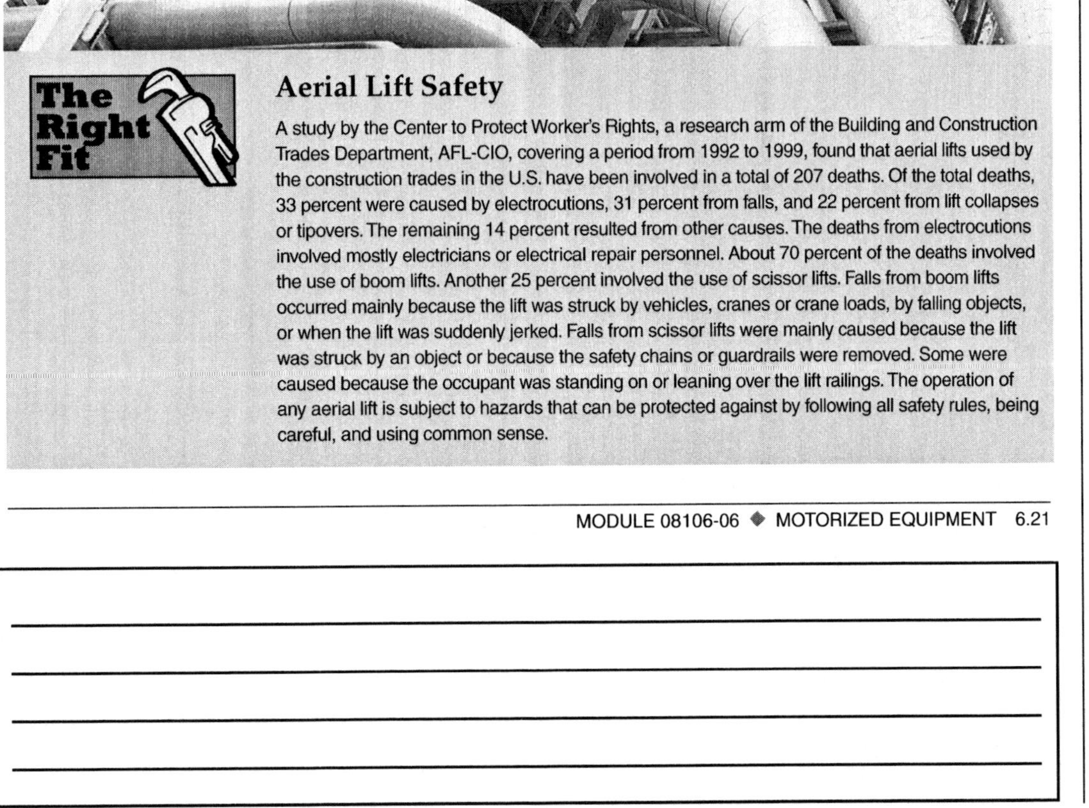

Aerial Lift Safety

A study by the Center to Protect Worker's Rights, a research arm of the Building and Construction Trades Department, AFL-CIO, covering a period from 1992 to 1999, found that aerial lifts used by the construction trades in the U.S. have been involved in a total of 207 deaths. Of the total deaths, 33 percent were caused by electrocutions, 31 percent from falls, and 22 percent from lift collapses or tipovers. The remaining 14 percent resulted from other causes. The deaths from electrocutions involved mostly electricians or electrical repair personnel. About 70 percent of the deaths involved the use of boom lifts. Another 25 percent involved the use of scissor lifts. Falls from boom lifts occurred mainly because the lift was struck by vehicles, cranes or crane loads, by falling objects, or when the lift was suddenly jerked. Falls from scissor lifts were mainly caused because the lift was struck by an object or because the safety chains or guardrails were removed. Some were caused because the occupant was standing on or leaning over the lift railings. The operation of any aerial lift is subject to hazards that can be protected against by following all safety rules, being careful, and using common sense.

MODULE 08106-06 ◆ MOTORIZED EQUIPMENT 6.21

Mechanical						
Structural damage/welds (1)	✓					✓
Parking brakes (2)	✓					✓
Tires and wheels (1)(2)(3)	✓					✓
Guides/rollers/slides (1)	✓					✓
Railings/entry chain/gate (2)(3)	✓					✓
Bolts and fasteners (3)	✓					✓
Rust (1)			✓			✓
Wheel bearings (2) King pins (1)(8)	✓					✓
Steer cylinder ends (8)				✓		✓
Electrical						
Battery fluid level (1)	✓					✓
Control switches (1)(2)	✓					✓
Cords and wiring (1)	✓					✓
Battery terminals (1)(3)	✓					✓
Terminals and plugs (3)	✓					✓
Generator and receptacle (2)	✓					✓
Limit switches (2)	✓					✓
Hydraulic						
Hydraulic oil level (1)	✓					✓
Hydraulic leaks (1)	✓					✓
Lift/lowering time (10)				✓		✓
Hydraulic cylinders (1)(2)		✓				✓
Emergency lowering (2)	✓					
Lift capacity (7)			✓			✓
Hydraulic oil/filter (9)					✓	✓
Miscellaneous						
Labels (1)(11) Manual (12)	✓					✓

Notes:

(1) Visually inspect. (3) Check tightness. (5)(6) N/A. (8) Lubricate.
(2) Check operation. (4) Check oil level. (7) Check relief valve setting. Refer to serial number nameplate. (9) Replace.

(10) General specifications. (12) Proper Operating Manual *must* be in the manual tube.
(11) Replace if missing or illegible.

*Record Inspection Date

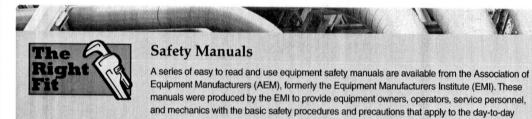

106F19.EPS

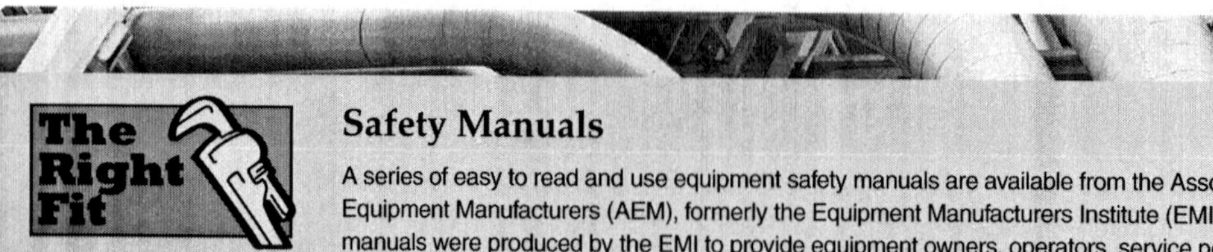

Safety Manuals

A series of easy to read and use equipment safety manuals are available from the Association of Equipment Manufacturers (AEM), formerly the Equipment Manufacturers Institute (EMI). These manuals were produced by the EMI to provide equipment owners, operators, service personnel, and mechanics with the basic safety procedures and precautions that apply to the day-to-day operation and maintenance of the specific equipment. These safety manuals are generic and are to be used in conjunction with the operator/user manuals provided by manufacturers for a specific piece of equipment.

Instructor's Notes:

6.22 PIPEFITTING ◆ LEVEL ONE

7.0.0 ◆ COMPACTION EQUIPMENT

Compactors are used to eliminate soil settlement. If the soil under a structure is not compacted, it will eventually settle and cause damage to the structure. This settling will cause the structure to deform. If the settling occurs more at one side or corner, cracks or structural failures can occur.

Compactors come in a wide variety of different styles and sizes. Each type of compactor is recommended for differing soil conditions. Compactors can also be installed on other equipment as an attachment. *Figure 20* shows two common types of compactors.

The compactors commonly used for various soil conditions are as follows:

- *Cohesive soils (silts and clay)* – Use an upright tamp or sheepsfoot vibrator roller.
- *Granular soils (sand or gravel)* – Use a flat plate tamp or smooth drum vibrator roller, or an upright tamp.
- *Mixed soil* – Use an upright tamp, sheepsfoot vibrator roller, smooth drum vibrator roller, or plate tamp.

This module covers one type of flat plate tamp compactor commonly used by pipefitters. All compactors have an operator's manual with the equipment. Refer to the operator's manual before operating any compactor.

7.1.0 Compaction Equipment Assemblies

The flat plate compactor shown in *Figure 21* includes the following components:

- Vibrator plate
- Engine bed
- Pulley cover
- Operating handle
- Clutch control lever
- Throttle lever
- Engine guard

7.2.0 Compaction Equipment Operator Qualifications

Only trained, authorized persons are permitted to operate a compactor. Personnel who have not been trained in the operation of compactors may only operate them for the purpose of training. The training must be conducted under the direct supervision of a qualified trainer.

7.3.0 Typical Compaction Equipment Controls

Compactors have operator controls for the engine and compaction plate. This section discusses typical controls associated with a flat plate tamp. *Figure 21* shows typical operator controls.

Controls associated with the compactor include the following:

- *Operating handle* – Provides the operator with a handle for controlling the compactor and mounting other controls.
- *Clutch control lever* – Used to engage the clutch for the compacting plate.
- *Throttle lever* – Controls engine speed.

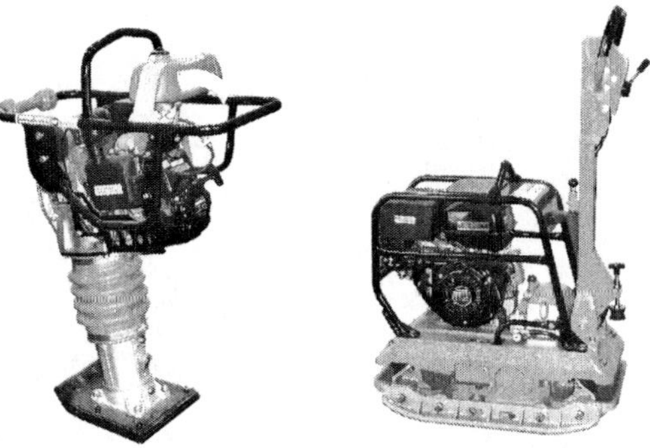

Figure 20 ◆ Compactors.

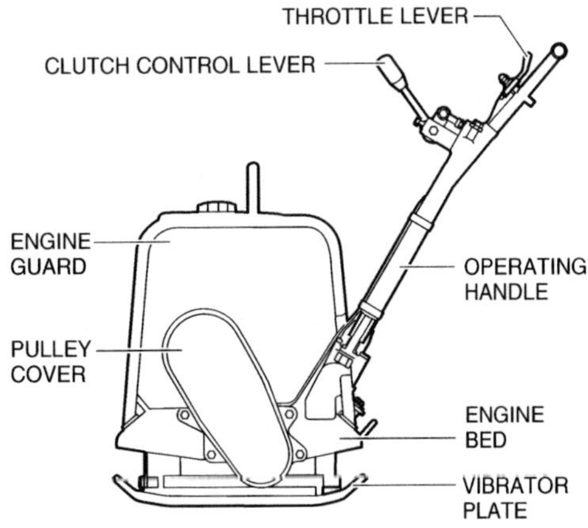

Figure 21 ◆ Flat plate compactor components.

Discuss the compactors commonly used for various soil conditions.

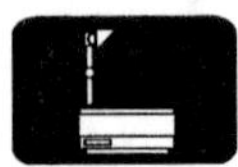

Show Transparency 14 (Figure 21). Describe the major components of a flat plate compactor.

Describe the qualifications needed to operate a compactor.

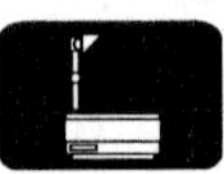

Show Transparency 14 (Figure 21) again. Describe the operator controls of a flat plate compactor.

Emphasize the need for safety when using a compactor.

Emphasize the importance of reading the operator's manual when using a compactor.

Describe typical preoperational checks that must be performed before using a compactor.

Provide an operator's manual for a compactor for the trainees to examine.

7.4.0 Compaction Equipment Safety Precautions

The safe operation of the soil compactor is the operator's responsibility. The following are general safety precautions that are applicable to compactors. Remember that the other safety precautions, already discussed in this module, also apply. Refer to the manufacturer's manual for specific safety precautions.

- After performing maintenance, make certain all guards are correctly installed and all safety devices are functioning.
- Be familiar with the safety devices and instructions for the compactor.
- Do not perform any maintenance work on the compactor unless authorized and qualified to do so.

7.5.0 Compaction Equipment Operation

The following discussion covers generic operation of a compactor. Manufacturers provide detailed operating instructions with their equipment. Refer to the operator's manual provided with the compactor for specific instructions. The general outline written here is intended to help you understand the operator controls. The operator should perform a complete inspection of the compactor and correct any deficiencies before beginning operation.

7.5.1 Compaction Equipment Preoperation Checks

Each manufacturer provides specific inspection procedure(s) for the machine being used in the operator's manual.

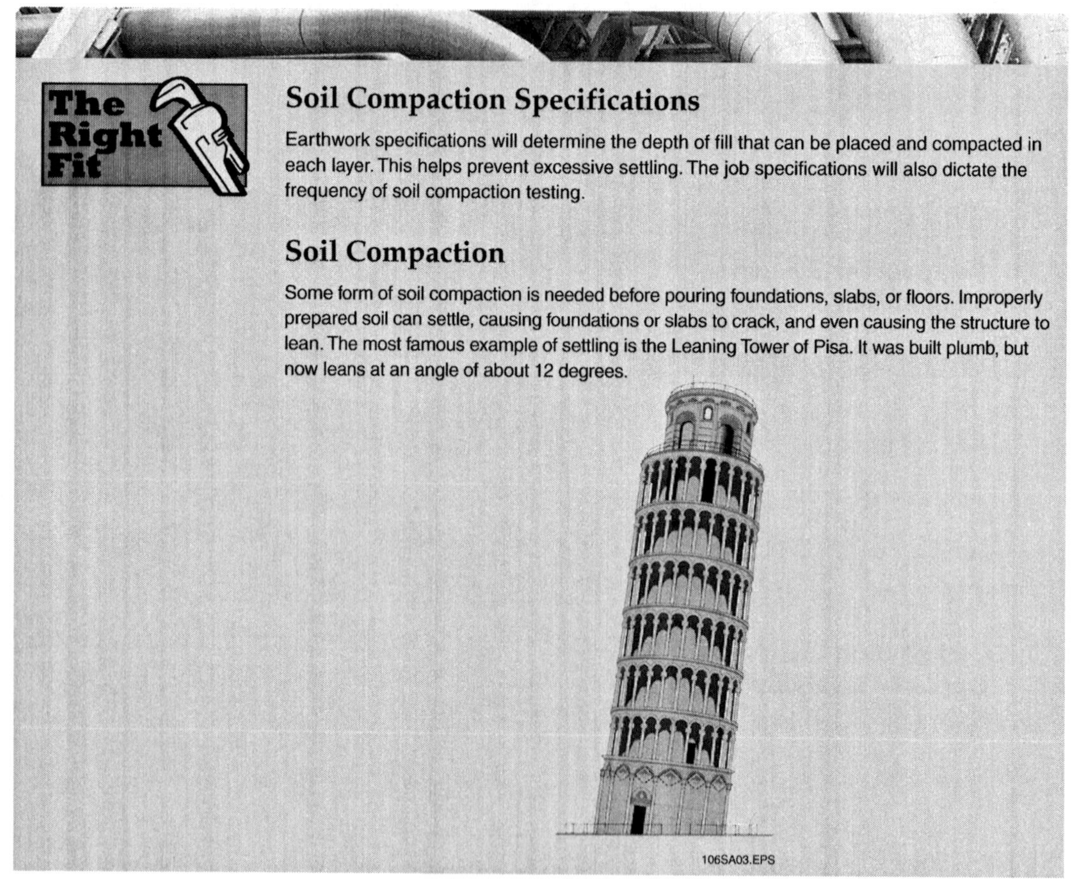

Soil Compaction Specifications

Earthwork specifications will determine the depth of fill that can be placed and compacted in each layer. This helps prevent excessive settling. The job specifications will also dictate the frequency of soil compaction testing.

Soil Compaction

Some form of soil compaction is needed before pouring foundations, slabs, or floors. Improperly prepared soil can settle, causing foundations or slabs to crack, and even causing the structure to lean. The most famous example of settling is the Leaning Tower of Pisa. It was built plumb, but now leans at an angle of about 12 degrees.

106SA03.EPS

Instructor's Notes:

6.24 PIPEFITTING ◆ LEVEL ONE

Before beginning operation, the operator must perform a preoperational check. Each manufacturer will provide a checklist with the equipment. The checklist normally includes the following types of items:

- Inspect the machine to verify that all guards and safety items are properly installed.
- Check the fuel level and add fuel, if required.
- Check the engine oil level and add oil, if required.
- Check the engine air filter and clean or replace it in accordance with the manufacturer's procedures.

7.5.2 Starting, Operating, and Stopping Compaction Equipment

Before beginning operation, the operator must perform a startup. Each manufacturer provides detailed startup instructions with the equipment. The instructions normally include the following types of items:

Step 1 Verify that the clutch control lever is in the disengaged position.

Step 2 Engage the choke if the engine is cold.

Step 3 Pull on the recoil starter handle until the engine starts. Allow the engine to run until warm, then release the choke.

Step 4 Use the throttle lever to control the engine's rpm during operation.

Each manufacturer provides detailed instructions for operation of the equipment. The instructions normally include the following types of items:

Step 1 Adjust the throttle until the desired engine speed is reached.

Step 2 Slowly engage the clutch using the clutch control lever.

Step 3 Use the operating handle to control the direction and movement of the compactor.

Each manufacturer provides detailed instructions for shutting down the equipment. The instructions normally include the following types of items:

Step 1 Use the throttle lever to decrease the engine's speed.

Step 2 Use the clutch control lever to disengage the clutch.

Step 3 Press the engine shutdown control to turn off the engine.

Emergency shutdown of the compactor is accomplished by pressing the engine shutdown control to stop the engine.

7.6.0 Compactor Operator's Maintenance Responsibility

The manufacturer's manual provides a schedule and procedures to be followed when performing maintenance on the compactor. Before performing any maintenance, the operator must be trained, authorized, and have the proper tools to perform any procedures. The following are examples of the types of maintenance that could be performed on a flat plate tamp compactor:

- Changing the engine oil
- Changing the engine air filters
- Lubricating parts
- Inspecting all guards and safety devices

8.0.0 ◆ FORKLIFTS

Forklifts are used to move, unload, and place material at required locations on the job site. Forklifts are also sometimes called **powered industrial trucks**. In this module, the term forklift will be used. This module introduces you to safe forklift operation. Forklifts are often misused. A firm foundation in forklift safety ensures greater productivity and a safer working environment.

8.1.0 Forklift Assemblies

Many different types of forklifts are available to meet different needs. Most forklifts used in construction are diesel-powered and fall into one of two broad categories: fixed mast and telescoping boom. Within each of these categories, forklifts are further differentiated by their drive train, steering, and capacity.

There are two types of fixed-mast forklifts: rough terrain and warehouse. Rough terrain forklifts are by far the most common type of fixed-mast forklift used in construction (see *Figure 22*). These forklifts are made for outside use. They have higher ground clearances than warehouse forklifts, larger tires, and leveling devices. Warehouse forklifts are designed for inside use. They have hard rubber tires that are usually of the same size.

When the term fixed-mast forklift is used in this module, it is assumed to refer to the rough terrain forklift.

Explain how to start, operate and stop a compactor.

Describe the maintenance requirements for a compactor.

Have trainees review Sections 8.0.0–11.5.0.

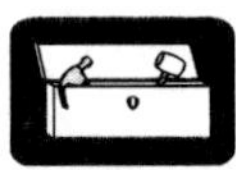

Ensure that you have everything required for teaching this session.

Describe different types of forklifts used on construction sites. Discuss the difference between warehouse and rough terrain forklifts.

Discuss the qualifications necessary to operate a forklift.

Explain that forklifts have either two- or four-wheel drive.

MODULE 08106-06 ◆ MOTORIZED EQUIPMENT 6.25

See the Teaching Tip for Section 8.0.0 at the end of this module.

Discuss the operation of two- and four- wheel drive forklifts.

Describe a fixed-mast forklift.

Describe a telescoping-boom forklift.

Provide an operator's manual for a forklift for the trainees to examine.

8.2.0 Forklift Operator Qualifications

Only trained, authorized persons are permitted to operate a powered forklift. Operators of powered forklifts must be qualified as to visual, auditory, physical, and mental ability to operate the equipment safely. Personnel who have not been trained in the operation of forklifts may only operate them for the purpose of training. The training must be conducted under the direct supervision of a qualified trainer.

8.3.0 Forklift Typical Controls

Forklifts have either two-wheel drive or four-wheel drive. The distinction is the same as that used with other vehicles.

A two-wheel-drive forklift has a drive train that transmits power to the front wheels. Many times, the front wheels are larger than the rear wheels and do not steer the forklift. These forklifts are good for general use, but may lack power in rough terrain.

A four-wheel-drive forklift has a drive train that transmits power to all four wheels. These forklifts are well suited to rough terrain and other conditions that might require additional traction.

The rear wheels of a forklift with two-wheel steering move when the steering wheel is turned. This feature is standard on most forklifts and provides increased maneuverability. The rear wheels are generally smaller than the front wheels.

When the four-wheel steering is engaged on a forklift with this feature, all the wheels steer. However, some forklifts offer the following three options:

- The rear wheels may be locked to allow the front wheels to steer.
- All wheels may move in the same direction. This is called crab steering or oblique steering.
- The front and rear wheels may move in opposite directions. This is called articulated steering or four-wheel steering.

8.3.1 Fixed-Mast Forklifts

The upright member along which the forks travel is called the mast. A typical fixed-mast forklift has a mast that may tilt as much as 19 degrees forward and 10 degrees rearward. These types of forklifts are suitable for placing loads vertically and traveling with loads, but their horizontal reach is limited to how close the machine can be driven to the pick-up or landing point. For example, a fixed-mast forklift cannot place a load of pipe beyond the edge of a trailer because of its limited reach.

106F22.EPS

Figure 22 ◆ Fixed-mast rough terrain forklifts.

106F23.EPS

Figure 23 ◆ Telescoping-boom forklift.

Instructor's Notes:

6.26 PIPEFITTING ◆ LEVEL ONE

8.3.2 Telescoping-Boom Forklifts

Telescoping-boom forklifts (*Figure 23*) provide more versatility in horizontal and vertical placement than a fixed-mast forklift. A telescoping-boom forklift is really a combination of a telescoping-boom crane and a forklift.

The mast can be either two-stage or three-stage. This designation refers to the number of telescoping channels built in the mast. A two-stage mast has one telescoping channel and a three-stage mast has two telescoping channels. The purpose of the telescoping channels is to provide greater lift height.

Some models of telescoping-boom forklifts have a level-reach fork carriage, which allows the fork carriage to be moved horizontally while the boom remains in a stationary position.

8.4.0 Forklift Safety Precautions

Safe operation is the responsibility of the operator. Operators must develop safe working habits and recognize hazardous conditions to protect themselves and others from death or injury. Always be aware of unsafe conditions to protect the load and the forklift from damage. Be familiar with the operation and function of all controls and instruments before operating a forklift. Before operating any forklift, read and fully understand the operator's manual.

The following safety rules are specific to forklift operation. Remember the other safety precautions already discussed in this module.

- Never put any part of your body into the mast structure or between the mast and the forklift.
- Never put any part of your body within the reach mechanism.
- Understand the limitations of the forklift.
- Do not permit passengers to ride unless a safe place to ride has been provided by the manufacturer.
- Never leave the forklift running unattended.

Safeguard pedestrians at all times by observing the following rules:

- Always look in the direction of travel.
- Do not drive the forklift up to anyone standing in front of an object or load.
- Make sure that personnel stand clear of the rear swing area before turning.
- Exercise particular care at blind spots, cross aisles, doorways, and other locations where pedestrians may step into the travel path.
- The use of a spotter is recommended when landing an elevated load with a telescoping-boom forklift.

8.5.0 Forklift Operation

This section contains general guidelines for operation of a forklift. The most important factor to consider when using any forklift is its capacity. Each forklift is designed with an intended capacity, and this capacity must never be exceeded. Exceeding the capacity jeopardizes not only the equipment but also the safety of everyone on or near the equipment. Each manufacturer supplies a capacity chart for each forklift model. Be sure to read and follow the capacity chart.

8.5.1 Picking Up a Load

Some forklifts are equipped with a sideshift device that allows the operator to horizontally shift the load several inches in either direction with respect to the mast. A sideshift device enables more precise placing of loads, but it changes the center of gravity and must be used with caution. If the forklift being used is equipped with a sideshift device, be sure to return the fork carriage to the center position before attempting to pick up a load.

Step 1 Check the position of the forks with respect to each other. They should be centered on the carriage. If the forks have to be moved, check the operator's manual for the proper procedure. Usually, there is a pin at the top of each fork that, when lifted, allows each fork to be slid along the upper backing

Review the safety precautions that must be followed when operating a forklift.

Explain that the most important consideration in operating a forklift is capacity.

Explain the steps for using a forklift to pick up a load.

Review the Case History and discuss the dangers in leaving equipment running unattended.

CASE HISTORY

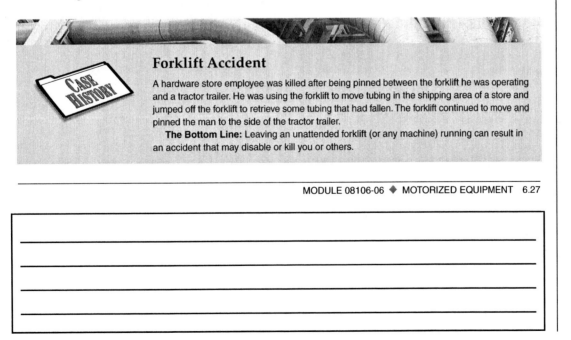

Forklift Accident

A hardware store employee was killed after being pinned between the forklift he was operating and a tractor trailer. He was using the forklift to move tubing in the shipping area of a store and jumped off the forklift to retrieve some tubing that had fallen. The forklift continued to move and pinned the man to the side of the tractor trailer.

The Bottom Line: Leaving an unattended forklift (or any machine) running can result in an accident that may disable or kill you or others.

Explain how to travel with a load.

Explain how to place a load with a forklift.

Discuss special consideration needed when placing an elevated load using a forklift.

Discuss special considerations for traveling with long loads, such as pipe, on a forklift.

Review the safety precautions that must be followed when operating a forklift.

plate until the fork centers over the desired notch.

Step 2 Travel to the area at a safe rate of speed. Always keep the forks lowered when traveling with a forklift.

Step 3 Before picking up a load with a forklift, make sure the load is stable. If it looks like the load might shift when picked up, secure the load. Knowing the center of gravity is crucial, especially when picking up tapered sections. Make a trial lift, if necessary, to determine and adjust the center of gravity.

Step 4 Approach the load so that the forks straddle the load evenly. It is important that the weight of all loads be distributed evenly on the forks. Overloading one fork at the expense of the other can damage the forks.

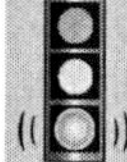

NOTE

If there is a load on the forks, the operator may not leave the seat.

In some cases, it may be advisable to measure the load and mark its center of gravity.

Step 5 Drive up to the load with the forks straight and level. If the load being picked up is on a pallet, be sure the forks are low enough to clear the pallet boards.

Step 6 Move forward until the leading edge of the load rests squarely against the back of both forks. If you cannot see the forks engage the load, ask someone to signal for you. This prevents expensive damage and injury.

Step 7 Raise the carriage, then tilt the mast rearward until the forks contact the load. Raise the carriage until the load safely clears the ground. Then tilt the mast fully rearward to cradle the load. This minimizes the chance that the load may slip during travel.

8.5.2 Traveling with a Load

Always travel with a load at a safe rate of speed. Never travel with a raised load. Keep the load as low as possible, and be sure the mast is tilted rearward to cradle the load.

As you travel, keep your eyes open and stay alert. Watch the load and the conditions ahead of you, and alert others of your presence. Avoid sudden stops and abrupt changes in direction. Be careful when downshifting, because sudden deceleration can cause the load to shift or topple. Be aware of front and rear swing when turning.

If you are traveling with a telescoping-boom forklift, be sure the boom is fully retracted.

If you have to drive on a slope, keep the load as low as possible and back down the slope. Do not drive across steep slopes. If you have to turn on an incline, make the turn wide and slow. If the load is large enough to obstruct your view, operate in reverse.

8.5.3 Placing a Load

Position the forklift at the landing point so that the load can be placed where you want it. Be sure everyone is clear of the load.

The area under the load must be clear of obstructions and must be able to support the weight of the load. If you cannot see the placement, use a signaler to guide you.

With the forklift in the unloading position, lower the load and tilt the forks to the horizontal position. When the load has been placed and the forks are clear from the underside of the load, back away carefully to disengage the forks or retract the boom on variable-reach units.

8.5.4 Placing Elevated Loads

Special care needs to be taken when placing elevated loads. Some forklifts are equipped with a leveling device that allows the operator to rotate the fork carriage to keep the load level during travel. When placing elevated loads, it is extremely important to level the machine before lifting the load.

One of the biggest potential safety hazards during elevated load placement is poor visibility. There may be workers in the immediate area who cannot be seen. The landing point itself may not be visible. Your depth perception decreases as the height of the lift increases. To be safe, use a signaler to help you spot the load.

Use tag lines to tie off long loads.

Drive the forklift as close as possible to the landing point with the load kept low. Set the parking brake. Raise the load slowly and carefully while maintaining a slight rearward tilt to keep the load cradled. Under no circumstances should the load be tilted forward until the load is over the landing point and ready to be set down.

If the forklift's rear wheels start to lift off the ground, stop immediately, but not abruptly. Lower the load slowly and reposition it, or break it down into smaller components if necessary. If surface conditions are bad at the unloading site, it may be necessary to reinforce the surface conditions to provide more stability.

Instructor's Notes:

6.28 PIPEFITTING ◆ LEVEL ONE

8.5.5 *Traveling with Long Loads*

Traveling with long loads presents special problems, particularly if the load is flexible and subject to damage. Traveling multiplies the effect of bumps over the length of the load. A stiffener may be added to the load to give it extra rigidity.

To prevent slippage, secure long loads to the forks. This may be done in one of several ways. A field-fabricated cradle may be used to support the load. While this is an effective method, it requires that the load be jacked up.

The forklift may be used to carry pieces of rigging equipment. This method requires slings and a spreader bar.

In some cases, long loads may be snaked through openings that are narrower than the load itself. This is done by approaching the opening at an angle and carefully maneuvering one end of the load through the opening first. Avoid making quick turns because abrupt maneuvers cause the load or its center of gravity to shift.

8.5.6 *Using the Forklift to Rig Loads*

The forklift can be a very useful piece of rigging equipment if it is properly and safely used. Loads can be suspended from the forks with slings, moved around the job site, and placed.

All the rules of careful and safe rigging apply when using a forklift to rig loads. Be sure not to drag the load or let it swing freely. Use tag lines to control the load.

Never attempt to rig an unstable load with a forklift. Be especially mindful of the load's center of gravity when rigging loads with a forklift.

When carrying cylindrical objects, such as oil drums, keep the mast tilted rearward to cradle the load. If necessary, secure the load to keep it from rolling off the forks.

8.6.0 Forklift Operator's Maintenance Responsibility

This section of the module is intended to familiarize you with some of the preventive maintenance procedures required of forklift operators, including:

- Check the operator's manual for lubrication points and suggested lubrication periods. These are given in terms of the number of hours of operation.
- Always check all fluid levels and use the recommended fluids when refilling.
- Because forklifts often operate in dusty and dirty environments, it is important to keep the moving parts of the machine well greased. This reduces wear, prolongs the life of the machine, and ensures safe operation. Follow the manufacturer's recommendations for lubricants.

Forklifts are also provided with a daily checklist from the manufacturer. The checklist covers all the items that must be checked on the forklift every day. *Figure 24* shows an example of a daily checklist and the items that are usually checked. These may differ from unit to unit.

9.0.0 ◆ BACKHOE/LOADER

A backhoe/loader (*Figure 25*) is a dual-purpose, highly maneuverable machine used to dig trenches, foundations, and similar excavations. It is also used to move dirt, crushed stone, gravel, and other materials around the job site.

Backhoes/loaders are equipped with either a gasoline or diesel engine. The backhoe bucket and loader bucket attachments are hydraulically operated under the control of a single operator who performs all backhoe and loader operations.

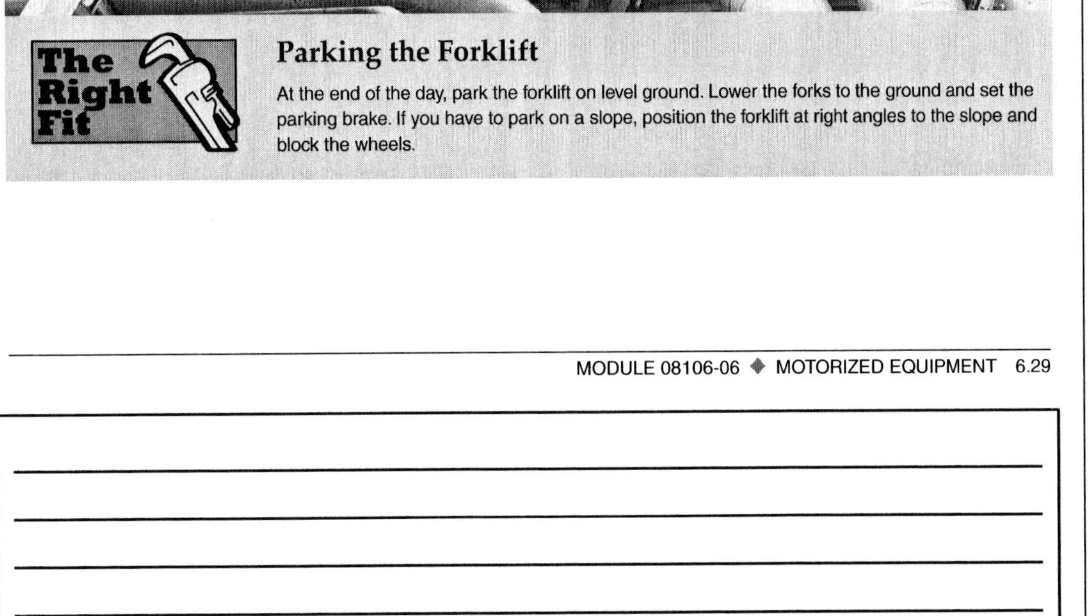

Parking the Forklift

At the end of the day, park the forklift on level ground. Lower the forks to the ground and set the parking brake. If you have to park on a slope, position the forklift at right angles to the slope and block the wheels.

Explain how to use the forklift to rig a load.

Discuss the maintenance requirements for operating a forklift.

Show Transparency 15 (Figure 24). Review the forklift operator's daily inspection checklist.

Show trainees how to identify forklift trucks and recognize the hazards involved in working around them.

Have trainees practice identifying forklift trucks and recognizing the hazards involved in working around them. Note the proficiency of each trainee. This laboratory corresponds to Performance Task 7.

Describe a backhoe/loader. Discuss the qualifications needed to operate a backhoe.

Discuss general safety considerations for operating a backhoe.

MODULE 08106-06 ◆ MOTORIZED EQUIPMENT 6.29

OPERATORS' DAILY CHECKLIST

Check Each Item Before Start Of Each Shift   Date:__________

Check One: ☐ Gas/LGP/Diesel Truck ☐ Electric Sit-down ☐ Electric Stand-up ☐ Electric Pallet

Truck Serial Number:__________ Operator:__________ Supervisor's OK: __________

Hour Meter Reading: __________

Check each of the following items before the start of each shift. Let your supervisor and/or maintenance department know of any problem. DO NOT OPERATE A FAULTY TRUCK. Your safety is at risk.

After checking, mark each item accordingly. Explain below as necessary.

Check boxes as follows: ☐ OK ☐ NG, needs attention, or repair. Circle problem and explain below.

OK	NG	Visual Checks
		Tires/Wheels: wear, damage, nuts tight
		Head/Tail/Working Lights: damage, mounting, operation
		Gauges/Instruments: damage, operation
		Operator Restraint: damage, mounting, operation, oily, dirty
		Warning Decals/Operators' Manual: missing, not readable
		Data Plate: not readable, missing adjustment
		Overhead Guard: bent, cracked, loose, missing
		Load Back Rest: bent, cracked, loose, missing
		Forks: bent, worn, stops OK
		Engine Oil: level, dirty, leaks
		Hydraulic Oil: level, dirty, leaks
		Radiator: level, dirty, leaks
		Fuel: level, leaks
		Battery: connections loose, charge, electrolyte low
		Covers/Sheet Metal: damaged, missing
		Brakes: linkage, reservoir fluid level, leaks
		Engine: runs rough, noisy, leaks

OK	NG	Visual Checks
		Steering: loose/binding, leaks, operation
		Service Brake: linkage loose/binding, stops OK, grab
		Parking Brake: loose/binding, operational, adjustment
		Seat Brake (if equipped): loose/binding, operational, adjustment
		Horn: operation
		Backup Alarm (if equipped): mounting, operation
		Warning Lights (if equipped): mounting, operation
		Lift/Lower: loose/binding, excessive drift, leaks
		Tilt: loose/binding, excessive drift, "chatters," leaks
		Attachments: mounting, damaged, operation, leaks
		Battery Test (electric trucks only): indicator in green
		Battery: connections loose, charge, electrolyte low while holding full forward tilt
		Control Levers: loose/binding, freely return to neutral
		Directional Control: loose/binding, find neutral OK

Explanation of problems marked above: __________

106F24.EPS

Instructor's Notes:

6.30 PIPEFITTING ◆ LEVEL ONE

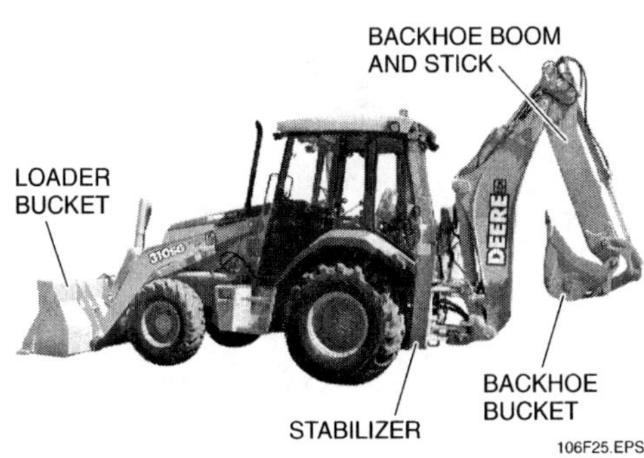

Figure 25 ◆ Backhoe/loader.

9.1.0 Backhoe/Loader Operator Qualifications

Before operating a backhoe/loader, the following requirements must be met:

- The operator must successfully complete a training program that includes actual operation of the backhoe/loader.
- The operator is responsible for reading and understanding the safety manual and operator's manual provided with the equipment.
- The operator must know the safety rules and regulations for the job site.

9.2.0 Backhoe/Loader Safety Precautions

The following is a list of general safety rules specific to the operation of a backhoe/loader. Remember that the other safety precautions, already discussed in this module, also apply.

- Mount the backhoe/loader only at points that are equipped with steps and/or handholds.
- Face the backhoe/loader while entering and leaving the operator's compartment.
- Do not mount a backhoe/loader while carrying tools or supplies.
- Do not mount a moving backhoe/loader.
- Do not use controls as handholds when entering the operator's compartment.
- Do not obstruct your vision when traveling or working.
- Never lift, move, or swing a load over a truck cab or over workers.
- When traveling, operate at speeds slow enough so that you have complete control of the backhoe/loader at all times, especially when traveling over rough or slippery ground and when on hillsides. Never place the transmission in neutral to allow the backhoe/loader to coast.
- Never approach overhead power lines with any part of the backhoe/loader unless you are in strict compliance with all local, state, and federal required safety precautions.
- Make sure that you know the underground location of all gas and water pipelines and of all electrical or fiber-optic cables.

When operating the backhoe the following safety precautions apply:

- Never enter or allow anyone to enter the backhoe swing pivot area.
- Operate the backhoe from the correct backhoe operating position. Never operate the backhoe controls from the ground. Never allow riders in or on the backhoe.
- Do not dig under the backhoe or its stabilizers. This is prohibited in order to prevent cave-ins and the chance of the backhoe falling into the excavation.
- When operating the backhoe on a slope, swing to the uphill side to dump the load, if possible. If required to dump downhill, swing only as far as required to dump the bucket.
- Always dump the soil far enough away from the trench to prevent cave-ins.

When operating the loader the following safety precautions apply:

- Carry the bucket low for maximum stability and visibility.

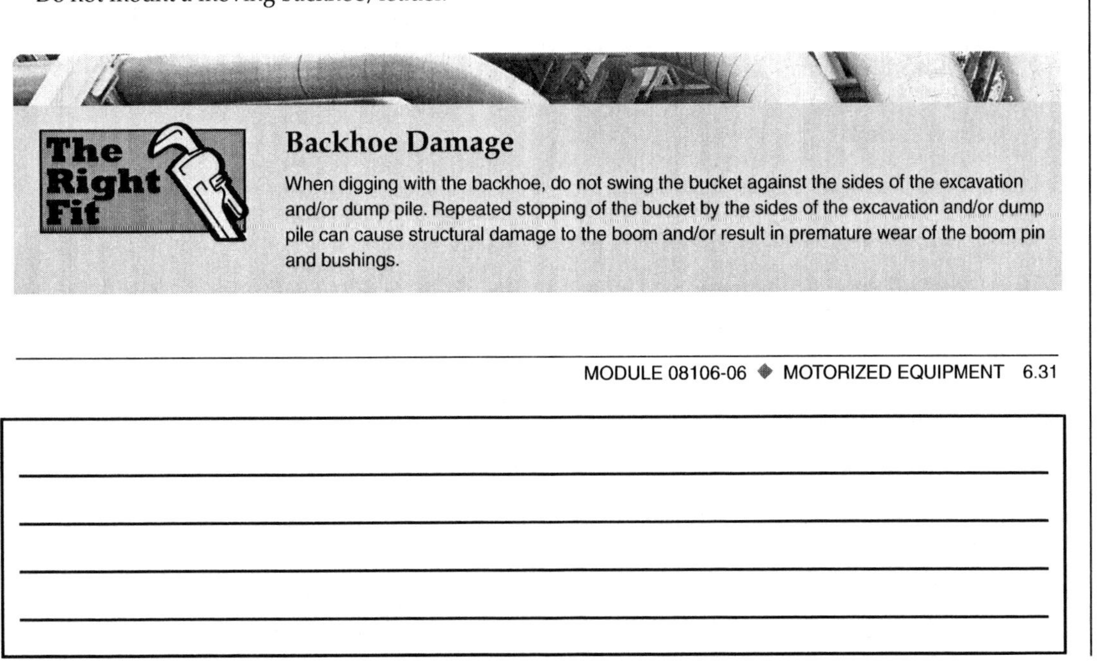

Backhoe Damage

When digging with the backhoe, do not swing the bucket against the sides of the excavation and/or dump pile. Repeated stopping of the bucket by the sides of the excavation and/or dump pile can cause structural damage to the boom and/or result in premature wear of the boom pin and bushings.

Review the safety guidelines for operating a backhoe.

Review the safety guidelines for operating a loader.

Describe different types of trenchers.

Provide operator's manuals for a backhoe and a trencher for the trainees to examine.

MODULE 08106-06 ◆ MOTORIZED EQUIPMENT 6.31

Describe the major components of a trencher.

Review the qualifications needed to operate a trencher.

Explain that the controls vary on different trenchers and emphasize the importance of reading the operator's manual.

Show Transparency 16 (Figure 27). Describe the controls used to operate a trencher.

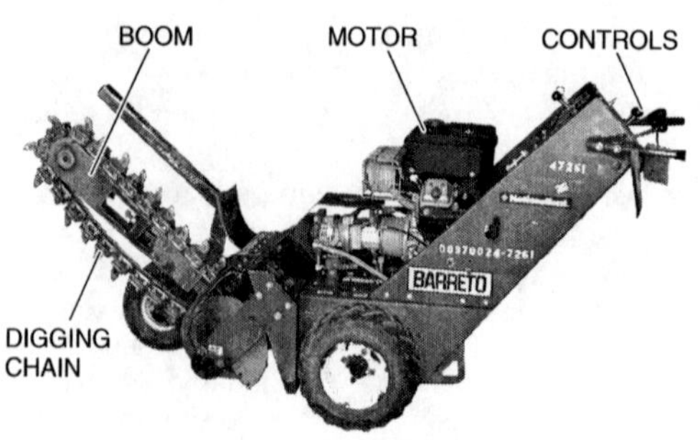

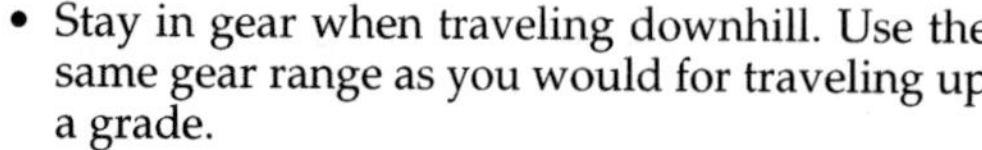

Figure 26 ◆ Trenchers.

- Stay in gear when traveling downhill. Use the same gear range as you would for traveling up a grade.
- When on a steep slope, drive up or down the slope; do not drive across the slope. If the bucket is loaded, drive with the bucket facing uphill. If empty, drive with the bucket pointed downhill.
- When operating the loader, make sure that the backhoe is in the transport lock position to prevent backhoe movement.
- When working at the base of a bank or overhang, never undercut a high bank and/or operate the loader close to the edge of an overhang or ditch.

10.0.0 ◆ TRENCHERS

Trenchers are used to dig trenches for the installation of underground services. There are different styles of trenchers available depending upon the scale of the work to be performed. This module will discuss two types of trenchers and provide examples of operating procedures for each type (*Figure 26*). The two styles of trenchers are a pedestrian (walk-behind) and compact (ride-on). Trenchers use either a gasoline or diesel engine to drive a hydraulic system that provides power to the drive wheels and the digging boom.

10.1.0 Trencher Assemblies

The major assemblies of the trenchers are discussed in this section. The two trenchers each have a boom. The boom provides the path that the digging chain follows when in operation. The boom is raised, lowered, and moved from side to side using hydraulic cylinders. The size of the boom determines the maximum depth of the trench. Digging chains have replaceable digging teeth that remove material from the ditch. Both trenchers use an engine to power hydraulic system components and move the trencher. The pedestrian machine uses a gasoline engine, and the compact trencher has either a gasoline or diesel engine. The engine provides power to move the machine around the work area and to move the trencher forward as the materials are removed from the trench. Both machines have an operator control area. These areas and controls differ because of the way each of the trenchers is operated. The compact trencher requires a certified operator.

The pedestrian machine is used by the operator walking along behind the trencher. The compact trencher is a ride-on machine with associated controls for driving the trencher. It is also equipped with a blade for backfilling the trench when work is complete. The compact trencher requires a certified operator.

10.2.0 Trencher Operator Qualifications

Only trained, authorized persons are permitted to operate a trencher. Operators of trenchers must be qualified as to visual, auditory, physical, and mental ability to operate the equipment safely. Personnel who have not been fully trained in the operation of trenchers may only operate them for the purpose of training. The training must be conducted under the direct supervision of a qualified trainer.

10.3.0 Typical Trencher Controls

This section covers the typical controls associated with pedestrian-style and compact trenchers. This discussion is for explanation only and may not include all controls that could be found on all trenchers. The operator must read and fully understand the operator's manual provided by the equipment manufacturer. The operator's manual

Instructor's Notes:

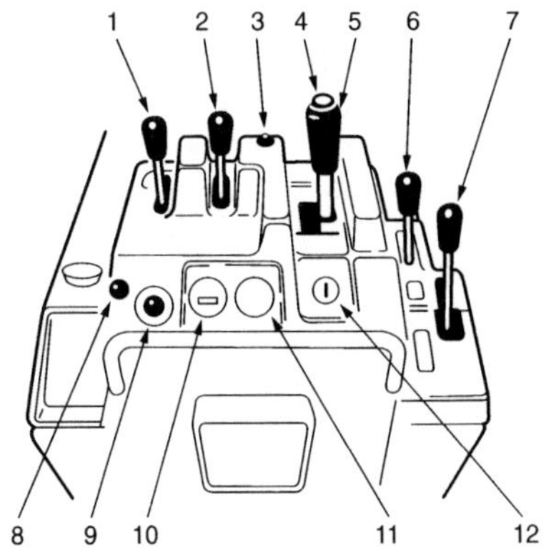

1. AXLE LOCK CONTROL
2. BOOM DEPTH CONTROL
3. ENGINE OIL PRESSURE LIGHT
4. OPERATOR PRESENCE SWITCH
5. SPEED/DIRECTION CONTROL
6. PUMP CLUTCH
7. DIGGING CHAIN CONTROL
8. THROTTLE LEVER
9. CHOKE
10. HOUR METER
11. VOLTMETER
12. IGNITION SWITCH

106F27.EPS

Figure 27 ◆ Pedestrian trencher control panel.

will contain specific instructions for the operation of the trencher.

Figure 27 shows a typical pedestrian trencher control panel. The panel contains the controls and indicators the operator uses when moving or digging with the trencher.

The following text identifies the controls and indicators, including a brief explanation of the purpose of each component.

- *Axle lock control* – This control shifts the trencher in and out of two-wheel drive. Two-wheel drive is used for digging, driving over rough ground, loading and unloading the trencher from a trailer, and when the trencher is parked or is being left unattended. Trencher maneuverability is reduced in two-wheel-drive mode.
- *Boom depth control* – This control raises and lowers the digging boom.
- *Engine oil pressure light* – This light indicates when engine oil pressure has dropped below an acceptable level. If the indicator lights during operation, the engine should be turned off and the oil level checked.
- *Operator presence (deadman) switch* – This switch must be depressed whenever the speed and direction control is moved from the neutral position or the digging chain control is moved from the stop position. Releasing the switch causes the engine to shut off.
- *Speed/direction control* – This control selects the speed and direction of machine travel and provides steering control. The speed/direction control must be in neutral when starting the trencher.
- *Pump clutch* – The pump clutch controls the application of engine power to the hydraulic pump and should be disengaged when starting the engine. The pump clutch must be engaged before using the speed/direction control, boom depth control, or digging chain control.
- *Digging chain control* – This control starts and stops rotation of the digging chain. It must be in the stop position before starting the engine. If not in the stop position, the engine will not start.
- *Throttle lever* – The throttle lever regulates engine speed.
- *Choke* – The choke helps to start a cold engine.
- *Hour meter* – The hour meter registers the engine operating time. It is used in the scheduling of lubrication and maintenance.
- *Voltmeter* – The voltmeter shows battery charge. With the ignition switch in the on position, but with the engine not running, it should read near 12 volts.
- *Ignition switch* – This is generally a three-position (stop/on/start) switch with a removable key.

10.4.0 Trencher Safety Precautions

The safe operation of a trencher is the operator's responsibility. The following are general safety precautions that must be observed when operating a trencher. The other safety precautions already discussed in this module also apply to the operation of a trencher.

- Incorrect procedures could result in death, injury, or property damage. Learn how to use the equipment correctly.
- Moving digging teeth can kill you or cut off an arm or a leg. Stay away from the boom when it is operating.
- The trencher may move when the chain starts to dig. Allow 3′ between the end of the chain and any obstacles.
- Keep the digging boom low when operating on a slope.
- If there is any doubt about the classification of a job site, or if the job site could contain electrical utilities, classify and treat the job site as electrical.

Discuss safety precautions that must be followed when operating a trencher.

Review safety precautions needed when working around gas lines and procedures that must be followed when a gas line is hit.

Discuss safety procedures that must be followed if an electrical line is hit.

Emphasize the importance of following the operating procedures in the operator's manual.

MODULE 08106-06 ◆ MOTORIZED EQUIPMENT 6.33

Emphasize the importance of inspecting a trencher before operating it. Review the items that must be checked or inspected.

Review the steps for starting a trencher.

The following are general work site precautions to be followed by the operator. When practical, stay upwind of the trencher. This reduces the amount of dust and dirt blown toward you. It may also reduce the amount of gas around you if you hit a gas line. If a gas line is hit, do the following to minimize any danger to personnel:

- Quickly turn off the ignition switch key.
- Leave the machine as quickly as possible.
- Contact the gas company to shut off the gas.
- Do not return to the area until given permission by the gas company.

In the event an underground electrical cable is hit, do the following to minimize any danger to personnel:

- If on the ground, stay where you are and do not touch any equipment.
- Warn other personnel that a strike has occurred and that they should stay away from any equipment and the immediate area.
- Contact the utility company to shut off the power.
- Do not return to the area until given permission from the utility company.

10.5.0 Pedestrian Trencher Operation

This section covers generic operation of a trencher. Each manufacturer provides detailed operating instructions with the equipment. Refer to the operator's manual provided with the trencher for specific instructions. The general guidelines given here are intended to help you understand the operator controls. You should perform a complete inspection of the trencher and correct any deficiencies before beginning operation.

A careful inspection of the equipment must be performed before beginning work. Each manufac-

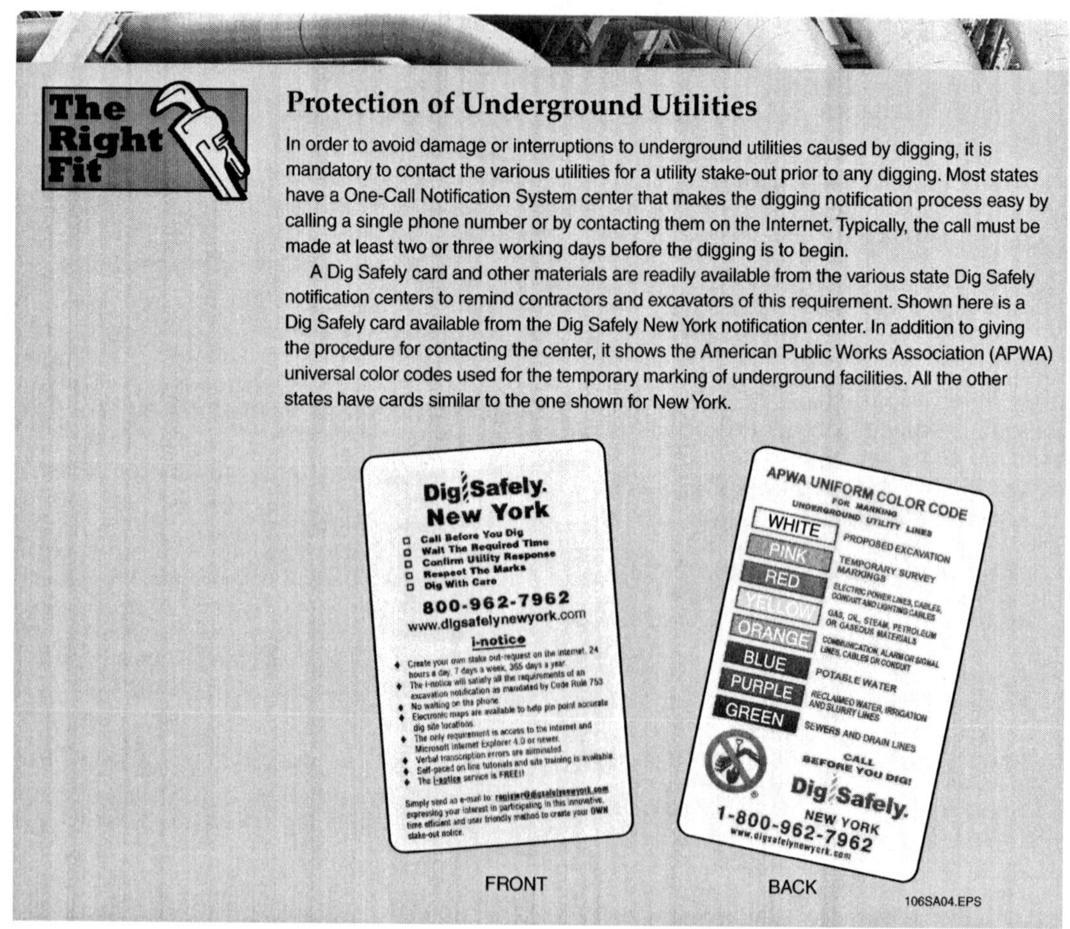

Protection of Underground Utilities

In order to avoid damage or interruptions to underground utilities caused by digging, it is mandatory to contact the various utilities for a utility stake-out prior to any digging. Most states have a One-Call Notification System center that makes the digging notification process easy by calling a single phone number or by contacting them on the Internet. Typically, the call must be made at least two or three working days before the digging is to begin.

A Dig Safely card and other materials are readily available from the various state Dig Safely notification centers to remind contractors and excavators of this requirement. Shown here is a Dig Safely card available from the Dig Safely New York notification center. In addition to giving the procedure for contacting the center, it shows the American Public Works Association (APWA) universal color codes used for the temporary marking of underground facilities. All the other states have cards similar to the one shown for New York.

Instructor's Notes:

6.34 PIPEFITTING ◆ LEVEL ONE

turer provides specific inspection procedures in the operator's manual.

The checklist will normally include the following types of items:

- Check the engine oil level, hydraulic oil level, fuel level, fuel filter, hoses and clamps, air filter system, and all of the lubrication points.
- Inspect tires for proper pressure and general condition.
- Check all control function positions. Verify that all guards and shields are in place and secure. Inspect the condition of drive belts. Inspect the digging chain for tension and condition.
- Inspect the condition of the battery and cables.
- Verify that all safety warning signs are in place and readable.

10.5.1 Starting the Trencher

The procedures for starting the trencher are contained in the operator's manual. A typical starting procedure is as follows:

Step 1 Begin by disengaging the pump clutch. Place the speed/directional control in neutral, and move the digging chain control to stop. If the engine is cold, use the choke.

Step 2 Insert the key in the ignition switch and turn to the on position. Turn the key to start; when the engine starts, release the key. Idle the engine five to ten minutes before moving to the job site.

Step 3 Engage the pump clutch slowly while warming up the engine to warm the hydraulic oil.

Step 4 While the engine is idling, check the operation of the speed/directional control, boom depth control, and digging chain control.

10.5.2 General Operation

The procedures for driving a trencher around the job site when not digging, called **mobiling** the trencher, are contained in the operator's manual. Mobiling normally proceeds as follows:

Step 1 Using the boom depth control, raise the boom to the transport position. On rough terrain, keep the boom as low as possible to help with stability.

Step 2 Engage or disengage the axle lock to match ground conditions. On level ground, the axle lock should be disengaged to increase maneuverability. On rough ground, the axle lock should be engaged to improve traction. The axle lock should always be in the engaged position when the machine is left unattended.

Step 3 Press the operator presence switch down. Use the speed/directional control to move the machine in the desired direction.

Explain how to move the trencher around the job site and how to stop it.

Review the steps for digging a trench with a trencher.

Review the Case History and discuss safety precautions that can save lives.

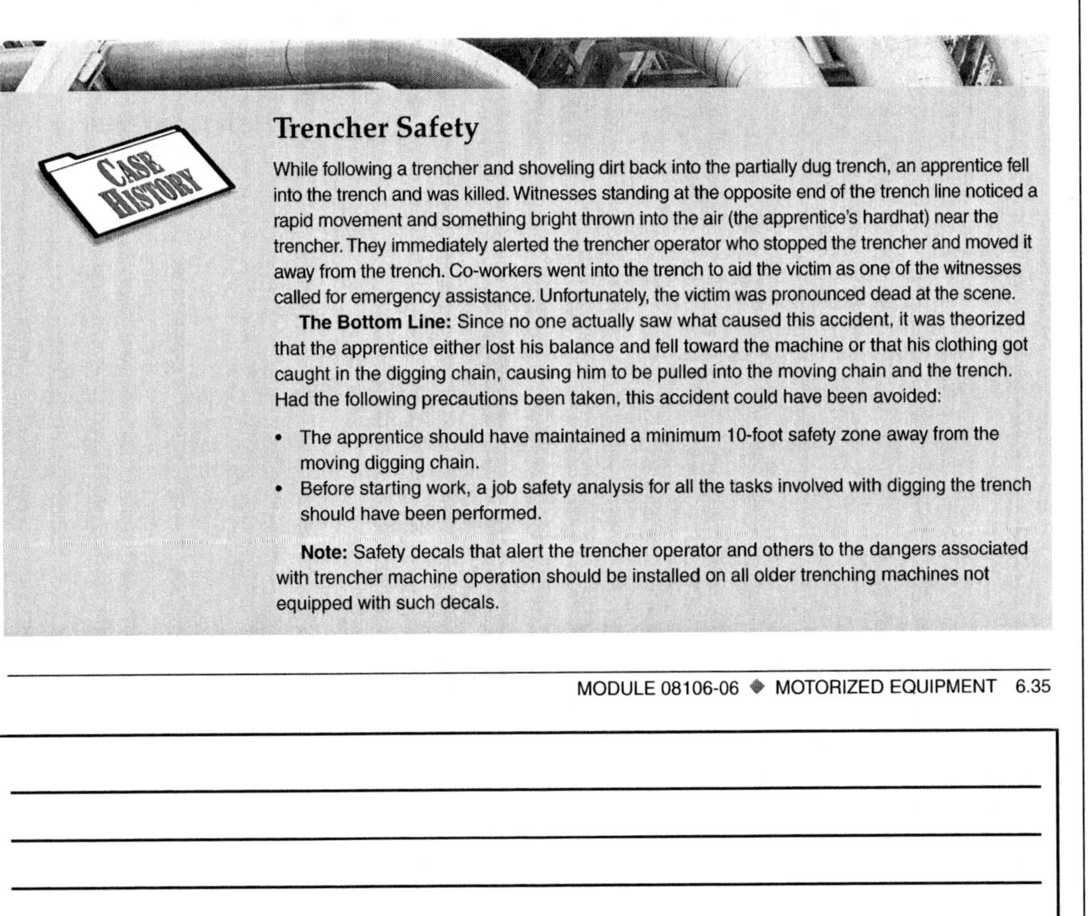

Trencher Safety

While following a trencher and shoveling dirt back into the partially dug trench, an apprentice fell into the trench and was killed. Witnesses standing at the opposite end of the trench line noticed a rapid movement and something bright thrown into the air (the apprentice's hardhat) near the trencher. They immediately alerted the trencher operator who stopped the trencher and moved it away from the trench. Co-workers went into the trench to aid the victim as one of the witnesses called for emergency assistance. Unfortunately, the victim was pronounced dead at the scene.

The Bottom Line: Since no one actually saw what caused this accident, it was theorized that the apprentice either lost his balance and fell toward the machine or that his clothing got caught in the digging chain, causing him to be pulled into the moving chain and the trench. Had the following precautions been taken, this accident could have been avoided:

- The apprentice should have maintained a minimum 10-foot safety zone away from the moving digging chain.
- Before starting work, a job safety analysis for all the tasks involved with digging the trench should have been performed.

Note: Safety decals that alert the trencher operator and others to the dangers associated with trencher machine operation should be installed on all older trenching machines not equipped with such decals.

MODULE 08106-06 ◆ MOTORIZED EQUIPMENT 6.35

Explain how to stop trenching operations.

Discuss an operator's maintenance responsibilities when operating a trencher.

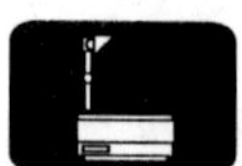

Show Transparency 17 (Figure 28). Review a typical trencher maintenance schedule.

Describe how cranes are used on construction sites.

Describe a boom truck.

Describe a cherry picker.

Use Your Senses

During normal operation of the trencher, listen for thumps, bumps, rattles, squeaks, squeals, or other unusual sounds. Smell for odors such as burning insulation, hot metal, burning rubber, or hot oil. Feel for any changes in the way the trencher is operating. Look for problems with wiring and cables, hydraulic connections, or other equipment. Correct anything you hear, smell, feel, or see that is different from what is expected, or that seems to be unsafe.

Step 4 Steer the machine with the speed/directional control.

The stop mobiling procedure is also contained in the operator's manual. A typical procedure is as follows:

Step 1 Move the speed/direction control to the neutral position.

Step 2 Release the operator presence switch and place the throttle in the idle position.

Step 3 Using the boom depth control, lower the digging boom to the ground. Engage the axle lock and disengage the pump clutch.

Step 4 Turn the ignition key to off.

If you are leaving the machine, remove the key from the ignition switch.

10.5.3 Trench Digging

The trenching procedure is contained in the operator's manual. A typical procedure is as follows:

Step 1 Begin digging the trench by slowly engaging the pump clutch. Drive the trencher to a point in line with the intended trench. Digging is done with the trencher moving away from the digging boom. Align the trencher with the boom pointing at the starting end of the trench.

Step 2 Place the speed/directional control in neutral. Engage the axle lock. Lower the digging chain to within 1 inch of the ground using the boom depth control.

Step 3 Place the digging chain control in the rotate position. The digging chain will begin to move. Slowly lower the digging boom to the desired digging depth using the boom depth control.

Step 4 When the digging boom reaches the desired depth, adjust the engine throttle for optimum digging speed. Do not overload the engine at low rpm. Use the speed/directional control to begin trenching and to regulate the ground travel speed. Find the best ground travel speed for digging without causing the engine to lug down. Minor, slow direction changes can be made using the speed and direction control.

The stop trenching procedure is also contained in the operator's manual. A typical procedure is as follows:

Step 1 To stop digging, place the speed/direction control in neutral. Using the boom depth control, raise the boom to ground level.

Step 2 Place the digging chain control in the stop position. Raise the boom all the way up using the boom depth control.

Step 3 Use the speed/direction control to drive away from the trench. On level ground, disengage the axle lock and shut down the machine.

10.6.0 Trencher Operator's Maintenance Responsibility

Figure 28 shows a typical maintenance schedule for a trencher. The interval hours column shows the number of operating hours between each required maintenance function. The reference number column provides a general location on the machine where the maintenance is to be performed. The description column provides the operator with the maintenance task that needs to be performed. Some charts have a lube column that tells the operator the type of lubricant to use and a point column that provides the total number of points at which the lubricant is applied.

Examples of the types of maintenance required include changing the engine oil, lubricating the digging chain clutch, and changing the hydraulic oil filter. Proper maintenance will extend the life and performance of the machine.

Instructor's Notes:

6.36 PIPEFITTING ◆ LEVEL ONE

REF.	HOURS	DESCRIPTION
1	10	CHECK TIRE PRESSURE
2	10	ADJUST DIGGING CHAIN TENSION
3	25	WASH & OIL AIR FILTER PRECLEANER
4	50	CHECK BATTERY
5	50	CHECK BELT TENSION
6	100	REPLACE AIR FILTER PAPER ELEMENT & PRECLEANER
7	100	CHECK DRIVE WHEEL END PLAY AND LUG NUTS
8	100	CHECK FUEL FILTER
9	100	CLEAN COOLING FINS ON CYLINDER HEAD AND BARREL
10	200	REPLACE FUEL FILTER
11	AS NEEDED	RE-TORQUE HEADSHAFT SPROCKET BOLTS
12	AS NEEDED	CHECK LEVEL IN FUEL TANK

106F28.EPS

Describe a carry deck.

Describe lattice boom cranes. Describe a lattice boom crawler crane.

11.0.0 ◆ CRANE TYPES AND USES

The primary function of a mobile crane is to lift and swing loads. A mobile crane consists of a rotating superstructure, operating machinery, and a boom mounted on a vehicle such as a truck, crawler, or railcar. Since their introduction in the 1800s, mobile cranes have gone through constant change, from steam power to air-cooled diesel engines. The manufacturers of these mobile cranes continuously strive to meet the needs of an ever-changing industry. Because of the wide variety of uses for cranes, many different types of cranes have been developed.

106F29.EPS

11.1.0 Boom Trucks

Boom trucks are unlike any of the other mobile cranes in that they are mounted on carriers that are not solely designed for crane service. These cranes are mounted onto a commercial truck chassis that has been strengthened to carry the crane. The boom truck can also be used to haul loads and be driven on the highway system when the boom is in the transport position. Boom trucks typically range in capacity from 6 to 22 tons. *Figure 29* shows a typical boom truck.

11.2.0 Cherry Pickers

Cherry pickers are rough-terrain mobile cranes. They have oversized tires that allow them to move across the rough terrain of construction sites and other broken ground. Two types of cherry pickers are the fixed cab and the rotating cab. *Figure 30* shows two cherry pickers.

11.3.0 Carry Decks

A carry deck is a small hydraulic mobile crane. Carry decks are primarily used in industrial applications where working surfaces are significantly better than those found on most construction sites. The compact size of the carry deck (15 feet long by 7 feet wide) allows it to be maneuvered easily about the job site. Most carry decks offer up to 8½ tons of lifting capacity with the outriggers in place and up to 7½ tons for lifting and transporting work. Most carry decks come with water-cooled diesel engines,

MODULE 08106-06 ◆ MOTORIZED EQUIPMENT 6.37

Describe lattice boom truck cranes.

Discuss mobile crane safety.

Show trainees how to identify types of hydraulic cranes and recognize the hazards involved in working around them.

Have trainees practice identifying hydraulic cranes and recognizing the hazards involved in working around them. Note the proficiency of each trainee. This laboratory corresponds to Performance Task 8.

but they are also available with gasoline-powered and propane (LP) engines. The carry deck also has a transport area that is designed to lock down and carry up to 15,000 pounds of materials. *Figure 31* shows a carry deck.

11.4.0 Lattice Boom Cranes

Lattice boom cranes are capable of much greater boom lengths than hydraulic boom cranes. Site preparation and erection of lattice boom cranes may take anywhere from a day for a smaller crane to weeks for larger cranes. The length of setup time depends on the height of the boom and the lift attachments to be used. The newer lattice boom cranes are powered by computer-controlled hydraulic systems. They have redundant fail-safe devices to assure operational safety. These devices consist of, but are not limited to: automatic braking systems, load moment measuring devices, mechanical system monitoring devices, and function lockout systems that stop operation when a system fault occurs. These controls provide for better control and accuracy than older friction-operated devices.

106F30.EPS

11.4.1 Lattice Boom Crawler Cranes

Lattice boom crawler cranes are specifically designed for the extreme duty associated with the use of a crawler crane. Lattice boom crawler cranes are used to pick up a heavy load and track with it to a nearby location. The reliability and versatility of the crawler-mounted lattice boom crane makes it the most widely applied crane design in use today. Besides lifting heavy loads to great heights, these cranes are ideal for applications with a high duty cycle. *Figure 32* shows a lattice boom crawler crane.

11.4.2 Lattice Boom Truck Cranes

Lattice boom truck cranes provide the mobility of a truck crane with the extreme lifting capacity of a lattice boom crane. Depending on the size of the crane and its gross vehicle weight, some components, such as counterweights and outriggers, may have to be removed before highway travel to meet local, state, and federal weight restrictions. *Figure 33* shows a lattice boom truck crane.

11.5.0 Mobile Crane Safety

Safely lifting and moving machinery and equipment is not automatic. It involves making several decisions regarding the object being moved and the lifting and rigging equipment to be used. Safely lifting and moving equipment is the responsibility of everyone involved. Everyone must be alert, responsible, and aware of all the factors involved to use mobile cranes safely. The following guidelines should be followed when working with or around mobile cranes:

106F31.EPS

Figure 31 ◆ Carry deck.

Instructor's Notes:

6.38 PIPEFITTING ◆ LEVEL ONE

106F32.EPS

Figure 32 ◆ Lattice boom crawler crane.

106F33.EPS

Figure 33 ◆ Lattice boom truck crane.

MODULE 08106-06 ◆ MOTORIZED EQUIPMENT 6.39

Have the trainees complete the Review Questions, and go over the answers prior to administering the Module Examination.

Review Questions

1. The operator should check for leaks in a pressurized hydraulic system by _____.
   a. checking the hydraulic oil level in the supply tank
   b. feeling all the hoses and connectors
   c. using a piece of wood or cardboard
   d. using a flashlight and an inspection mirror

2. Before disconnecting any hydraulic system hoses, the operator should shut down the engine and _____.
   a. check the hydraulic fluid level
   b. cycle the hydraulic controls
   c. set the parking brake
   d. chock the wheels

3. When refueling a piece of equipment, keep the nozzle or fuel funnel in contact with the filler neck to _____.
   a. prevent the possibility of sparks
   b. lower the possibility of a fuel spill
   c. protect you from breathing the fuel fumes
   d. make it easier to see when the fuel tank is full

4. When a battery is frozen, _____.
   a. disconnect the negative lead before attaching the charger
   b. do not charge it or attempt to jump-start it
   c. jump-start the equipment following the manufacturer's instructions
   d. replace the battery and charge the old battery

5. A _____ should be consulted about the grounding requirements of a generator.
   a. fellow worker
   b. qualified operator
   c. licensed electrical contractor
   d. representative from OSHA

6. If you find evidence of arcing on the control panel of a tow-behind generator, _____.
   a. continue with operation of the generator
   b. replace the control panel
   c. locate and repair the problem before operating
   d. call the manufacturer for additional information

7. When a portable air compressor is used along with another compressed air source, make sure _____.
   a. there is a check valve installed at the service valve
   b. all air vents are clear
   c. the engine and air compressor air filters are clean
   d. the compressor is on level ground

8. The type of pump that would be used with high-viscosity liquids is the _____ pump.
   a. trash
   b. mud
   c. submersible
   d. centrifugal

9. The components of an aerial lift include a _____.
   a. base, auger, and bucket
   b. base, platform, and lifting mechanism
   c. platform, boom, and compactor
   d. compactor, control panel, and forks

10. The _____ allows hydraulic fluid to flow and allows the aerial lift to be pushed by hand.
    a. emergency lowering valve
    b. free wheeling valve
    c. stabilizer hardware
    d. steer controller

Instructor's Notes:

6.40 PIPEFITTING ◆ LEVEL ONE

Review Questions

11. The _____ are major components of a plate-style compactor.
    a. boom and chain
    b. platform and base
    c. vibrator plate and engine bed
    d. fixed-mast and telescoping boom

12. The mast of a typical fixed-mast forklift can be tilted _____ degrees forward.
    a. 19
    b. 39
    c. 59
    d. 79

13. For a backhoe/loader equipped with all-wheel steering, the mode of steering where the operator can choose independent rear-axle operation is called _____ steering.
    a. rear-axle
    b. two-wheel
    c. circle
    d. all-wheel

14. When first starting a trencher, _____ of clearance should be allowed between the digging chain and any obstacles.
    a. 1'
    b. 2'
    c. 3'
    d. 4'

Figure 1

106RQ01.EPS

15. The machine shown in *Figure 1* is a _____.
    a. cherry picker
    b. carry deck
    c. boom truck
    d. lattice boom crane

MODULE 08106-06 ◆ MOTORIZED EQUIPMENT 6.41

Summarize the major concepts presented in the module.

Summary

Motorized equipment is another useful tool to make the pipefitter's job easier. This module has introduced you to some of the various types of motorized equipment you will use throughout your career. You should always remember to read and follow the manufacturer's operating procedures and preventive maintenance schedules before attempting to use any type of motorized equipment. If you follow the manufacturer's instructions, you can obtain a long life of safe service from the motorized equipment on the job site.

Keep in mind that anyone operating motorized equipment must be trained, and in many cases certified, to operate the specific equipment. The purpose of this module is to introduce you to the motorized equipment used in pipefitting work. Whether or not you are actually trained to use some or all of this equipment will depend on the policies of your school and/or employer.

Notes

Instructor's Notes:

Trade Terms Quiz

Fill in the blank with the correct trade term that you learned from your study of this module.

1. Electrical current is measured in _________.
2. Move bulk items around the job site with a(n) _________.
3. A(n) _________ may be part of a circuit or may simply hold a spare fuse.
4. The machine used to supply air to pneumatic tools is called a(n) _________.
5. A machine that moves liquid using rotation is a(n) _________.
6. A machine that moves liquid using hydraulic pressure is a(n) _________.
7. A(n)_________ is used to provide electricity on a job site.
8. The speed or power of an engine can be automatically controlled by a(n) _________.
9. Another term for forklift is_________.
10. To prevent settling of soil after construction, a(n) _________ is used.
11. The rotating part of a centrifugal pump is called the _________.
12. Personnel, tools, and materials can be raised to overhead work areas using a(n) _________.
13. Two types of electrical connectors are _________ and _________.
14. The _________ is designed to protect circuits from overloads.
15. A truck-mounted lift designed for indoor or outdoor use is known as a(n) _________.
16. A liquid is pushed outward from the center of rotation by _________.
17. Electrical potential is measured in _________.
18. A _________ tool uses compressed air to do work.
19. A machine that uses a boom and chain to dig is called a(n) _________.
20. A(n) _________ is a unit of power equal to one joule per second.
21. In some pumps, the fluid is moved back and forth by a thin wall called a(n) _________.
22. When driving, or _________, a trencher on rough terrain, keep the boom as low as possible.
23. The speed at which fluid is ejected from a pump is called the _________.
24. A control that increases speed according to the movement of the control is a(n) _________.

Trade Terms

Aerial lift
Amperes (amps)
Centrifugal force
Centrifugal pump
Cherry picker
Circuit breaker
Compactor
Compressor
Diaphragm
Diaphragm pump
Forklift
Fuse holder
Generator
Governor
Impeller
Mobiling
Pneumatic
Powered industrial truck
Proportional control
Straight blade duplex connector
Trencher
Twist-lock connector
Velocity
Volt
Watt

Have the trainees complete the Trade Terms Quiz, and go over the answers prior to administering the Module Examination.

Administer the Module Examination. Record the results on Craft Training Report Form 200, and submit the results to the Training Program Sponsor.

Administer the Performance Test, and fill out Performance Profile Sheets for each trainee. If desired, trainee proficiency noted during laboratory sessions may be used to complete the Performance Test. Record the results on Craft Training Report Form 200, and submit the results to the Training Program Sponsor.

MODULE 08106-06 ◆ MOTORIZED EQUIPMENT 6.43

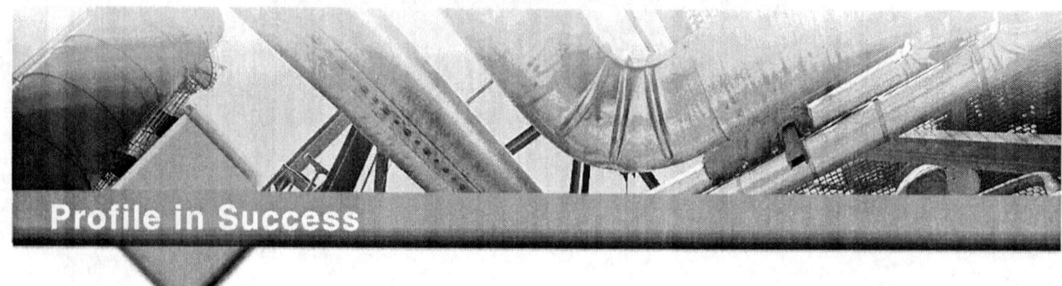

Profile in Success

Tina Goode

Becon Construction
Senior General Foreman

Tina began her career as a pipefitter in the early 80's—a time when females weren't as accepted in what were considered male-only jobs. She recalls a point early in her career where a fellow fitter and she were working on top of a boiler. During their work, he threw all the bolts to the ground, making Tina climb down each time to pick them up. She conquered these early trials, however, and worked her way up through the ranks to Foreman, General Foreman, and now to Senior General Foreman with Becon Construction.

How did you choose a career in the pipefitting field?
Well, I didn't start there. I actually started working for my dad in ironworking in the early 80's. One of my dad's friends worked for Don Love and he helped to get me a job in the plant. My first job was as a hole watch, then as a fire watch, but while doing these jobs I'd always watch what the pipefitters were doing. It looked like real interesting work. So I looked into training in the area. I went through my schooling at Fluor Daniel and was placed with AJ Mundy as a helper afterwards.

What types of training have you been through?
The training through Fluor Daniel was a formal apprenticeship. After I completed the program, I worked for several years as a helper. It really wasn't until I had a foreman who really believed in me that I realized my potential. After becoming a full-fledged pipefitter, it wasn't long until I was foreman. I was a foreman first with J.E. Merit and then with Becon and now I'm the senior general foreman with Becon.

What kinds of work have you done in your career?
There's a wide variety of work in pipefitting. I've hydrotested, pipefitted, performed underground pipe fabrication, pipe maintenance, and installed pipe supports. Of all the work, I enjoy the support details the most. My crew and I helped Bechtel become the best in the world with their supports. Pipe supports make the job.

Tell us about your present job.
As the senior general foreman with Becon, I supervise the other foremen, I perform the scheduling, the planning, and make sure everything that's needed to perform the job is on the job. I have nine foremen and approximately 100 craftspeople working for me.

What factors have contributed most to your success?
Persistence. Taking advantage of training wherever it is being offered. I have really studied the NCCER curriculum and have applied what I've learned in the field. I'm a certified craft instructor and a performance evaluator with NCCER as well as a subject matter expert on the Pipefitting authoring committee. I set goals for myself and never stop learning. After senior general foreman, there's a superintendent's job ahead. But more than anything, having people who believed in me helped me to get where I am today.

What advice would you give to those new to the pipefitting field?
Work hard and don't give up. Take advantage of school, anything anyone offers. Be sure this is the trade for you. You have to like what you're doing. It can be hot sometimes, and miserable and cold, but if you enjoy the job, you'll stay there. I'd encourage more women to get into the trade. They typically pay more attention to detail and that's a real asset in pipefitting. There's also a huge need out there for skilled pipefitters and companies are competing fiercely to get them on their job sites. There are good opportunities to advance if you apply yourself. A young woman was just hired at Austin Industrial as a lead pipefitter after her training with NCCER because she showed us she really wanted to learn and to work. Initiative and drive get noticed, no matter who or where you are.

Instructor's Notes:

Trade Terms Introduced in This Module

Aerial lift: A mobile work platform designed to transport and raise personnel, tools, and materials to overhead work areas.

Ampere (amp): A unit of electrical current.

Centrifugal force: The force that moves a substance outward from a center of rotation.

Centrifugal pump: A machine that moves a liquid by accelerating the liquid radially from the center of an impeller in motion.

Cherry picker: A truck-mounted or trailer-mounted lift designed for both indoor and outdoor use.

Circuit breaker: A device designed to protect circuits from overloads and be reset after tripping.

Compactor: A machine used to compact soil to prevent the settling of soil after construction.

Compressor: A motor-driven machine used to supply compressed air for pneumatic tools.

Diaphragm: A thin, flexible separating wall that flexes back and forth to move a fluid.

Diaphragm pump: A pump that uses a diaphragm to isolate the liquid being pumped from the operating parts in a mechanically driven pump or from the hydraulic fluid in a hydraulically actuated pump.

Forklift: A machine designed to facilitate the movement of bulk items around the job site.

Fuse holder: A device used to hold a fuse. It may be part of a circuit or may simply hold a spare fuse.

Generator: A machine used to generate electricity.

Governor: A device used to provide automatic control of speed or power for an internal combustion engine.

Impeller: A rotating part inside a pump that increases the speed of the fluid moving through the pump.

Mobiling: Term used for driving a trencher around the job site when not digging.

Pneumatic: Run by or using compressed air.

Powered industrial truck: See *forklift*.

Proportional control: A control that increases speed in proportion to the movement of the control.

Straight blade duplex connector: An electrical connector or style of outlet.

Trencher: A small machine used for digging trenches.

Twist-lock connector: A type of electrical connector.

Velocity: The speed at which a fluid is ejected from a pump.

Volt: A unit of electrical potential and electromotive force.

Watt: A unit of power that is equal to one joule per second.

MODULE 08106-06 ◆ MOTORIZED EQUIPMENT 6.45

Resources & Acknowledgments

Additional Resources

This module is intended to be a thorough resource for task training. The following reference works are suggested for further study. These are optional materials for continued education rather than for task training.

Construction Equipment Guide, Latest Edition. New York, NY: John Wiley & Sons.

Machinery's Handbook, Latest Edition. Erik Oberg, Franklin D. Jones, Holbrook L. Horton, and Henry H. Ryffel. New York, NY: Industrial Press, Inc.

Figure Credits

Ingersoll-Rand Co., 106F01A, 106F07

Topaz Publications, Inc., 106F01B, 106F02, 106F06, 106F25, 106F26A, 106SA04

Wacker Corporation, 106F09, 106F10, 106F12

Miller Electric Mfg. Co., 106SA01

Fluor Corporation, 106F11, 106F13

JLG Industries, Inc., 106F14, 106SA02

John Hoerlein, 106F20

Sellick Equipment Ltd., 106F22

Manitou North America, 106F23

DitchWitch®, 106F26B, 106E02

Terex Cranes, 106F29, 106F30B

Manitowoc Crane Group, 106F30A, 106F31, 106RQ01

Link-Belt Construction Equipment Co., 106F32, 106F33

Zachary Construction Corporation, Title Page

Instructor's Notes:

6.46 PIPEFITTING ◆ LEVEL ONE

MODULE 08106-06 — TEACHING TIPS

The following are suggested activities or instructional methods to help you teach the material in this AIG.

General

When you call on someone to answer a question, the rest of the class relaxes or even tunes out because they expect that the question and answer will take place only between you and the trainee you called on. Instead, use this technique to involve more trainees in answering questions and to keep them on their toes.

1. Ask the trainees to define a term or explain a concept.
2. After one trainee has answered, ask a trainee seated nearby if the answer is right. Then ask whether a trainee in the back of the room agrees.
3. Ask the trainees to explain why they think an answer is right or wrong.
4. Use the session to clear up incorrect ideas and encourage the trainees to learn from their mistakes.

Section 6.0.0 *Aerial Lifts*

Pipefitters often use aerial lifts to perform their job. This exercise will familiarize trainees with various types of aerial lifts. Trainees will need appropriate personal protective equipment, pencils, and paper. Arrange for a manufacturer's representative to give a presentation on the different types of lifts available. Topics to include are load capacity and safety issues. Allow 20 to 40 minutes for this exercise.

1. Introduce the topic and the speaker. Have the speaker give a presentation on the various types of lifts available, their capacities, and safety issues.
2. Have the trainees take notes and write down questions during the tour or presentation.
3. Ask the presenter to spend some time after the tour to answer any questions the trainees may have.

Section 8.0.0 *Forklifts*

Pipefitters often use forklifts to perform their job. This exercise will familiarize trainees with various types of forklifts. Trainees will need appropriate personal protective equipment, pencils, and paper. Arrange for a manufacturer's representative to give a presentation on different types of forklifts available. Topics to include are load capacity and safety issues. Allow 20 to 40 minutes for this exercise.

1. Introduce the topic and the speaker. Have the speaker give a presentation on the various forklifts available, their capacities, and safety issues.
2. Have the trainees take notes and write down questions during the tour or presentation.
3. Ask the presenter to spend some time after the tour to answer any questions the trainees may have.

MODULE 08106-06 — ANSWERS TO REVIEW QUESTIONS

Answer	Section
1. c	2.3.0
2. b	2.3.0
3. a	2.4.0
4. b	2.5.0
5. c	3.5.0
6. c	3.5.1
7. a	4.5.2
8. b	5.2.0
9. b	6.1.0
10. b	6.3.0
11. c	7.1.0, Figure 20
12. a	8.3.1
13. d	9.2.1
14. c	10.4.0
15. b	11.4.0

Answers to Trade Terms Quiz

1. Amperes (amps)
2. Forklift
3. Fuse holder
4. Compressor
5. Centrifugal pump
6. Diaphragm pump
7. Generator
8. Governor
9. Powered industrial truck
10. Compactor
11. Impeller
12. Aerial lift
13. Straight blade duplex connectors, twist-lock connectors
14. Circuit breaker
15. Cherry picker
16. Centrifugal force
17. Volts
18. Pneumatic
19. Trencher
20. Watt
21. Diaphragm
22. Mobiling
23. Velocity
24. Proportional control

CONTREN® LEARNING SERIES — USER UPDATE

NCCER makes every effort to keep these textbooks up-to-date and free of technical errors. We appreciate your help in this process. If you have an idea for improving this textbook, or if you find an error, a typographical mistake, or an inaccuracy in NCCER's Contren® textbooks, please write us, using this form or a photocopy. Be sure to include the exact module number, page number, a detailed description, and the correction, if applicable. Your input will be brought to the attention of the Technical Review Committee. Thank you for your assistance.

Instructors – If you found that additional materials were necessary in order to teach this module effectively, please let us know so that we may include them in the Equipment/Materials list in the Annotated Instructor's Guide.

Write: Product Development and Revision
National Center for Construction Education and Research
P.O. Box 141104, Gainesville, FL 32614-1104

Fax: 352-334-0932

E-mail: curriculum@nccer.org

Craft ______ Module Name ______

Copyright Date ______ Module Number ______ Page Number(s) ______

Description ______

______

______

______

(Optional) Correction ______

______

______

(Optional) Your Name and Address ______

______

______

Glossary of Trade Terms

Apprenticeship Training, Employer and Labor Services (ATELS): The U.S. Department of Labor office that sets the minimum standards for training programs across the country.

Aerial lift: A mobile work platform designed to transport and raise personnel, tools, and materials to overhead work areas.

Ampere (amp): A unit of electrical current.

Assured equipment grounding conductor program: A detailed plan specifying an employer's required equipment inspections and tests and a schedule for conducting those inspections and tests.

Backfire: A loud snap or pop as a torch flame is extinguished.

Base plates: Flat discs or rectangles under scaffold legs that are used to evenly distribute the weight of the scaffold. Base plates come in different sizes for different scaffold heights.

Bevel: (1) An angle cut or ground on the end of a piece of solid material. (2) A cut made at an angle.

Burr: A sharp, ragged edge of metal usually caused by cutting pipe.

Carburizing flame: A flame burning with an excess amount of fuel; also called a reducing flame.

Casters: Wheels that are attached to the bottoms of scaffold legs instead of base plates. Most casters come with brakes.

Centrifugal force: The force that moves a substance outward from a center of rotation.

Centrifugal pump: A machine that moves a liquid by accelerating the liquid radially from the center of an impeller in motion.

Chamfer: An angle cut or ground only on the edge of a piece of material.

Cherry picker: A truck-mounted or trailer-mounted lift designed for both indoor and outdoor use.

Chuck: The part of a machine that holds a piece of work tightly in the machine. A chuck is normally used only when the pipe or cutter will be rotated.

Circuit breaker: A device designed to protect circuits from overloads and be reset after tripping.

Compactor: A machine used to compact soil to prevent the settling of soil after construction.

Compressor: A motor-driven machine used to supply compressed air for pneumatic tools.

Conduit: A round raceway, similar to pipe, that contains conductors.

Coupling pin: A steel pin used to line up and join sections of a scaffold.

Cross braces: Steel pieces used to connect and support the vertical uprights of a scaffold.

Diaphragm pump: A pump that uses a diaphragm to isolate the liquid being pumped from the operating parts in a mechanically driven pump or from the hydraulic fluid in a hydraulically actuated pump.

Diaphragm: A thin, flexible separating wall that flexes back and forth to move a fluid.

Die grinder: A tool used to make male threads on a pipe or a bolt.

Die: A tool used to make male threads on a pipe or bolt.

Drag lines: The lines on the kerf that result from the travel of the cutting oxygen stream into, through, and out of the metal.

Dross: The material (oxidized and molten metal) that is expelled from the kerf when cutting using a thermal process.

Duty rating: American National Standards Institute (ANSI) rating assigned to ladders. It indicates the type of use the ladder is designed for (industrial, commercial, or household) and the maximum working load limit (weight capacity) of the ladder. The working load limit is the maximum combined weight of the user, tools, and any materials bearing down on the rungs of a ladder.

Fabrication: The act of putting together component parts to form an assembly.

Female threads: Threads on the inside of a fitting.

Ferrous metals: Metals containing iron.

Flare: A pipe end that has been forced open to make a joint with a fitting.

Flashback: The flame burning back into the tip, torch, hose, or regulator, causing a high-pitched whistling or hissing sound.

Forklift: A machine designed to facilitate the movement of bulk items around the job site.

Fuse holder: A device used to hold a fuse. It may be part of a circuit or may simply hold a spare fuse.

Glossary of Trade Terms

Generator: A machine used to generate electricity.

Governor: A device used to provide automatic control of speed or power for an internal combustion engine.

Ground fault circuit interrupter (GFCI): A fast-acting circuit breaker that senses small imbalances in the circuit caused by current leakage to ground and, in a fraction of a second, shuts off the electricity.

Guardrails: Protective rails attached to a scaffold. The top rail is 42 inches above the scaffold floor, and the middle rail is halfway between the top rail and the toeboard.

Harness: A device that straps securely around the body and is connected to a lifeline. It is part of a personal fall arrest system.

Hinge pin: A small pin inserted through a leg and base plate or caster wheel and used to hold these parts together.

Horsepower (hp): A unit of power equal to 745.7 watts or 33,000 foot-pounds per minute.

Impeller: A rotating part inside a pump that increases the speed of the fluid moving through the pump.

Kerf: The edge of the cut.

Ladder: A wood, metal, or fiberglass framework consisting of two parallel side pieces (rails) connected by rungs on which a person steps when climbing up or down. Ladders may either be of a fixed length that is permanently attached to a building or structure, or portable. Portable ladders have either fixed or adjustable lengths and are either self-supporting or not self-supporting.

Leveling jack: A threaded, adjustable screw located between the legs and the base plates or caster wheels of a scaffold. It is used to raise or lower parts of a scaffold to level it on uneven surfaces.

Male threads: Threads on the outside of a pipe.

Mobiling: Term used for driving a trencher around the job site when not digging.

Neutral flame: A flame burning with correct proportions of fuel gas and oxygen.

Occupational Safety and Health Administration (OSHA): The federal government agency established to ensure a safe and healthy environment in the workplace.

On-the-job training (OJT): Job-related learning acquired while working.

Outside diameter: A measurement of the outside width of a pipe.

Oxidizing flame: A flame burning with an excess amount of oxygen.

Pierce: To penetrate through metal plate with an oxyfuel cutting torch.

Pipe fitting: A unit attached to a pipe and used to change the direction of fluid flow, connect a branch line to a main line, close off the end of a line, or join two pipes of the same size or of different sizes.

Pneumatic: Run by or using compressed air.

Powered industrial truck: See forklift.

Proportional control: A control that increases speed in proportion to the movement of the control.

Ratchet: A device that allows a tool to rotate in only one direction.

Revolutions per minute (rpm): The number of complete revolutions an object will make in one minute.

Scaffold floor: A work area platform made of metal or wood.

Scaffold: An elevated work platform for workers and materials.

Scaffolding: A temporary built-up framework or suspended platform or work area designed to support workers, materials, and equipment at elevated or otherwise inaccessible job sites.

Slag: A nonmetallic product resulting from the mutual dissolution of flux and nonmetallic impurities in some welding and brazing processes.

Soapstone: Soft, white stone used to mark metal.

Straight blade duplex connector: An electrical connector or style of outlet.

Sweat: A method of joining pipe in which solder is applied to the joint and heated until the solder flows into the joint.

Tack welds: Short welds used to hold parts in place until the final weld is made.

Thread gauge: A tool used to determine how many threads per inch are cut in a tap, die, bolt, nut, or pipe. Also called a pitch gauge.

Glossary of Trade Terms

Toeboard: A 4-inch railing attached around the scaffold floor to prevent tools and materials from falling off the scaffold.

Trencher: A small machine used for digging trenches.

Twist-lock connector: A type of electrical connector.

Velocity: The speed at which a fluid is ejected from a pump.

Vertical upright: The end section of a scaffold, also known as a buck, made of welded steel. It supports other vertical uprights or upper end frames and the scaffold floor.

Volt: A unit of electrical potential and electromotive force.

Watt: A unit of power that is equal to one joule per second.

Index

Index

X

Y

Z